Statistics for Sensory and Consumer Science

Statistics for Sensory and Consumer Science

TORMOD NÆS

Nofima Mat, Norway

and

PER B. BROCKHOFF

Danish Technical University, Denmark

and

OLIVER TOMIC

Nofima Mat, Norway

A John Wiley and Sons, Ltd., Publication

This edition first published 2010

Registered office
John Wiley & Sons Ltd, The Atrium, Southern Gate, Chichester, West Sussex, PO19 8SQ, United Kingdom

For details of our global editorial offices, for customer services and for information about how to apply for permission to reuse the copyright material in this book please see our website at www.wiley.com.

Library of Congress Cataloging-in-Publication Data

Næs, Tormod.
Statistics for sensory and consumer science / Tormod Næs and Per B. Brockhoff and Oliver Tomic.
p. cm.
Summary: "As we move further into the 21st Century, sensory and consumer studies continue to develop, playing an important role in food science and industry. These studies are crucial for understanding the relation between food properties on one side and human liking and buying behaviour on the other. This book by a group of established scientists gives a comprehensive, up-to-date overview of the most common statistical methods for handling data from both trained sensory panels and consumer studies of food. It presents the topic in two distinct sections: problem-orientated (Part I) and method orientated (Part II), making it to appropriate for people at different levels with respect to their statistical skills. This book succesfully makes a clear distinction between studies using a trained sensory panel and studies using consumers. Concentrates on experimental studies with focus on how sensory assessors or consumers perceive and assess various product properties. Focuses on relationships between methods and techniques and on considering all of them as special cases of more general statistical methodologies. It is assumed that the reader has a basic knowledge of statistics and the most important data collection methods within sensory and consumer science. This text is aimed at food scientists and food engineers working in research and industry, as well as food science students at master and PhD level. In addition, applied statisticians with special interest in food science will also find relevant information within the book"– Provided by publisher.
Summary: "This book will describe the most basic and used statistical methods for analysis of data from trained sensory panels and consumer panels with a focus on applications of the methods. It will start with a chapter discussing the differences and similarities between data from trained sensory and consumer tests"– Provided by publisher.
Includes bibliographical references and index.
ISBN 978-0-470-51821-2 (hardback)
1. Food–Sensory evaluation. 2. New products. I. Brockhoff, Per B. II. Tomic, Oliver. III. Title.
TX546.N34 2010
664′.07–dc22

2010016197

A catalogue record for this book is available from the British Library.

ISBN (Hbk) 9780470518212

Typeset in 10/12pt Times by Aptara Inc., New Delhi, India.

Contents

Preface

Sensory and consumer studies play an important role in food science and industry. They are both crucial for understanding the relation between food properties on one side and human liking and buying behaviour on the other. These studies produce large amounts of data and without proper data analysis techniques it is impossible to understand the data fully and to draw the best possible conclusions.

This book aims at presenting a comprehensive overview of the most important and frequently used statistical methods for handling data from both trained sensory panels and consumer studies of food. A major target group for this book is food scientists and food engineers working in research and industry. Another important target group is food science students at master and PhD level at universities and colleges. Applied statisticians with special interests in food science will also find important material in the book. The goal is to present the reader with enough material to make him/her able to use the most important methods and to interpret the results safely.

The book is organised in two main parts, Part I has a quite different focus to Part II. Part I has a problem oriented focus and presents a number of typical and important situations in which statistical methods are needed. Part II has a method oriented perspective and all the methods discussed in Part I are described more thoroughly here. There is a strong link between Part 1 and Part II through extensive cross-referencing. The structure of the book is also presented in Figure 1.1.

The book will have focus on relationships between methods and techniques and on considering all of them as special cases of more general statistical methodologies. In this way we will avoid, as far as possible, using "local dialect" for each of the themes discussed. Conjoint analysis is an example of an area which has developed into a separate discipline with a particular terminology and culture. In our approach conjoint analysis will be considered and presented as an example of an intelligent use of experimental design and analysis of variance which are both classical disciplines in statistics.

It will be assumed that the reader has a basic knowledge of statistics and also the most important data collection methods within sensory and consumer science. For some of the more advanced parts of Part II, an elementary knowledge of matrix algebra will make reading easier.

Tormod Næs, Per B. Brockhoff and Oliver Tomic

Acknowledgments

We would like to thank Nofima Mat and DTU for having made it possible for us to write this book. We would also like to thank colleagues for important discussions and support related to the examples used for illustration.

Tormod Næs: A large part of the book was written while I was visiting scientist at Dipartimento delle Biotechnologie Agrarie at University of Firenze (winter 2008-2009). The colleagues at University of Firenze are thanked for this opportunity. I would also like to thank my wife for staying faithfully and patiently with me during this period in our small house in Impruneta outside Firenze.

Oliver Tomic: I would like to dedicate this book to my daughter Nina, my wife Heidi and my parents Milica and Marko.

Per B. Brockhoff: My contribution to this book is dedicated to my wife Lene and my daughters Trine, Fie and Julie.

1
Introduction

Some of the most important aspects in food science and food industry today are related to human perception of the food and to enjoyment associated with food consumption. Therefore, very many activities in the food sector are devoted to improving already existing products and developing new products for the purpose of satisfying consumer preferences and needs. In order to achieve these goals one needs a number of experimental procedures, data collection techniques and data analysis methods.

1.1 The Distinction between Trained Sensory Panels and Consumer Panels

In this book we will, as usually done, make a clear distinction between studies using a trained sensory panel (Amerine *et al.*, 1965; O'Mahony, 1986; Meilgaard *et al.*, 1999) and studies using consumers (Lawless and Heyman, 1999). The former is either used for describing degree of product similarities and differences in terms of a set of sensory attributes, so-called sensory profiling, or for detecting differences between products, so-called sensory difference testing. For the various attributes, the measurement scale is calibrated and usually restricted to lie between a lower and an upper limit, for instance 1 and 9. A sensory panel will normally consist of between 10 and 15 trained assessors and be thought of as an analytical instrument. For consumer studies, however, the products are tested by a representative group of consumers who are asked to assess their degree of liking, their preference or their purchase intent for a number of products. These tests are often called hedonic or affective tests. While the trained sensory panel is only used for describing the products as objectively as possible, consumer studies are used for investigating what people or groups of people like or prefer. The number of consumers must be much higher than the number of assessors in a sensory panel in order to obtain useful and reliable information. Typically, one will use at least between 100 and 150 consumers in this type of studies.

Statistics for Sensory and Consumer Science Tormod Næs, Per B. Brockhoff and Oliver Tomic

Note that sometimes the term sensory science is used to comprise all types of tests where the human senses are used. This means that many consumer studies are also sensory tests. The difference between sensory consumer tests and sensory panel tests is the way they are used; sensory panels are used for describing the properties of products and sensory consumer tests are used for investigating the degree of liking. In this book it will be clear from the context and description of the situation which of these studies that is in focus.

Note also that one is often interested in relating the two types of data to each other for the purpose of understating which sensory attributes are important for liking. This is important both for product development studies, for developing good marketing strategies and also for the purpose of understanding more generally what are the opinions and trends in various consumer segments.

In this book, we will concentrate on experimental studies with focus on how sensory assessors or consumers perceive and assess various product properties. Consumer surveys of attitudes and habits will play a minor role here, even though some of the statistical methods treated may also be useful in such situations. In the examples presented main attention will be given to data that are collected by asking people about their opinion (stated acceptance or preference), but many of the same statistical methods can be used if data are obtained by monitoring of real behaviour (revealed acceptance or preference, see e.g. Jaeger and Rose, 2008).

For a broad discussion of possible problems and pitfalls when interpreting results from consumer studies we refer to Köster (2003).

1.2 The Need for Statistics in Experimental Planning and Analysis

In sensory and consumer science one is typically interested in identifying which of a number of potential factors that have an influence on the sensory attributes and/or the consumer liking within a product category. For obtaining such information, the most efficient experimental procedures can be found within the area of statistical experimental design (Box *et al.*, 1978). These are methods which describe how to combine the different factors and their levels in such a way that as much information as possible is obtained for the lowest possible cost. For optimising a product, one will typically need to work in sequence, starting out by eliminating uninteresting factors and ending up with optimising the most important ones in a limited experimental region. In all phases it is more efficient to consider series of experiments where all factors are varied instead of investigating one factor at a time. Important building blocks in this tradition are the factorial designs, fractional factorial designs and central composite designs (Chapter 12). The concepts of randomisation and blocking, for systematically controlling uninteresting noise factors, are also important here. Another important point is representativity, which means that the objects and assessors are selected in such a way that they represent the situation one is interested in the best possible way. For instance, if a whole day's production of a product is to be investigated, one should not investigate consecutive samples, but rather pick them at random throughout the day.

The data sets that emerge from sensory and consumer experiments are typically quite large and the amount of information available about relations between them is limited. It is therefore important to have data analysis methods that in a pragmatic way can handle such situations. Focus in the present book will be on analysis of variance (ANOVA) and regression based methods (Chapter 13, and 15) and methods based on PCA for data compression (Chapter 14, 16, 17). An important aspect of all these methods is that they are versatile and can be used in many practical applications. We will be interested in significance testing for indicating where the most important information is and plotting techniques for visual inspection of complex relations. Validation of models by the use of empirical data will also be important. The main philosophy in the exposition will be simplicity, transparency and practical usefulness.

It is important to emphasise that in order to get the most out of statistical design and analysis methods, one must use as much subject matter knowledge as possible. It is only when statistical and subject matter knowledge play well together that the best possible results can be obtained. It is also worth mentioning that although the book is primarily written with a focus on applications within food science, many of the same methods can be used also for other applications where sensory and consumer aspects are involved.

Other books that cover some of the same topics as discussed here are Gacula *et al.* (2009), Mazzocchi (2008), Næs and Risvik (1996), and Meullenet *et al.* (2007).

1.3 Scales and Data Types

In most of the book we will consider sensory panel data and also consumer rating data as continuous interval scale data. This means that the differences between two different values will be considered meaningful, not only the ordering of the data. One of the advantages of taking such a perspective is that a much larger set of methods which are easy to use and understand are made available. It is our general experience that such an approach is both reasonable and useful.

If the data are collected as rank data or as choice/preference data, it is necessary to use methods developed particularly for this purpose. For choice based conjoint experiments one will typically treat the data as nominal categorical data with a fixed set of outcomes and analyse with for instance generalised linear models (see e.g. Chapter 15). The same type of methods can also be used for rank data, but here other options are also available (see Chapter 17).

1.4 Organisation of the Book

This book is organised in two parts, Part I (Chapters 1–10) and Part II (Chapters 11–17). The first part is driven by applications and examples. The second part contains descriptions of a number of statistical methods that are relevant for the application in Part I. In Part I we will refer to the relevant methodologies presented in Part II for further details and discussion. In Part II we will refer to the different chapters in Part I for typical applications of the methods described. The more practically oriented reader may want to focus on Part I and look up the various specific methods in Part II when needed. The more statistically oriented reader may prefer to do it the other way round. The structure of the book is illustrated in Figure 1.1.

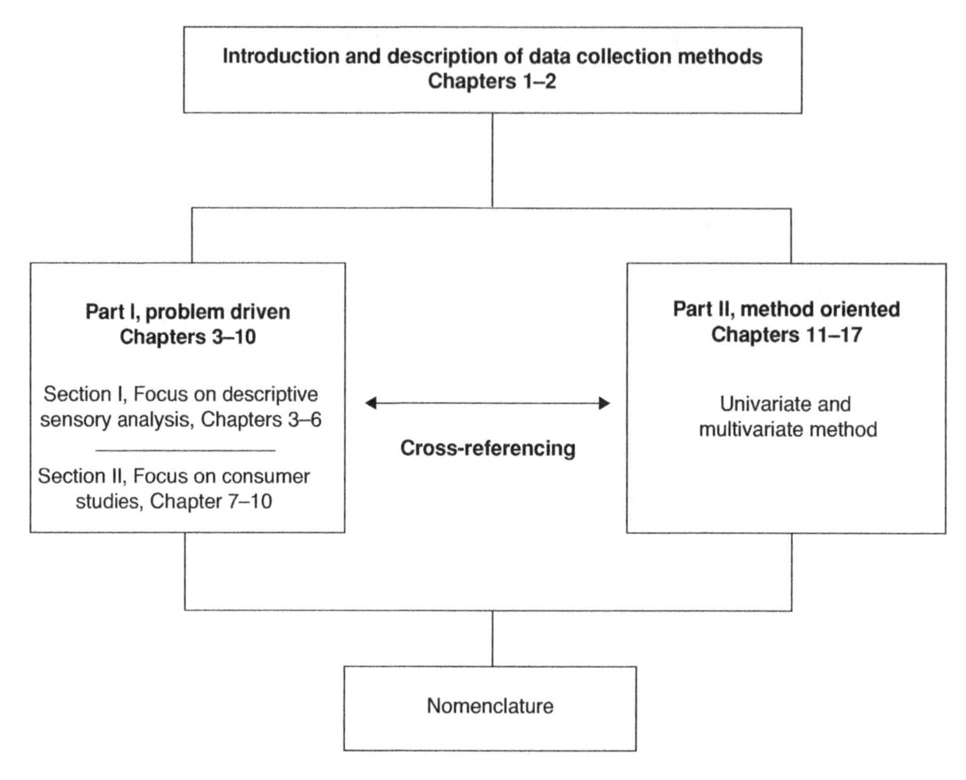

Figure 1.1 Description of how the book is organised.

References

Amerine, M.A., Pangborn, R.M., Roessler, E.B. (1965). *Principles of Sensory Evaluation of Food*. New York: Academic Press.

Box, G.E.P., Hunter, W., Hunter, S. (1978). *Statistics for Experimenter*. New York: John Wiley & Sons, Inc.

Gacula, M.C. Jr., Singh, J., Bi, J., Altan, S. (2009). *Statistical Methods in Food and Consumer Science*. Amsterdam, NL: Elsevier.

Jaeger, S.R., Rose, J.M. (2008). Stated choice experimentation, contextual influences and food choice. A case study. *Food Quality and Preference* 10, 539–64.

Köster, E.P. (2003). The psychology of food choice. Some often encountered fallacies. *Food Quality and Preference* 14, 359–73.

Lawless, H.T., Heymann, H. (1999). *Sensory Evaluation of Food: Principles and Practices*. New York: Chapman & Hall.

Mazzocchi, M. (2008). *Statistics for Marketing and Consumer Research*. Los Angeles: Sage Publications.

Meilgaard, M., Civille, G.V., Carr, B.T. (1999). *Sensory Evaluation Techniques* (2nd edn). Boca Raton, Florida: CRC Press, Inc.

Meullenet, J-F, Xiong, R., Findlay, C.J. (2007). *Multivariate and Probabilistic Analysis of Sensory Science Problems*. Ames, USA: Blackwell Publishing.

Næs, T., Risvik, E. (1996). *Multivariate Analysis of Data in Sensory Science*. Amsterdam: Elsevier.

O'Mahony, M. (1986), *Sensory Evaluation of Food, Statistical Methods and Procedures*. New York: Marcel Dekker, Inc.

2

Important Data Collection Techniques for Sensory and Consumer Studies

This chapter gives a brief description of some of the most important methodologies for collecting data in sensory and consumer science. For more detailed and comprehensive presentations of these and related methods we refer to Amerine *et al.* (1965), Lawless and Heyman (1999) and O'Mahony (1986).

2.1 Sensory Panel Methodologies

2.1.1 Descriptive Sensory Analysis

Descriptive sensory analysis or so-called sensory profiling is probably the most important method in sensory analysis and also the one that will be given the most attention here. This is a methodology which is used for describing products and differences between products by the use of trained sensory assessors. The main advantage of sensory analysis as compared to for instance chemical methods is that it describes the properties of a product in a language that is directly relevant for people's perception, for instance degree of sweetness, hardness, colour intensity etc. The sensory panel used this way is thought of and used as an analytical instrument.

Typically, a sensory panel consists of between 10 and 15 trained assessors, usually recruited according to their ability to detect small differences in important product attributes. Before assessment of a series of products, the assessors gather to decide on the attributes to use for describing product differences. In some cases, certain attributes may also be given prior to this discussion. Usually, one will utilise some of the products for the purpose of calibrating the scale to be used, if not calibrated by other means. In some cases the assessors

Statistics for Sensory and Consumer Science Tormod Næs, Per B. Brockhoff and Oliver Tomic

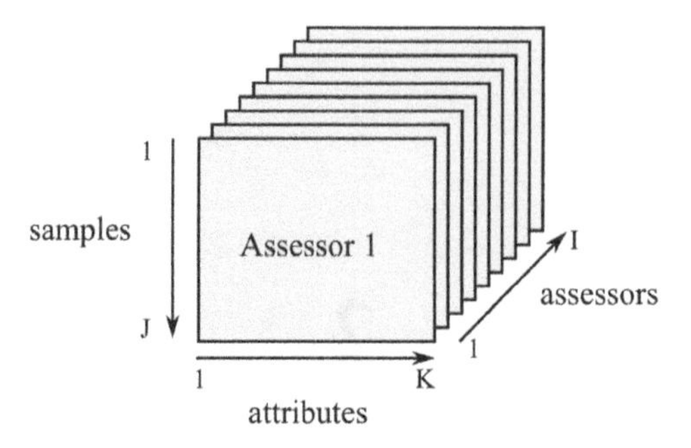

Figure 2.1 Three-way sensory data. *This illustration depicts the general structure of sensory profiling data. For each assessor (i), there is a data table consisting of measurements (scores) for a number of attributes (k) and a number of products (j). If there are replicate (r) in the data set, these can be added as additional rows.*

are allowed to use their own vocabulary (free choice profiling, FCP, see e.g. Arnold and Williams, 1987), but this type of analysis will not be given attention here. In most cases, the number of attributes will be between 10 and 20, but this depends on the complexity of the products and also the scope of the study. In the actual testing session, all assessors are given the products in random order and without knowing anything about the product differences, labelling, brands etc.; so-called blind tasting. The intensity scores for the different products are either given as numbers between a lower and an upper limit or as indications on a line, either on a computer screen or on paper. Descriptive sensory analysis produces data that can be presented in a three-way data structure as indicated in Figure 2.1. In most cases, each sample is tested in duplicate or triplicate. Within the three-way framework of Figure 2.1 replicates can be accommodated by representing them as new samples or products.

Different ways of presenting the samples to the assessors are possible. One possibility is to have standard sample(s) present during the whole session for the purpose of providing a stable calibration of the panel. In most cases, however, the samples are just presented to the assessor without any of these additional tools. For the purpose of the statistical analysis of the data, this aspect has little influence.

In this book sensory analysis data will be treated in several of the chapters. In some of the chapters the only focus is on sensory analysis itself and how it can be used to detect and understand sensory differences between products (for instance Chapter 5). In other cases, it will be used in combination with consumer data for the purpose of understanding better the consumer preferences and acceptances (Chapter 9). A separate chapter will be devoted to quality control of sensory panels (Chapter 3). This is of particular importance for the assessment of panel reliability and for panel improvement.

The main reasons for having several assessors, instead of one or only a few, in a panel are that this gives more precise assessments of product attributes, it provides an automatic internal quality control of the panel and that individual differences can be detected and analysed. The latter can in some cases be very important for assessing differences in use of the scale, for detecting differences in thresholds etc.

2.1.2 Discrimination Tests

Another type of sensory test is the class of so-called discrimination or difference tests, which are suitable for testing whether the assessors, either consumers or sensory panellists, are able to detect small differences between products tested. An important example is for recruiting assessors for a sensory panel. In such cases potential candidates are typically asked to distinguish between samples with very similar intensities of bitterness, sweetness, sourness, umami and saltiness. Another important example is for analysing product substitutability (Ennis, 1993a), which is typically of interest when introducing a new and cheaper ingredient. The triangle test is probably the most commonly used discrimination test. This is based on giving each assessor a so-called triangle consisting of two identical and one different sample. He/she is then asked to identify the one which is different. The random probability for this is equal to 1/3 if the products are identical. These standard tests for such hypotheses are usually based on the binomial distribution. More advanced methods based on Thurstonian modelling (see Chapter 7) can, however, be used to obtain more insight.

2.2 Consumer Tests

2.2.1 Hedonic or Affective Tests

Sensory analysis by trained sensory panels describes the products as objectively as possible, but in order to obtain information about what people like, various types of consumer studies are needed. Linking the two types of analyses is of particular importance since one is often interested in understanding what are the main drivers for food choice or liking. For instance, is the acceptance of a certain product related to sweetness or another sensory attribute or have extrinsic attributes like various types of information or packaging a larger impact? While a sensory panel is primarily selected based on the assessors ability to detect and measure sensory aspects of products, a consumer study will normally be based on results from consumers that are randomly drawn from a certain population. In some cases one may decide to select consumers that are consumers of a particular product or one may decide to do a more systematic sampling for the purpose of for instance ensuring a certain distribution of a demographic variable.

In this book, main emphasis will be given to experimental consumer studies with a hedonic response, which are typical and frequently used both in an industrial and in a research context. Both sensory product-related attributes as well as extrinsic attributes related to information and packaging will be in focus. Important examples to be discussed are conjoint analysis and preference mapping studies. Large surveys based on questionnaires related to habits, attitudes etc. will only be touched upon briefly, but many of the methods considered in this book will also be useful in such a context. Other methods that will not be covered in this book are methods based on deeper interviews and discussions with consumers.

Experimental consumer tests may be conducted in central locations or labs, in homes or via internet. For the purpose of this book, which is mainly about statistical methods, a clear distinction between these techniques will not be made, because the data structures and statistical analysis techniques are usually the same for all. For a deeper discussion of all these aspects we refer to Lawless and Heymann (1999) and references therein.

2.2.2 Self-explicated Tests

The simplest types of tests that can be conducted are the self-explicated tests. These are tests in which the consumers are asked which of a series of attributes they put most emphasis on when they select food products or when they make choices. They may also be asked to rank the importance of the attributes. These tests can be useful and are not necessarily inferior to others (see Gustafsson *et al.* (2003)), but they also have a number of drawbacks. First of all they cannot be used for assessing interactions between the attributes, which can sometimes be a major concern. Secondly they require a mental processing which is not typical for a buying situation. Main emphases will therefore here be given to experimental strategies that combine various product attributes or contexts of interest using experimental design methodologies. The consumers are then asked to assess the different combinations of attributes varied.

2.2.3 Rating Based Studies

Rating based studies will be given main attention here. These are experimental studies where the consumers are asked to assess either their degree of liking, their degree of acceptance or their probability of buying the products for each of the combinations tested. In for instance purely sensory tests it is natural to ask about degree of liking while in more concept oriented tests including also several extrinsic attributes, purchase intent or purchase probability may sometimes be more relevant. Since the statistical analysis is usually the same regardless of which question is asked, we will not distinguish strictly between the different questions asked to the consumers, whether it is expected liking (Deliza and MacFie, 1996), actual liking of a real product or purchase probability. We refer to Lawless and Heymann (1999) and to Mela (2000) for discussion of various aspects related to the different types of consumer responses.

In some cases the different attributes tested are presented verbally or by using illustrations. Important examples are related to for instance health claims, brand name and packaging. In many cases, however, it is natural to bring in real products in order to assess the relative importance of sensory and extrinsic information as well as their possible interactions. The sensory perception is then brought in as an important aspect of the assessors' rating of the products.

The consumers will typically, as for sensory analysis, give a score between a lower and an upper value for each of the products regardless of which question they are asked to respond to. The scale is in many cases anchored with for instance 'like very little' and 'like very much', but this is not always done.

2.2.4 Ranking Tests

Another type of tests is the ranking test. In this type of test all possibilities (samples) are presented simultaneously to the consumers and they are asked to rank them according to for instance liking or purchase intent. If there are many combinations to be tested, the sorting can be done in sequence, by first splitting in two, then in two again etc. until all have been ranked. Note that this type of test is impossible or at least very difficult to use in context studies where it is impossible to present all alternatives at the same time. Ranking tests are useful, but the main drawback lies in the analysis. Since the data are ordinal and the sum of all assessments for each assessor sum to a constant, this limits the number of possible analysis methods considerably. Some possibilities exist as will be discussed later in this book (Chapter 7 and Chapter 17), but the methods are often more difficult to understand and the number of possibilities is limited.

2.2.5 Choice Tests

Choice studies (Louviere *et al.*, 2000) will also be discussed in this book (Chapter 8). These are tests for which the consumers are given a number of so-called choice sets and for each choice set they are asked to select the one they like best. A choice set is constructed by selecting systematically a number of products from the full design of possibilities. Some researchers claim that these tests are more realistic than rating and ranking tests since they fit better to a real buying situation. This is, however, questionable and studies exist which indicate that it does not always matter so much for the conclusions in which way the data are collected (see for instance Jaeger and Rose, 2008 for a discussion). As for the ranking studies, the number of analysis methods is more limited and they are more difficult to use and understand than for rating tests.

2.2.6 Auctions and Monitoring of Behaviour

Price is often an important factor to consider in probability-to-buy studies. A number of specific techniques have been developed for this purpose, for instance various types of auctions (Lange *et al.*, 2002) and willingness to pay tests. These methods will not be given special attention here, although several of the standard statistical methods treated in Part II of the book can be used also for this purpose. Price is, however, also easy to incorporate as a separate factor in conjoint analysis and this will be the main approach taken here.

Real consumer behaviour in buying situations is, as mentioned above, more relevant than what they say, but also much more difficult to measure. Different methods have, however, been developed based on auctions or monitoring of actual choices made in a cafeteria or in a supermarket. Consumers are for instance given an amount of money before the test and they are asked to spend the money the way they like under certain restrictions on choice within product categories. Another possibility is to organise a set of different options in a cafeteria, with the possibility of repeated experimentation. The data can be analysed by the methods presented in this book.

2.2.7 Additional Consumer Attributes and Characteristics

When consumer acceptance data have been collected and analysed, one will typically also be interested in understanding the individual differences between consumers in a better way. This can be done if additional consumer attributes have been collected during or after the actual testing session. This can be done using a questionnaire with questions related to demographic variables such as gender, age, family size and variables related to attitudes and habits, either on a general basis or related to the actual product studied. These consumer attributes can be related to the individual differences observed using regression or tabulation methods (Chapter 8). Segmentation is an important concept which comes in here

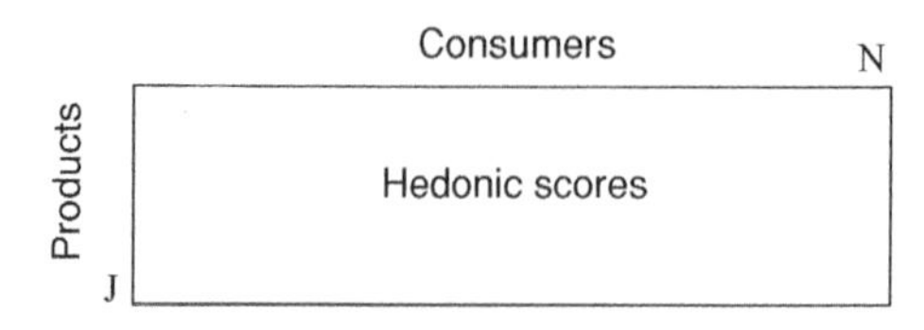

Figure 2.2 Two-way consumer data. *This illustration simply depicts the structure of a matrix of consumer hedonic scores for a number of products.*

(Chapter 10). This can be done either a priori based on the consumer attributes themselves or based on the consumer liking pattern with subsequent analysis or relations to consumer attributes.

The data set based on a hedonic consumer test with additional consumer attributes can typically be represented in data tables with simple structure as indicated in Figure 2.2.

References

Amerine, M.A., Pangborn, R.M., Roessler, E.B. (1965). *Principles of Sensory Evaluation of Food*. New York: Academic Press.

Arnold, G.M., Williams, A.A. (1987). The use of generalised procrustes techniques in sensory analysis. In J.R. Piggott (ed.), *Statistical Procedures in Food Research*. London: Elevier Science Publishers, 244–53.

Deliza, R., MacFie H.J.H. (1996). The generation of sensory expectation by external cues and its effect on sensory perception and hedonic ratings: A review. *Journal of Sensory Studies* 11, 103–28.

Ennis, D.M. (1993a). The power of sensory discrimination methods. *Journal of Sensory Studies* 8, 353–70.

Gustafsson, A., Herrmann, A., Huber, F (2003). *Conjoint Measurement. Methods and Applications*. Berlin: Springer.

Jaeger, S.R., Rose, J.M. (2008). Stated choice experimentation, contextual influences and food choice. A case study. *Food Quality and Preference* 10, 539–64.

Lange, C. Martin, C., Chabanet, C., Combris, P., Issanchou, S. (2002). Impact of the information provided to consumers on their willingness to pay for Champagne: comparison with hedonic scores. *Food Quality and Preference* 13, 597–608.

Lawless, H.T., Heymann, H. (1999). *Sensory Evaluation of Food: Principles and Practices*. New York: Chapman & Hall.

Louviere, J.J., Hensher, D.A. and Swait, J.D. (2000). *Stated Choice Methods: Analysis and Applications*. Cambridge: Cambridge University Press.

Mela, D. J. (2000). Why do we like what we like? *Journal of the Science of Food and Agriculture* 81, 10–16.

O'Mahony, M. (1986). *Sensory Evaluation of Food, Statistical Methods and Procedures*. New York: Marcel Dekker, Inc.

3

Quality Control of Sensory Profile Data

This chapter focuses on quality control of trained sensory panels. This subject is important for determining panel reliability, for being able to improve the panel and for better handling of the actual data. A number of different methods will be proposed and discussed together with a strategy for how the different methods can be used together to identify different types of problems. Some of the methods are multivariate and may be used to obtain an overview of all the attributes. Other methods are univariate and therefore suitable for more detailed studies of specific attributes. A discussion is provided regarding advantages and disadvantages of the different techniques. An example from analysis of apples will be used to illustrate the different methods and how they can be used together to reveal a number of different problems. Most plots presented here are made in the open source software package PanelCheck available at www.panelcheck.com

The chapter is strongly related to the remedies and panel improvement methods in Chapter 4 and is based on theory that can be found in Chapters 11, 13, 14 and 17.

3.1 General Introduction

This chapter is about quality control of sensory profile data. For such data to be meaningful, it is important that the assessors are calibrated and that they use the sensory attributes the same way (see Amerine *et al.* 1965). This is typically obtained through a discussion between the panel leader and the assessors prior to analysis. From the set of samples that are to be tested in the tasting session, usually a few are selected and used as a basis for discussion. These samples should preferably represent the extreme states of intensity for the attributes that are to be used to describe the product.

Regardless of how well the calibration is done, there will always be individual differences between the assessors in their way of assessing the samples. Some of these are related to differences in the assessors' sensitivity and cognitive processing of the sensory stimuli. Other differences are less basic and may for instance be related to such things as different use of the intensity scale or differences in ability to discriminate between the samples, due to for instance lack of concentration or poor sensory memory. The focus of this chapter will be on detecting the latter type of differences since these are usually considered to represent nuisance effects (Næs, 1990; Tomic *et al.*, 2007; Brockhoff, 2003b) and may be reduced by extended and targeted training. In some cases it may be difficult to distinguish between the two types of differences using data analysis only. Therefore, subject matter knowledge will as always play a central role. Some of the same methods as discussed here can also be used for comparing differences between panels instead of differences between assessors (see for instance Hunter and McEwan, 1998; McEwan, 1999; Lê *et al.*, 2008; Tomic *et al.*, 2010a).

If individual differences in performance are neglected, the final results may suffer from bias and imprecise conclusions. Provided that resources are available, one should always gather the panel for performance feedback and discuss possible reasons for the differences detected and in this way continuously improve the panel performance. In concrete cases, however, one will usually have to live with the data at hand and seek to make the best out of them even when the individual differences are large. In a worst case scenario, it might be necessary to discard certain assessors or attributes from the data set in order to eliminate large unwanted variability. This, however, is an undesirable approach that may raise economical and also ethical issues. A better solution could be to pre-process the data by one of the methods discussed in Chapter 4 (see also Romano *et al.* 2008) and possibly weigh down negative effects before further analysis. This may be achieved by computing for instance weighted product averages across assessors with weights determined from the assessors' individual performance. Another possibility is to try to model the individual differences explicitly in order to enhance the usefulness of the data (Brockhoff and Sommer, 2008).

All these remedy aspects will be discussed in more detail in Chapter 4. In the present chapter we will concentrate on techniques for *detecting and visualising* individual differences for quality control purposes. It should be mentioned, however, that these methods can also be important for obtaining improved knowledge about sensory analysis as a measurement technique and about important differences in assessors' capabilities as panel members.

The most important individual differences that one can find in sensory data are listed in Table 3.1. We also refer to Figure 3.1 for a graphical illustration of some of these effects. The first point in the Table 3.1 is related to how the assessors use the intensity scale. Note that this effect does not have any direct link to the quality of the assessor. Still, information about the use of scale is important because large differences may have an unfortunate effect on the average panel results if not accounted for in the data analysis. The second point is related to agreement or confusion among the assessors regarding the definition of an attribute. A well trained and calibrated panel should have a high degree of reproducibility across all assessors, i.e. they should score similarly on the tested products and have similar or identical sample ranking. The third point is directly related to the error variance or the assessors' ability to reproduce or repeat a similar intensity value for the same stimulus. The

Table 3.1 Some important types of individual differences in sensory panel data, see also Figure 3.1.

1. Use of scale: Differences in mean and variability/range of the scores
2. Agreement/reproducibility across the panel: Disagreement in ranking of the objects
3. Repeatability: Different level of precision – differences between independent replicates
4. Discrimination: Differences in ability to discriminate between products

fourth point is related to the third, but is more specific in the sense that here the only focus is on detecting differences between products. If an increased error variance for an assessor is accompanied with a larger span of the scale used, the ability of the assessor in detecting differences between products may be as good as for the rest (see below).

The presentation order of the tools in this chapter corresponds to the order in which they are normally used in practice. For a graphical illustration of the workflow we refer to the flow chart in Figure 3.2. The way the methods are used in practice will, however, also depend on focus, personal habits and preferences.

To obtain full insight into the performance of the individual assessors and the panel as a whole, one needs to combine information from more than only one method. The reason for this is that each method provides unique information covering only certain aspects of all possible performance issues. Some of the methods are related to general statistical techniques such as ANOVA and Procrustes analysis which are discussed in a broader framework in Part II of the book (Chapters 13 and 17), while other methods are specific for quality control of sensory profiling data.

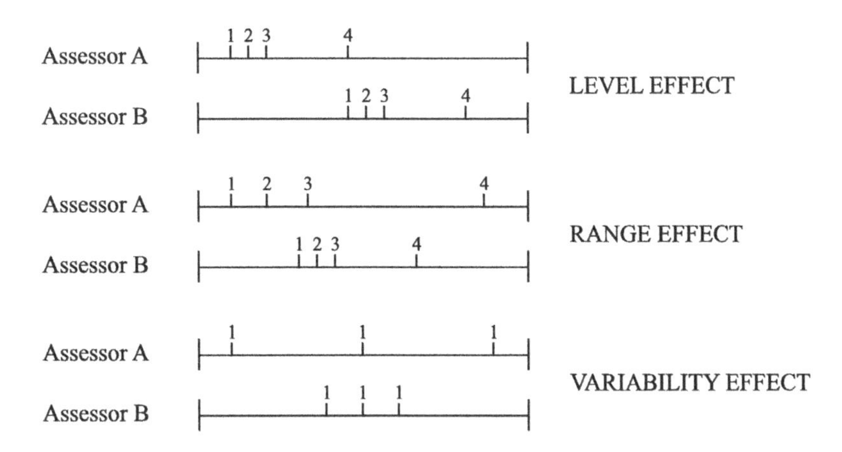

Figure 3.1 ***Illustration of individual differences in use of the scale.*** *The two upper lines correspond to two different assessors using the lower and the upper part of the scale respectively for assessing the differences between 4 products. The two lines in the middle illustrate two assessors who use the range very differently. The bottom lines show two assessors with very different replicate error.*

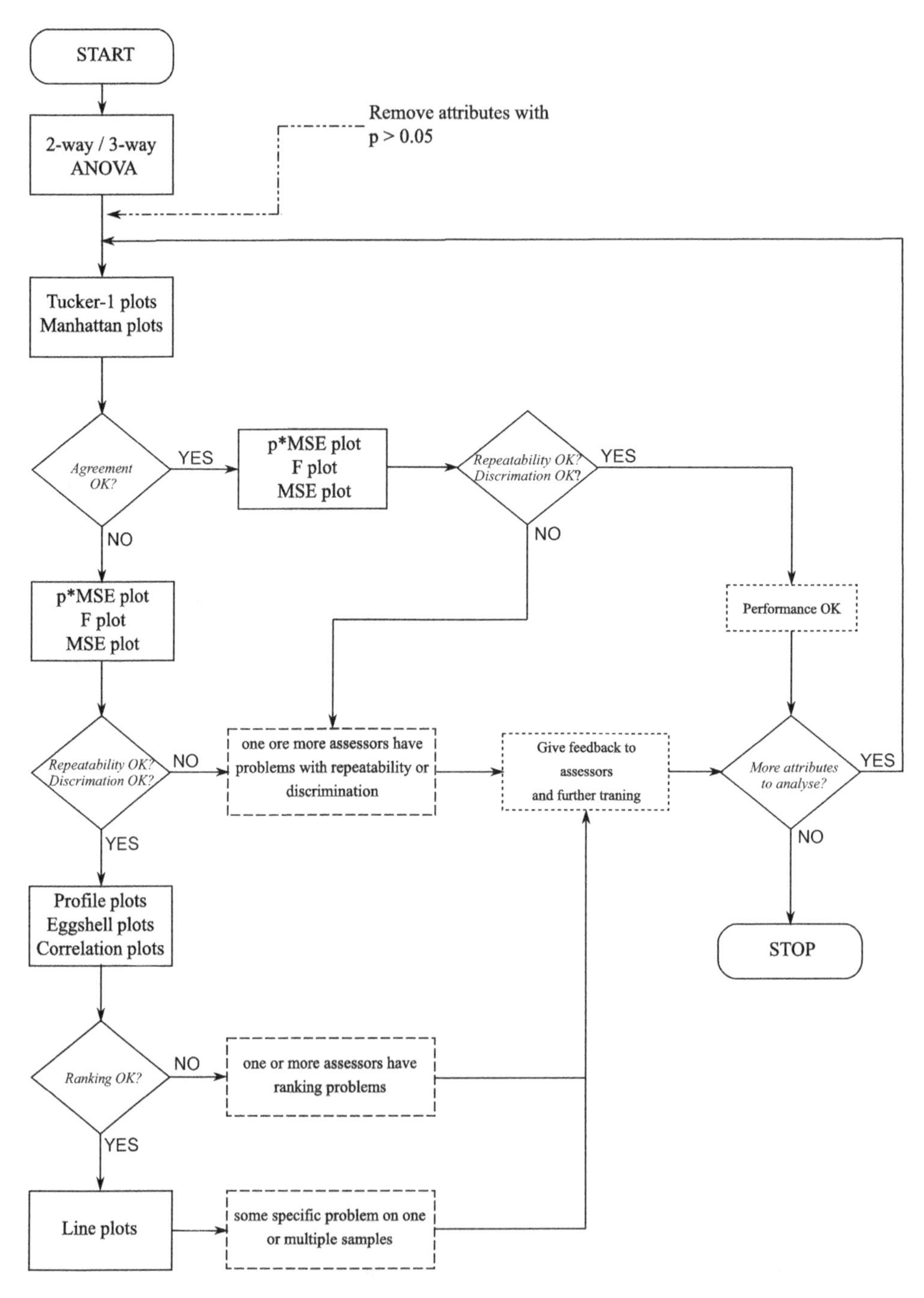

Figure 3.2 Flow chart for how to combine the tools discussed in the text. *Reprinted from European Food Research and Technology, 230, Tomic et al, Analysing sensory panel performance in a proficiency test using the PanelCheck software. 2009, with permission from Springer.*

Below we will start by discussing simple inspection of the raw data. This is always recommended since extreme outliers or errors should be detected and removed prior to further analysis. As a next step one will normally seek an overview of the panel performance using a mixed model ANOVA (Chapter 13) for all the attributes followed by multivariate analysis of all the assessors and attributes simultaneously using the Tucker-1 method (Chapters 14 and 17). The various problems detected in these analyses are then typically investigated using more detailed techniques such as the correlation plot and the profile plots. In other words, the working strategy starts with methods focussing on the overall aspects of the data and concludes with a more detailed analysis of specific problems detected.

Most methods discussed in this chapter assume a numeric scale for the intensity of the attributes. One of the methods presented (the eggshell plot, see below, Section 3.7.3) is, however, particularly developed for rank data.

The design of the sensory panel study may sometimes have an effect on some of the tools discussed here. We refer to Chapters 5 and 13 for a discussing of various ways of conducting sensory analysis and how this may lead to different replicate structure in the data. When relevant, these aspects will be highlighted in the discussion below.

For illustration, we will in this chapter use a data set from sensory analysis of apples which has the following dimensions: $J = 7$ apple varieties are tested using $I = 9$ assessors and $K = 20$ sensory attributes. There are two randomised replicates in the data set. The attributes used to describe the apple products were gloss, wax coat, grass odour, fruit odour, flower odour, grass flavour, honey flavour, fruit flavour, flower flavour, acidity, sour taste, sweet taste, bitter taste, skin toughness, hardness, chewing resistance, brittleness, meliness, juiciness and aftertaste.

3.2 Visual Inspection of Raw Data

Investigating and visualising the raw data prior to further analysis is always recommended. In this way obvious mistakes and outliers are eliminated from the analysis. Moreover, one can obtain an initial impression of the main structures of the data set and detect possible tendencies that may be of interest later on during the analysis. A number of simple tools are available for this purpose.

A straightforward and often used technique for getting an initial overview of the raw data is to use average values accompanied by standard deviations (see e.g. Chapters 11 and 5). These results can be presented for the complete data set or with focus on either individual assessors or specific attributes. An example is given in Figure 3.3 focusing on the attribute acidity with each column representing one specific assessor. The standard deviations are superimposed (both upwards and downwards) in order to indicate the variability of the observations. With a perfectly trained and calibrated sensory panel all assessors should have identical mean scores. In our example, however, one can easily see that mean scores and also the size of corresponding standard deviations vary across assessors. For instance, assessor B has a mean score equal to 6.2 with a standard deviation equal to 2.2 while for assessor E the corresponding values are 4.7 and.0.9. In this manner one can easily spot assessors that deviate strongly from the others.

Although plots showing means and standard deviations are useful for overview purposes, they are less suitable for detecting outliers in the data. A large standard deviation can for

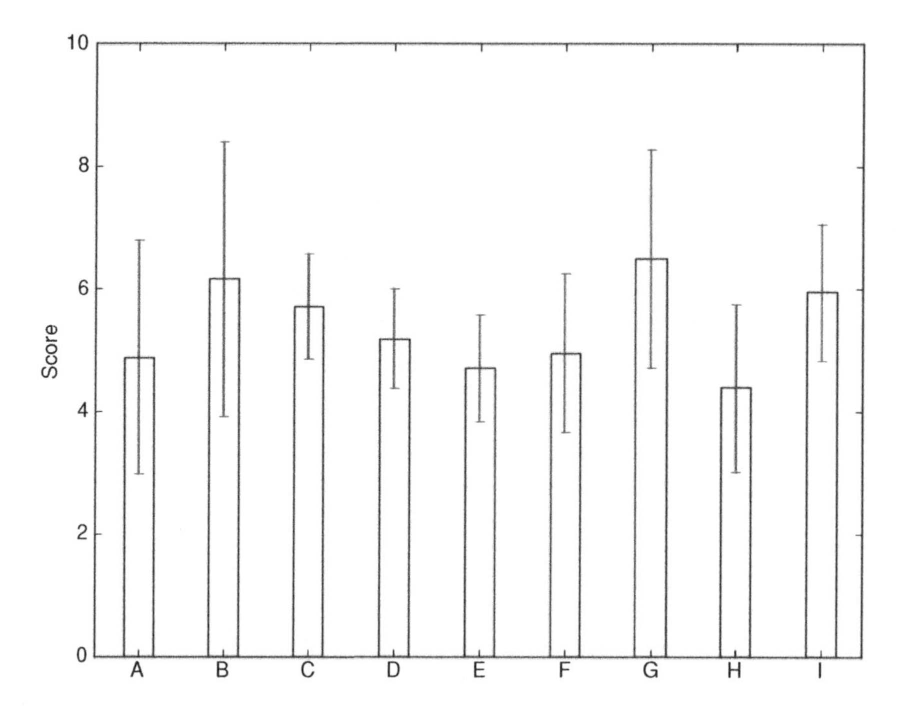

Figure 3.3 Mean and standard deviation for all assessors for the attribute acidity. *The means are plotted as columns and the standard deviations are superimposed both upwards and downwards on the corresponding column.*

instance be due to an outlier, but the outlier itself will not be visible. For this purpose, box plots or histograms are better suited because they highlight individual outlier values and provide more information about the distribution of the data. Examples of these types of plots are presented in Figure 3.4 and Figure 3.5, respectively. The Box plot (see Chapter 11) is presented for the same data as used in Figure 3.3 while the histogram is presented for one assessor only (attribute acidity). A disadvantage with histograms is that large numbers of them are required to allow for a complete overview over all assessors and attributes. A total of I^*K histograms of this type are available meaning that for the apple data described above, a total of $9^*20 = 180$ histograms are needed for a full evaluation. Therefore, histograms should mainly be used for cases where some assessors appear to be very different from the rest.

In order to obtain further details about the raw data, one may use so-called line plots, a way of visualising data that is highly relevant and has a shape familiar to the sensory scientist. These line plots show the product profiles averaged across assessors and replicates with the scores of individual assessors superimposed in the same plot. An example of this plot is given in Figure 3.6 for one of the products. The horizontal axis represents the attributes and the vertical axis represents the intensity score values. Attribute average scores are connected by straight lines giving the plot a characteristic pattern that visualises the main properties of the product. The intensity scores of each assessor are marked with the same

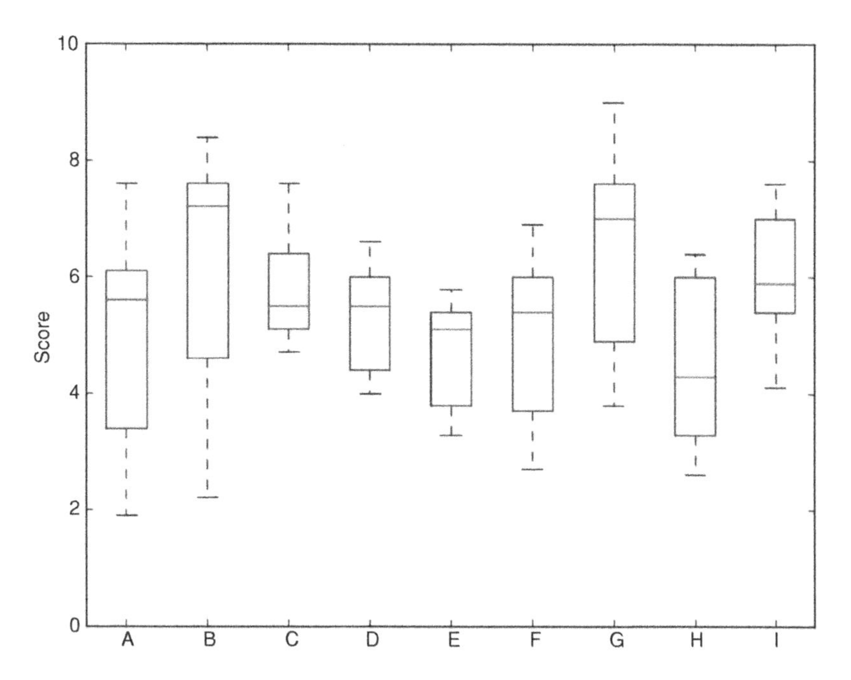

Figure 3.4 Box plot for the same data as presented in Figure 3.3.

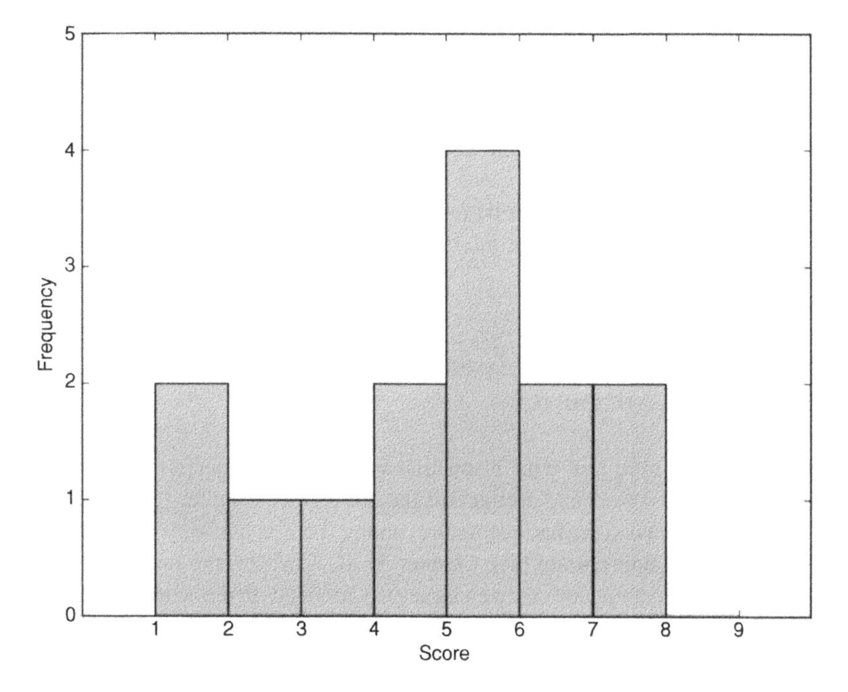

Figure 3.5 ***Histogram.*** *The histogram depicts the scores for one assessor and the attribute acidity.*

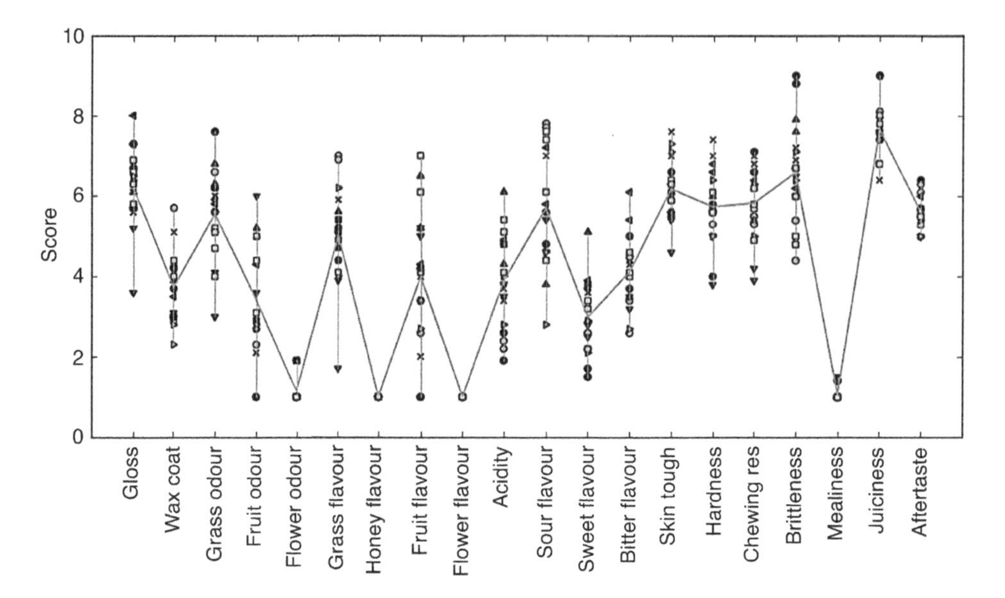

Figure 3.6 Line plot. *The plot shows the panel average and the individual scores for all assessors for one of the products (Pink Lady). Replicates for one assessor are shown with the same symbol.*

symbol or colour. In this way it is possible to spot large replicate differences. Furthermore, for every attribute a vertical line is added, indicating the range of scores that the panel has used. Ideally, this line should be as short as possible. As can be seen, there are quite large individual differences for most of he attributes. Similar plots will also be discussed further below in Chapter 3.8.

All the analyses in this section can be used regardless of replicate structure and design used in the sensory test, but should be interpreted accordingly.

3.3 Mixed Model ANOVA for Assessing the Importance of the Sensory Attributes

The first step after simple raw data inspection is typically to perform a simple mixed model ANOVA with assessors and products as the effects (see Chapter 13). If replicates are random replicates with no systematic structure among them, a two-way mixed model with interaction is the most appropriate (see Chapter 5). If, however, the replicate structure is systematic, it may be advantageous to use a three-way model with replicate as the third effect as discussed in Chapter 5. In such a case, all two-way interactions should be incorporated and the replicate effect treated as random. Mixed model ANOVA can also be conducted for unbalanced data, i.e. with different number of replicates for each assessor and product combination. With missing cells in the data set, i.e. with missing product and assessor combinations, one should, however, be more careful since the definition of interactions is not obvious in such cases.

The main reason for using this method here is for possibly eliminating unimportant attributes. If an attribute has no significant main product effects or interactions, it can be safely claimed that the panel as a whole is not able to distinguish between the products for this attribute. For most purposes nonsignificant attributes have no influence on the results and should be eliminated from further investigation. It may, however, be useful to take a quick look at one of the individual plots below, for instance the p*MSE plot, even for the nonsignificant attributes since one or a few very 'unreliable' assessors may possibly be the reason for the lack of significance. If nonsignificant attributes are considered important for the product profile and it is believed that differences between the products are really present, further training is required.

Two-way mixed model ANOVA results for the apple data are provided in Figure 3.7. Alternatively, the same results can be displayed in a table showing numbers instead of bars. Figure 3.7 contains three plots, one for the assessor effects, one for the product effects and one for assessor*product interaction. Each plot contains a number of bars representing the F-values for the tests (see Chapter 13). In addition, each bar may be coloured according to a set of significance levels: white ($p \geq 0.05$), grey ($0.01 \leq p < 0.05$), darker grey ($0.001 \leq p < 0.01$) and black ($p < 0.001$). The colouring of the bars may be used as a visual enhancement for easier identification of significance level.

As can be seen from Figure 3.7, the product effect is significant for all attributes, most of them at a very low level of p. This means that the panel as a group discriminates well between the samples for all attributes. There are, however, quite large differences between those attributes that distinguish the most (sweet flavour) and the least (honey flavour). Most attributes also have a significant assessor effect and some of them have significant interactions. These results are clear indications that the assessors use the scale differently. They have a different average and they also sometimes score the differences between the products differently (Chapter 4).

As always when using ANOVA in sensory analysis, one should keep an eye on the distribution of the residuals. One needs to check whether their distribution is reasonably normal and whether it contains any outliers. These and other aspects related to model assumptions are discussed in Chapters 13 and 15. In the example presented in Chapter 15 (Figure 15.7), the residuals have a distribution which is close to normal.

A couple of alternative ANOVA models proposed by Brockhoff and Skovgaard (1994) and Brockhoff and Sommer (2008) are also useful for attribute evaluation of this type. These models will be discussed later in this chapter (Section 3.5.3), Chapters 4 and 5.

3.4 Overall Assessment of Assessor Differences Using All Variables Simultaneously

After elimination of unimportant variables by the use of mixed model ANOVA, a logical next step is to carry out a multivariate analysis of all remaining attributes in order to obtain a simultaneous overview of the relation between attributes. Several such methods are discussed in Chapters 14 and 17, but here we will focus mainly on a methodology often

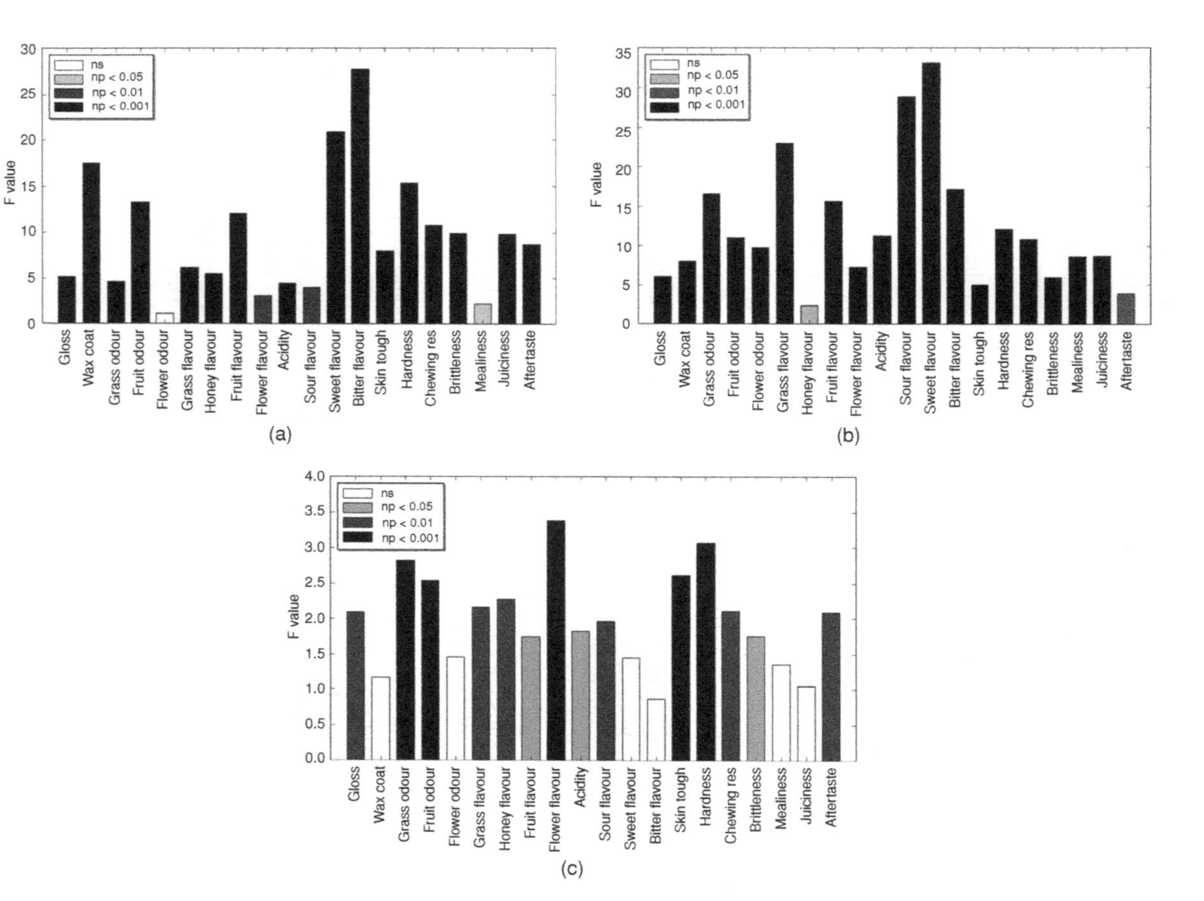

Figure 3.7 ***a, b,c. ANOVA results.*** *The F-values are presented for the a) assessor effect, b) the product effect and c) the interactions for all attributes measured.*

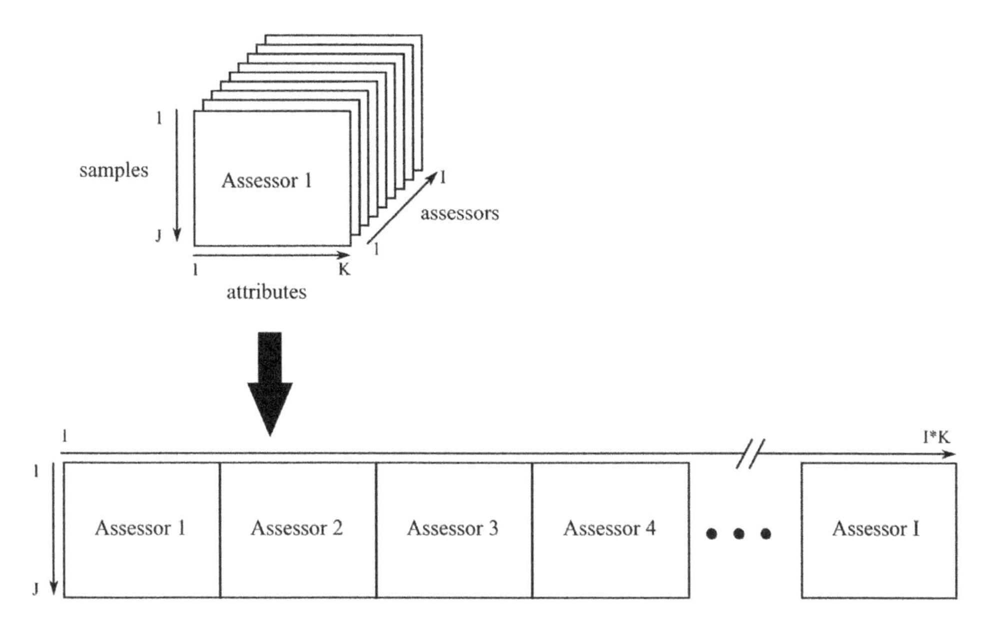

Figure 3.8 ***Unfolding.*** *The process of unfolding for the Tucker-1 method. The different "sheets" for all assessors of the three-way structure are put adjacent to each other.*

referred to as Tucker-1 (Consensus PCA (CPCA)). This method is based on simply using PCA on the horizontally unfolded matrix obtained from the three-way sensory data structure as shown in Figure 3.8. This new unfolded matrix then consists of J rows, where each row represents the average across replicates, and I^*K columns. If replicates are incorporated, the number of rows is extended to J^*R. The Tucker-1 methodology has shown to be useful for detecting assessors that differ from the rest and attributes that are affected by poorly performing assessors (Dahl *et al.*, 2008). The Tucker-1 method is used mainly as a screening tool that provides a quick overview of assessor performance and determines how to proceed with more detailed studies later.

An underlying assumption behind the use of this methodology is that it is possible to describe most of the information adequately by a relatively small number of common principal components or latent variables, for instance 2 or 3. Experience has shown that this is often possible in practice. The Tucker-1 model for the unfolded data can be written as

$$\mathbf{Y_i} = \mathbf{TP_i}^{\mathbf{T}} + \mathbf{E_i} \qquad (3.1)$$

where $\mathbf{Y_i}$ represents the data matrix of the i'th assessor, $\mathbf{T}$ is the matrix of common (consensus) principal component scores and the $\mathbf{P_i}^{\mathbf{T}}$ matrix represents the loadings for assessor number j. As can be seen, the scores $\mathbf{T}$ are common for all assessors, only the loadings are different. Hence, one can present the results of this analysis in two types of plots: a common scores plot displaying how the samples relate to each other and a loadings plot displaying the individual loadings for all assessor-variable combinations. Consider for instance the apple data set described earlier. Here the sensory panel consists of $I = 9$

assessors describing $J = 7$ samples using $K = 20$ sensory attributes. The common scores plot then shows 7 samples whereas the loadings plot will contain $9 * 20 = 180$ loading values. This is a very high number, but as shown below, this plot is generally used only with some of the points highlighted.

3.4.1 Correlation Loadings Plots

A useful way of presenting information about importance of the different variables is to use a correlation loadings plots as described in Chapter 14. For assessor and attribute combinations with low signal to noise ratio or for situations where the assessors interpret the attribute differently, the corresponding correlation loadings will generally be located closer to the centre (origin) than others, making it possible to identify variables with weak relation to the general underlying data structure.

Dahl *et al.* (2008) proposed to generate K identical plots highlighting correlation loadings of one single attribute at a time using a label to indicate the assessors. This series of plots represents a practical way of identifying assessor-attribute combinations which are different from the others in some way. The common scores mentioned above play a minor role in this context.

Some examples of Tucker-1 correlation loadings plots are presented in Figure 3.9 (of attributes sweetness and wax coating highlighted). The two ellipses (the plot can also be expanded in the vertical direction to show circles instead of ellipses, but for interpretation purposes this has no effect on the results) in the plot represent 50 % and 100 % explained variance. For a well-trained and calibrated panel the correlation loadings of the attribute under investigation should be close to the outer ellipse with all panellists clustered closely together.

It can be seen that the two correlation loadings plots are quite different. The first plot shows that for the attribute sweetness all assessors agree well with one another and that sweetness is strongly related to the first consensus dimension. For the attribute wax coating, one can see that all assessors are somewhat scattered over the upper part of the correlation loadings plot. This indicates that there is less agreement across the panel than it is the case for attribute sweetness. Three assessors clearly lie inside the inner circle (i.e. less than 50 % explained variance) and four others are only just outside of it. Only assessors A and I can be claimed to have high explained variance in their data for this attribute. Since most of the assessors are located in the upper part of the plot, this attributes seems to be mostly related to component 2.

The Tucker-1 analysis is usually done without standardising the variables (Chapter 14), but if it is obvious that some of the assessors or some of the attributes have very different variance than the rest, it may be advantageous to standardise prior to analysis (see Chapters 5 and 14). Note that since correlation loadings are used for this purpose, standardisation has less influence on the plot than when regular loadings are used.

It is important to mention that 3 is the minimum number of objects required for PCA to give a two-dimensional plot and that in this case all loadings will be located at the outer ellipse (or circle). Even with a slightly larger number of samples, the correlation loadings plot will have all or most of the assessor/attribute combinations close to the outer ellipse (or circle). In such cases it is important to also consider other plots, such as for instance the profile plot (see below, Section 3.7.2). In general, the Tucker-1 plot is most useful for a higher number of samples (for instance for 7 and higher).

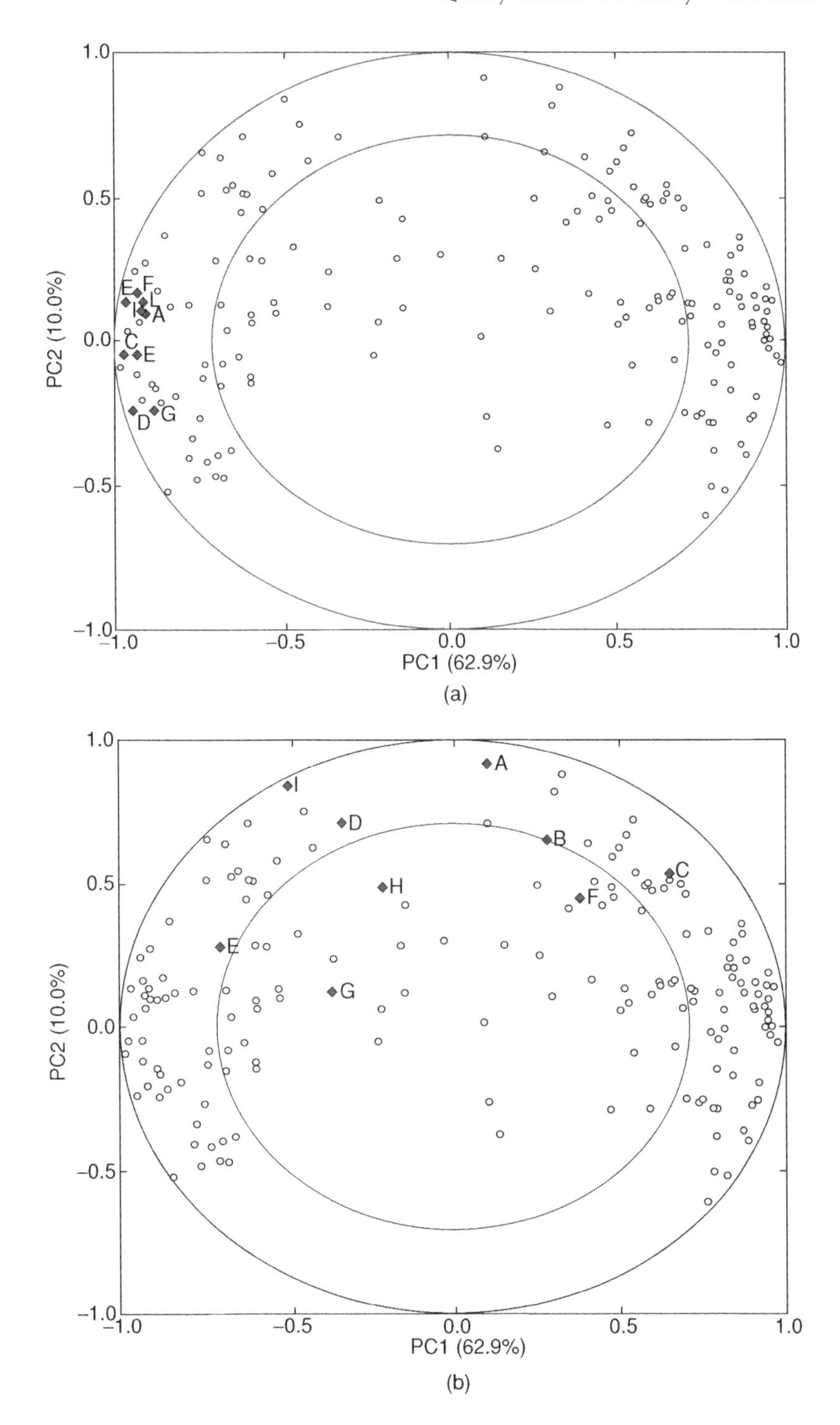

Figure 3.9 a, b. Tucker-1 correlation loading plots. *All assessors are highlighted for the attribute a) sweetness and b) wax coating.*

3.4.2 The Manhattan Plot

The Manhattan plot (Dahl *et al.*, 2008), Chapter 14) is another important option for providing information about differences between assessors. These plots are easy to look at and provide useful information for screening purposes. Manhattan plots can be used to visualise the explained variances for each assessor-attribute combination for Tucker-1 or for individual PCA analyses for the different attributes (the latter used for illustration below). The method is most useful when the explained variances for the different attributes are presented in separate plots.

Two examples of this are provided in Figure 3.10, one for attribute sweet taste and one for aftertaste. The horizontal axis represents the assessors and the vertical axis represents the principal components from the individual PCAs. The cumulative explained variances for each principal component are visualised with colours varying from black to white, representing 0 % and 100 % explained variance respectively. This implies that the colour in the plot is generally light when few components are needed.

The Manhattan plot for attribute sweet taste shows an example of good panel performance. Already for principal component one, high explained variance is achieved for each assessor, with six out of 9 assessors having more than 90 % variance explained. Assessor F has the lowest explained variance (only 70.4 %), which is considerably lower than for assessor B which has 97.4 % explained variance. Nevertheless, differences between the assessors appear to be relatively small in this case. The attribute aftertaste shows a situation with quite different performance across assessors. In general, one can see that the colours are much darker in this plot indicating that there is less systematic variance for the attribute. There are obvious differences between assessors G and H, with assessor G having an explained variance of 89.2 % for principal component one and assessor H only 0.1 %. After 4 principal components assessor G and H have 98.1 % and 5.0 % explained variance, respectively, indicating that there are great differences in the relation between attribute aftertaste and the remaining attributes for the different assessors.

The Manhattan plots provide no information on how the assessors rank the samples. This means that assessors may have explained variances that look very similar, but their sample ranking can be completely different. Other plots need to be consulted to obtain this information (for instance the profile plot described below in Section 3.7.2).

3.5 Methods for Detecting Differences in Use of the Scale

Although different use of scale (see point 1 in the Table 3.1) is not necessarily directly related to the quality of the assessors, it is still of interest to detect it and also correct for it. First of all, computing panel averages across assessors becomes a more natural and reliable practice when done on data with the same scale. Secondly, it has been shown by Romano *et al.* (2008) and Næs (1990) that correcting for scaling differences can reduce the interaction effects. Another important aspect is that this type of information may be useful for obtaining better calibrated panels in the future.

In general, different use of the scale has two important aspects; the level effect which corresponds to the positioning of the mean score along the intensity scale and the range effect which corresponds to the variability of the scores (see Figure 3.1). How to correct for scaling differences will be discussed in Chapter 4.

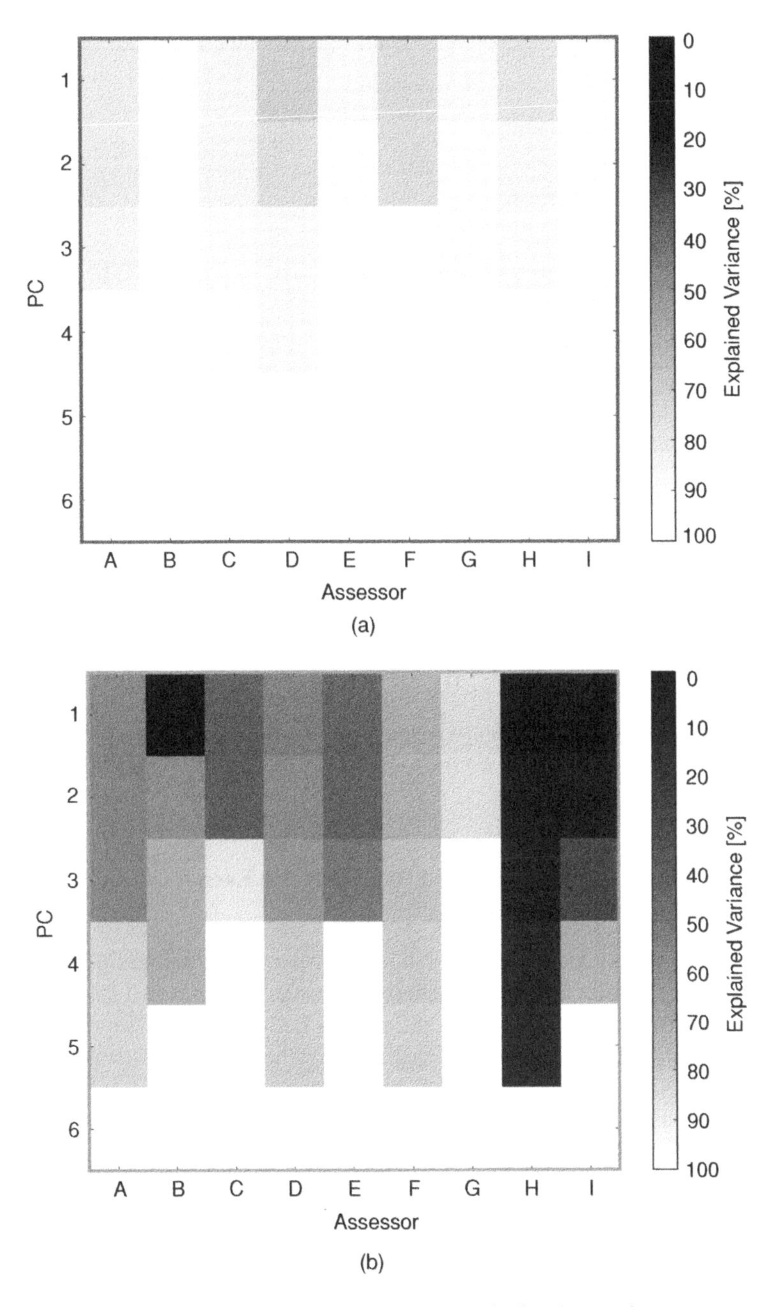

***Figure 3.10** **a, b. Manhattan plot.** Manhattan plot made for the attribute a) sweetness and b) aftertaste. The plot shows the explained variance for each assessor.*

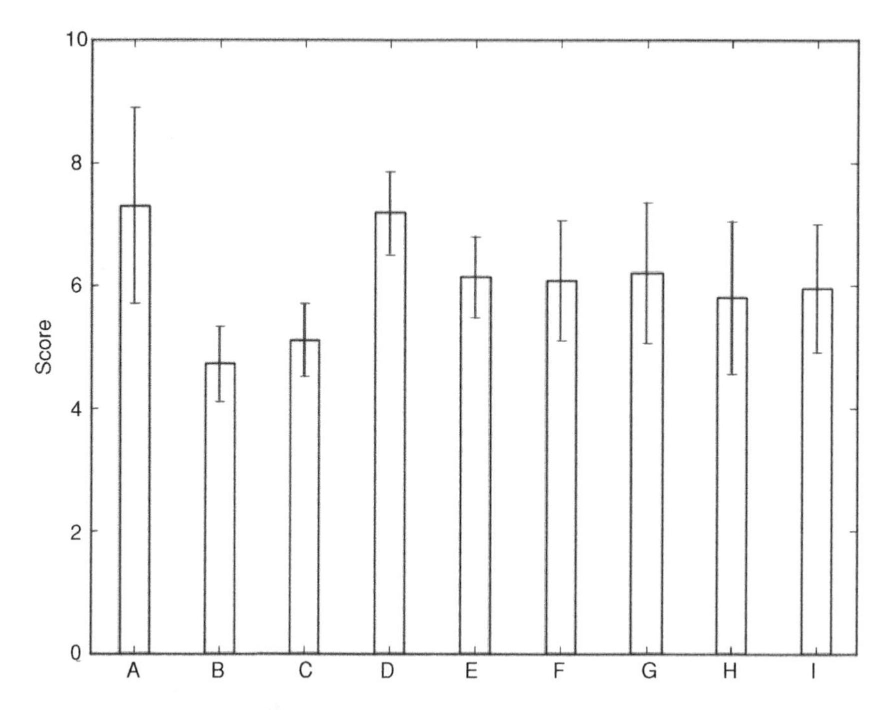

Figure 3.11 ***Scaling differences.*** *The plot shows differences in use of the scale. Both means and standard deviations (presented both upwards and downwards) are presented for each individual for the attribute brittleness.*

3.5.1 Computing Means and Standard Deviations for Each Assessor and Attribute Combination

The simplest way of detecting level and range differences is by the use of regular means and standard deviations for all attribute-combinations across all tested samples and their replicates. Figure 3.11 visualises scaling differences between the assessors by displaying the means and standard deviations for all assessors for the attribute brittleness of the apple data set.

As can be seen, assessor A, has a relatively high mean for brittleness and also a relatively large standard deviation. Assessor B, on the other hand, has a low mean and a small standard deviation. The difference in level between the two is as large as 2.6 units, which is relatively much considering that a 9 unit scale was used to describe the products. The standard deviation of assessor A is more than 2 and a half times larger than that of assessor B.

3.5.2 Minimising Differences between Assessor Scores for the Same Product and Attribute

Another method which is more sophisticated is the method proposed in Næs (1990) based on a mathematical technique developed by Ten Berge (1977). The average scores (over replicates) y_{ijk} for assessor i, product j and attribute k, are all multiplied by a constant c_{ik}

(independent of the samples j) which are optimised such that the scores for all products and attributes become as similar as possible. The optimisation is done for each attribute (k) separately using the criterion

$$Criterion(k) = \sum_{j=1}^{J} \sum_{i<i'} (c_{ik} y_{ijk} - c_{i'k} y_{i'jk})^2 \tag{3.2}$$

In order to obtain a nontrivial solution, a restriction is put on the c-values (see Næs (1990) for details). The algorithm for finding the solution is simple and based only on simple eigenvector calculations.

The advantage of this method as opposed to the method above is that it also offers diagnostic procedures. For instance, one can look at the individual contributions to the sum to identify assessors that do not fit to the rest. Negative constants c_{ik} are indications of attribute-assessor combinations for which the order of samples is reversed, thus indicating a confusion or very large random noise.

3.5.3 The Brockhoff Assessor Model

In the so-called Brockhoff assessor model, the interactions between product and assessor are modelled as a multiplicative effect (see Brockhoff and Skovgaard, 1994). The model takes into account different use of scale and also different individual variances. For each attribute k, the model can be written as

$$y_{ijkr} = \alpha_{ik} + \gamma_{ik} v_{jk} + \varepsilon_{ijkr}, \tag{3.3}$$

where $\varepsilon_{ijkr} \sim N(0, \sigma_{ik}^2)$, the v_j's represent the different product effects, the α_i's represent assessor levels and the γ_i's represent the individual scaling values. Assessors with large γ_i-value use a larger portion of the scale than the average assessor and assessor with a small value use a smaller range than the average. Systematic interaction effects not accounted for by the multiplicative terms are now in the error terms ε_{ijkr}. It is easy to show (see also Romano *et al.*, 2008) that the model can be presented in such a way that it also contains the main effects for products, but for the purpose of assessor evaluation, this aspect has no influence. The estimates of the parameters in the model are obtained using an iterative algorithm described in the original paper by Brockhoff and Skovgaard (1994).

The systematic scaling differences, both the level differences represented by the α's and the range differences estimated by the γ's, can be plotted for each attribute separately, for each assessor separately or for all of them at the same time (see also Romano *et al.*, 2008). Note that also this method can be used to reveal different/opposite ranking of the samples.

Which of the methods above that is the best has not been evaluated properly, but the latter two are generally preferable since they can also be used to reveal opposite ranking.

3.6 Comparing the Assessors' Ability to Detect Differences between the Products

A very important aspect of an assessor is his/her ability to detect differences between the products (see points 3 and 4 in Table 3.1). If we are only considering attributes that are significant (see above, Section 3.3) for the panel considered as a whole, an individual assessor who is not able to detect differences between the products requires attention from the panel leader. Possible reasons for a lack of significance are low sensitivity to an attribute, poor sensory memory or simply a misunderstanding.

The easiest way of investigating this type of individual differences is to compare, for each assessor independently, the differences between the product averages with a measure of the random error based on differences between the sensory replicates. The most standard way of doing this is to use one-way ANOVA (see Chapter 13) based on the model

$$y_{ijkr} = \mu_{ik} + \beta_{ijk} + \varepsilon_{ijkr} \tag{3.4}$$

for each assessor (i) and attribute (k) combination. Here the β 's represent the product effects and the ε's represent the random error or noise (see also Chapter 13). The analysis of this model is based on computing the so-called F-ratio, which is basically a ratio of sums of squares of the differences between average values of the products and the variance of the replicates (often called *MSE*, Chapter 13). The F-values are computed for each assessor and attribute and then compared to a table of so-called critical values (see Chapters 11, 13). Alternatively, one computes the p-value for the observed F-values and compares them with a reference, for instance the significance level 0.05. Assessor-attribute combinations with values of p higher than for instance 0.05 are considered to be nonsignificant. But as usual, there is no sharp limit as to where to distinguish the good performers from the poor.

The F-values or the *MSE* values can be presented in various ways, but here we focus on the plotting procedures described in Næs and Solheim (1991) and Lea *et al.* (1995). The *F's* and *MSE's* can for instance be plotted as bars and sorted according to either assessor or attribute. An example of this is given in Figure 3.12 where it can be seen that there are quite large differences between the assessors and between the attributes. There are two horizontal lines in the plot, representing F-values at 5 % (lower line) and 1 % (upper line) significance level. In Figure 3.12 (F-value plot sorted by assessor) one can observe for instance that assessor B has more than twice as many attributes significant at 5 % level as compared to assessor E, which means that assessor B is much more reliable as an assessor than assessor E for detecting product differences.

Another possibility is the *p-MSE* plot (Lea *et al.*, 1995) where the p-values are plotted along the vertical axis vs. *MSE* values along the horizontal axis. The best assessors are those with a low p-value and a corresponding low *MSE* value, meaning that they discriminate between the samples and at the same time are able to reproduce their own scores reliably. Note, however, that the p-value can be low even though the *MSE* value is relatively high if the assessor uses a large portion of the scale.

In p**MSE* plots one can plot p-values and *MSE* values of all possible assessor-attribute combinations and highlight the assessor or the attribute one is interested in. Again, each point in the p**MSE* plot represents an assessor-attribute combination, meaning that for the data set considered here, a total of 180 points are displayed (9 assessors * 20 attributes).

An example of a p**MSE* plot is given in Figure 3.13 for the attribute aftertaste. As can be seen, assessor A has a high p value (0.385) indicating that this assessor has the lowest ability to detect differences between samples. Furthermore, assessor A has the highest *MSE* value (0.76) of all assessors in the panel. Assessor A is clearly the weakest performer of all assessors regarding attribute aftertaste. In comparison, assessor B has a much lower p-value (0.002) and a MSE value equal to 0.04. Assessor B is clearly able to discriminate significantly between products and reproduce his/her own results with high precision.

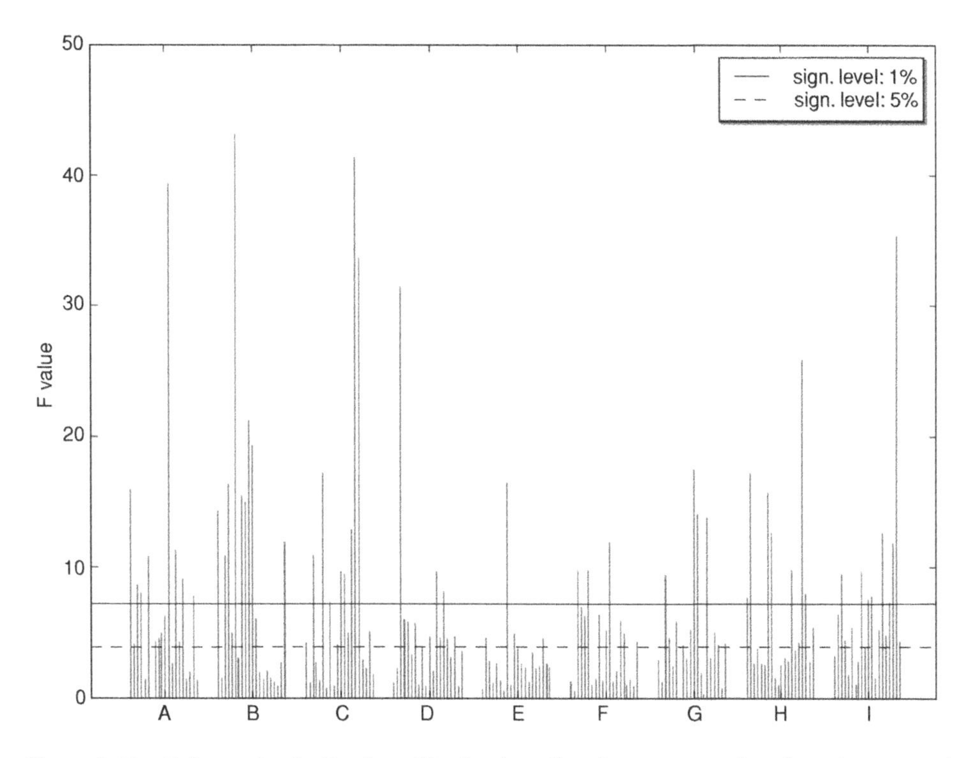

Figure 3.12 ***Column plot for* F*-values.*** *The* F*-values (for all assessors and attributes) are sorted by assessor and by attributes within assessor.*

Note that two or more replicates are required for doing this test. The number of replicates is, however, allowed to vary between the products and between the assessors. It is known that F-tests can be sensitive to large outliers, so for this situation it is important to check for outliers prior to analysis. A possible way of doing this is to compute the residuals from the model and plot them for instance using a q-q plot as indicated in Chapter 15.

If the number of attributes or assessors is high, one can decide to look only at plots which contain the combinations that are identified as different in the Tucker-1 plot. It should also be mentioned that none of the ANOVA based plots provides information about sample ranking. For instance, two assessors may appear to have very similar F, p and MSE values, but their sample ranking can be different from each other. For detecting this problem, other plots need to be consulted, for instance correlation plots.

If another replicate structure is used than the standard random one, there may be a need for incorporating an extra replicate effect in the model in the same way as shown in Chapter 5. Likewise, if other designs are used for setting up the sensory experiment, the additional systematic effects should be incorporated in the ANOVA model.

3.7 Relations between Individual Assessor Ratings and the Panel Average

This whole sub-chapter is devoted to point 2 in Table 3.1, namely reproducibility of sensory panels as defined at the end of the book (in: Nomeclature, symbols and abbreviations). The

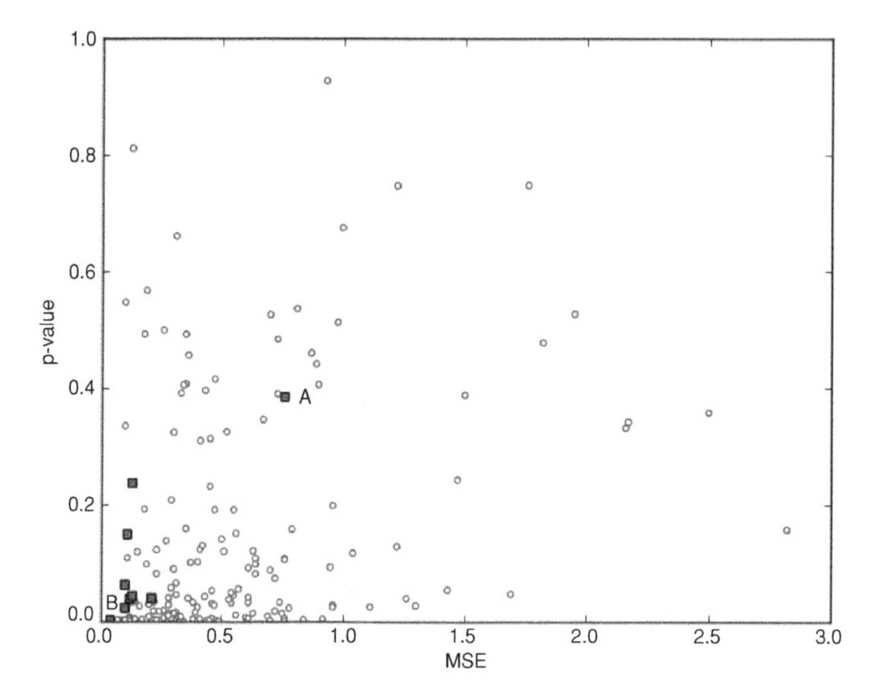

Figure 3.13** **p-MSE plot. *The points for the attribute aftertaste are highlighted.*

plots presented here are useful also for alternative replicate structures since they are usually based on averages over replicates.

3.7.1 Correlation Plots

A simple plot for investigating similarities between assessors is the correlation plot, where sample scores of one specific assessor are plotted versus the average scores for the whole panel. The plot is particularly useful for detecting misunderstandings, differences in ranking as well as outliers. Moreover, the plot visualises whether an assessor rates the tested samples over, under or at the same level as the panel. The plot can be generated for each replicate, but usually one will average over replicates before plotting. One can also compute the correlation coefficient for further interpretation and comparison of the performance of the assessors (see Chapter 11). With a perfectly calibrated panel, all assessor-panel combinations, represented by points, would lie on a 45 degrees straight line starting from the origin. In practice, however, this will never be the case.

When plotting all possible assessor-panel-combinations in the same plot, a total of I^*J points will be shown for each attribute. The points for the assessor under investigation may then be highlighted by using another colour or symbol.

Examples of correlation plots are given in Figure 3.14 for the attribute flower flavour. As mentioned above, the two plots (for two different assessors) are identical except for the highlighting of the points corresponding to the assessor under investigation. Assessor C (Figure 3.14a) has an average score of 1.0 for all samples except for sample Kanzi (1.2)

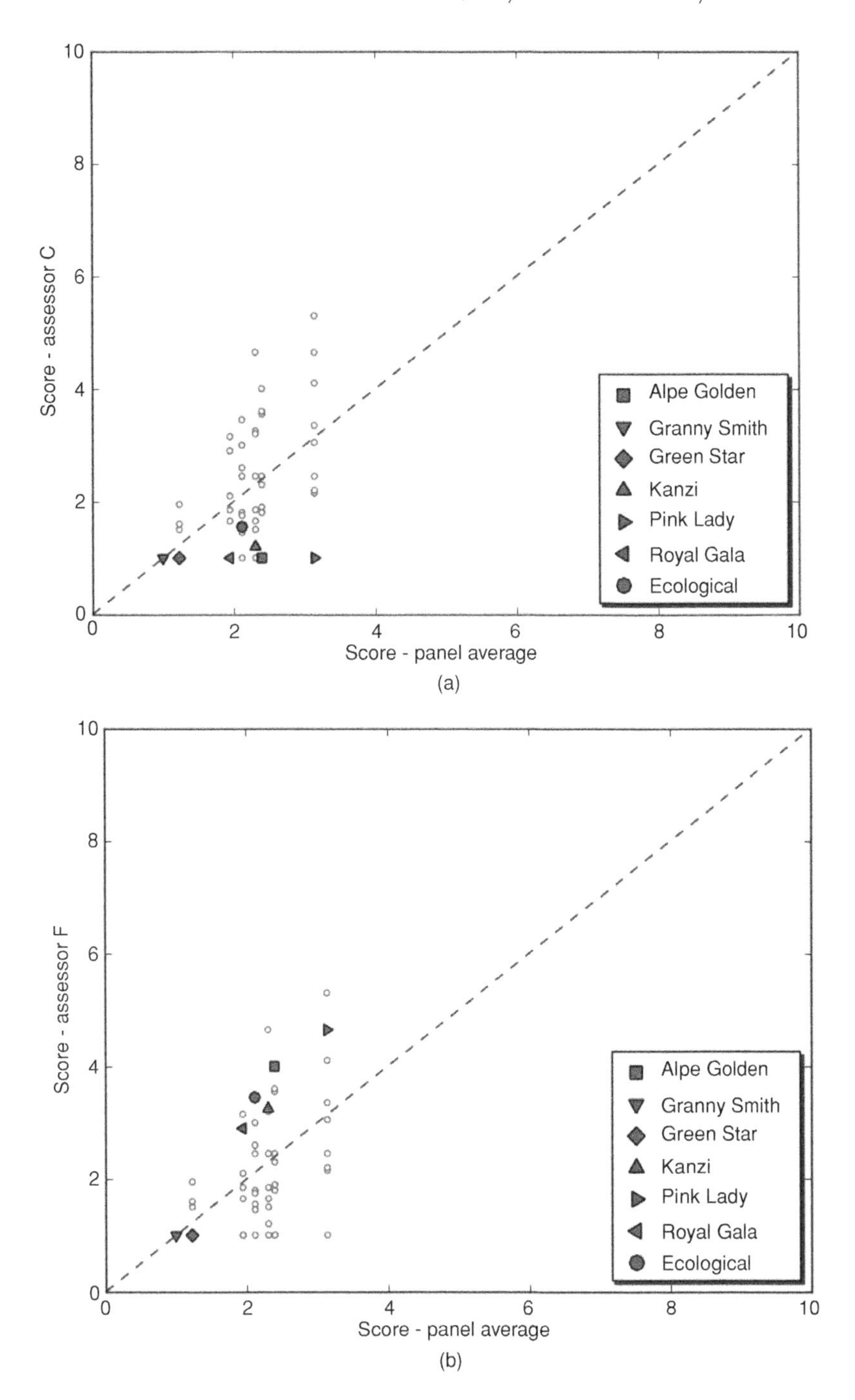

***Figure 3.14** **a, b. Correlation plot for assessors C an F.** The plots show the relation between the individual assessors and the panel average with one of the assessors highlighted in each of the plots. In both cases, the attribute flower flavour is considered.*

and Ecological (1.55). In contrast, the consensus/average sample score varies from 1.0 to 3.14 with only Granny Smith having an average of 1.0. This indicates that assessor C did not discriminate well between the samples regarding the attribute flower flavour and that he or she differed strongly from the consensus. Assessor F in Figure 3.14b used the intensity scale better and had average samples scores that varied between 1.0 and 4.65. The sample ranking for assessor F is almost identical to the consensus ranking.

3.7.2 Profile Plot

Another simple plot that allows for comparison of individual assessors against the panel consensus is the profile plot. One profile plot is generated for each attribute, which results in a total of K separate plots. Here the horizontal axis represents the tested samples and the vertical axis represents the intensity scores averaged over replicates. The samples along the horizontal axis are sorted according to consensus mean score. The scores of each assessor are linked by a straight line resulting in $I + 1$ lines including the line representing the consensus scores. The profile plot is best suited when the number of samples is low, i.e. from two up to about 10 samples. With a number of samples higher than that, one should consider using the eggshell plot as described below.

Figure 3.15 shows an example of a profile plot for the attribute sour taste. In the figure it can be seen that sample Pink Lady and Granny Smith have the lowest and highest consensus

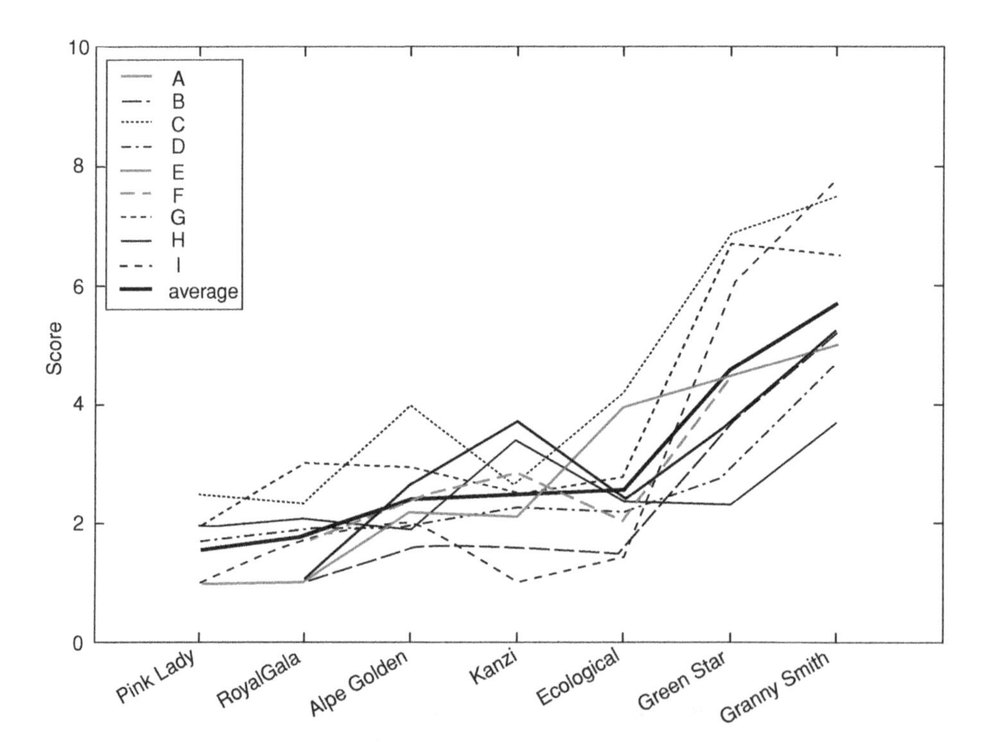

Figure 3.15 Profile plots for attribute sour taste. *The samples are sorted according to average score (solid line in the middle) and all the other lines represent he different assessors.*

intensity, respectively. In general it seems that assessors agree reasonably well regarding the ranking although there are also some clear individual differences.

It should be noted that the profile plot can also be used to visualise level and range differences between the assessors. In Figure 3.15 one can see that assessor C generally scores higher than the consensus on all samples. For assessor I the opposite is the case, with its scores constantly being lower than the consensus.

3.7.3 Eggshell Plot

This eggshell plot (Hirst and Næs, 1994; Næs *et al.*, 1994) is an alternative to the correlation plot and profile plot. It is only based on the ranking information computed from the original intensity scores, but it can also be used for rank data directly. The eggshell plot visualises consensus ranks vs. individual cumulative ranks thus highlighting ranking differences among assessors. The plot has its name from the resemblance to an eggshell and is most useful if the number of samples is higher than 10. But as is illustrated here it can also be useful for fewer samples.

The eggshell plot shares many similarities with the profile plot. As in the profile plot, each assessor is represented by its own line. Contrary to the profile plot, however, the consensus line in the eggshell plot is always located as a baseline below the other lines. Another similarity is that the samples along the horizontal axis are sorted by intensity of the averages (in this case the average ranks) in increasing order from left to right.

The information shown in the eggshell plot is based on first computing the consensus rank (the rank of the average of the ranks). The samples are then sorted according to the consensus as mentioned above. The cumulative (accumulation according to sorted order) ranks for the samples are then computed for each of the assessors. Thereafter, a constant depending on consensus sample rank is subtracted from the cumulative ranks in order to give the plot the characteristic shape that merits the name 'eggshell plot'.

An example of an eggshell plot is given in Figure 3.16 for the attribute yellow colour. As can be seen, there is a good agreement between the assessors since most assessors follow the consensus line quite closely, at least for the samples with the highest intensity. The only exception is assessor B who differs somewhat from the other assessors.

The area between the lower consensus line and one of the assessor lines can be interpreted as '1-correlation' between the two (Hirst and Næs, 1994). In other words, the area is inversely proportional to the correlation between the consensus and the assessor. The correlation used here is the Spearman rank correlation (Kendall, 1948). It is also possible to superimpose lines on the plot which correspond to given values of Spearman correlation. Further interpretations of the eggshell plot can be found in Hirst and Næs (1994).

The eggshell plot is usually made from averages over replicates, which means that it can be used for any of the replicate structures discussed in Chapter 5.

3.8 Individual Line Plots for Detailed Inspection of Assessors

An example of an individual line plot is presented in Figure 3.17 for sample Royal Gala and assessor A. The plot contains the grand average across the whole panel (dashed line) and the sample average across replicates for the particular assessor. In this way it is possible to compare the average of the assessor under investigation against the panel consensus

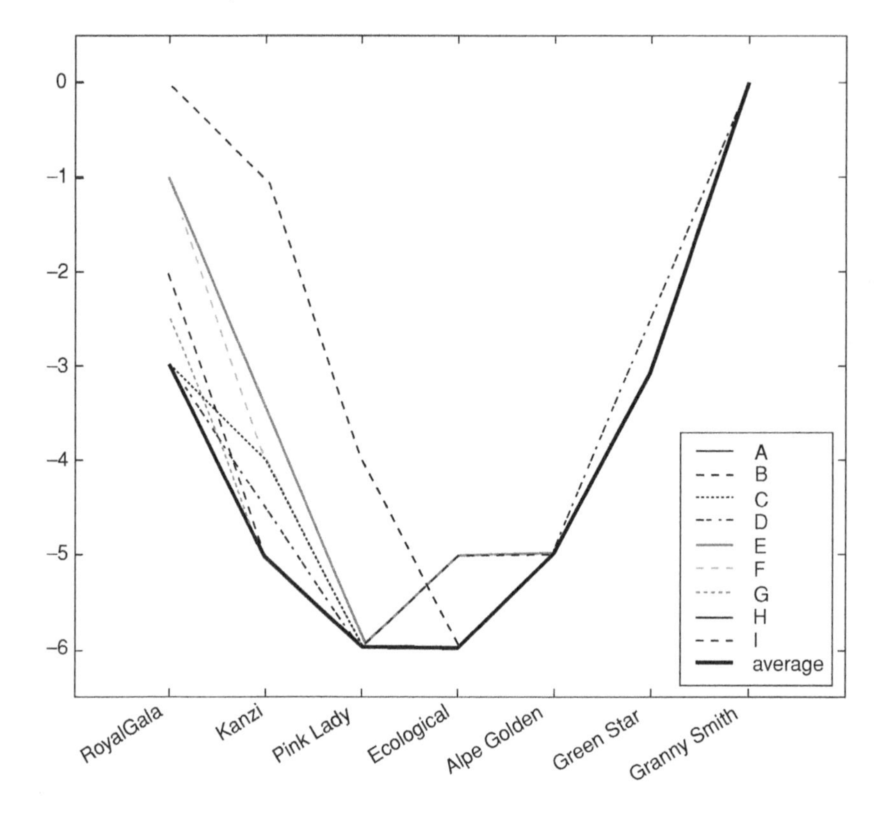

Figure 3.16 Eggshell plot for the attribute yellow colour. *The lower line corresponds to the consensus ranking and each of the other lines corresponds to individual assessors. The closer to the consensus an assessor is, the more similar he/she is to the panel average.*

revealing level effects across all attributes. In addition the vertical lines for each attribute indicate the range used for the replicates. Ideally, this range should be as small as possible. As can be seen, the performance for attribute acidity is extremely poor with exactly 4 score units difference between the two replicates. For the attributes sweet flavour and hardness, assessor A is systematically different from the grand average.

The main problem with this approach is that for a full study I^*J plots have to be considered. But as mentioned above, main attention at this point will typically be on a small number of problematic attributes discovered by the use of other plots. Since this plot is essentially a raw data plot, it can be used for any type of design as long as the results are interpreted accordingly.

3.9 Miscellaneous Methods

3.9.1 Consonance Analysis

The analysis of consonance (Dijksterhuis, 1995), i.e. consistency or agreement, is a useful method that seems to have been used relatively little in the literature. The method, which is

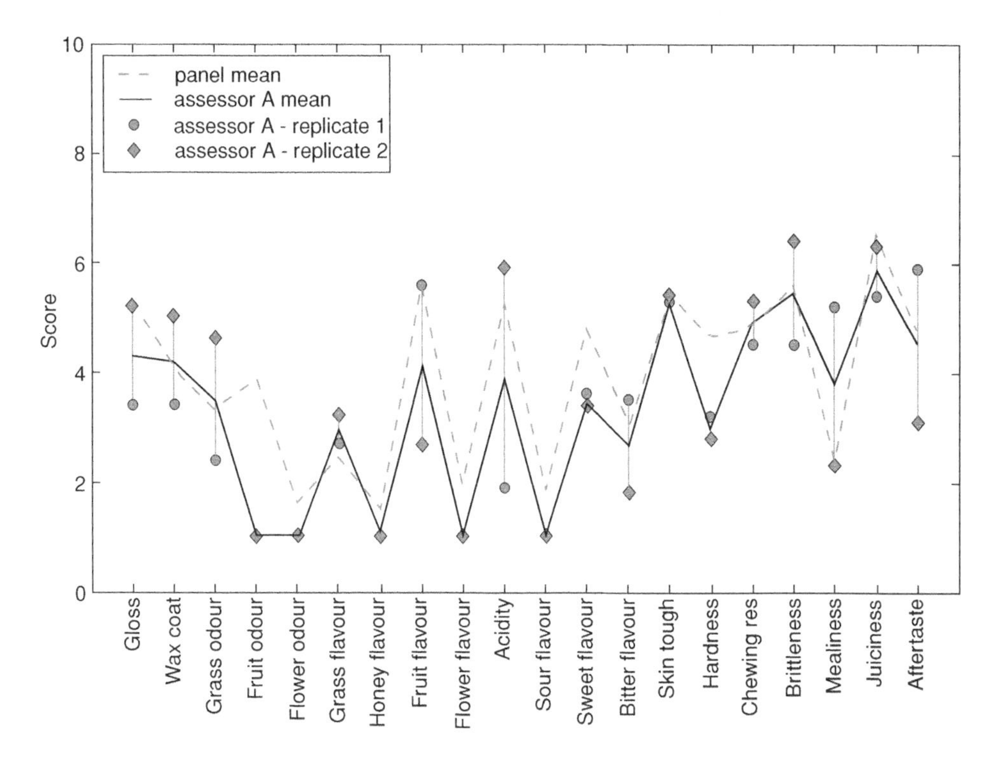

Figure 3.17 Individual line plots for single samples. *The plot shows the panel average (dotted line), the individual average (solid line) and the individual replicates for assessor A for one of the samples.*

simple and easy to interpret, is based on simply organising a data matrix for each attribute with the samples as rows and assessors and columns. A PCA is then used and the results are presented in a scores plot that shows how the samples are related and a loadings plot which displays how the assessors are related. The explained variance along principal component one should for a well trained panel be close to 100 %. Dijksterhuis (1995) also defines a performance index for the panel named 'Variance accounted for' or simply VAF. The VAF is computed by dividing the first eigenvalue by the sum of the remaining eigenvalues. This means that VAF can take values between 0 (no agreement at all) and 1 (perfect agreement) between the assessors.

Regarding datasets with multiple replicates, one can decide to either present the replicates as additional samples or as additional assessors. Looking at similarity between the replicates is an important tool for detecting those assessors that perform poorly. For both ways of organising the data, correlation loadings plots may be used instead of regular loadings plots.

3.9.2 Procrustes Analysis

Procrustes analysis (Chapter 17) has for a long time been used for detecting and correcting for individual differences in sensory analysis (Arnold and Williams, 1987; Dijksterhuis,

1996; Gower and Dijksterhuis, 2004). The model behind Procrustes analysis is based on allowing for individual differences in translation, isotropic scaling and rotation (Dijksterhuis, 1996; Gower, 1975). The former of these effects can be handled by subtracting the averages for each assessor and attribute. The other two effects are optimised by making the assessor scores as similar to each other as possible in a least squares sense. The assessors that are very different after rotation and scaling can be considered outliers. For the purpose of detecting abnormalities it is also possible to consider individual fit to the criterion used for optimisation.

For free choice profiling (FCP, Arnold and Williams, 1987) where one leaves the definition of attributes to each panellist, the method is very useful. In such cases it may be necessary, if the assessors use different number of attributes, to fill in some of the assessor matrices with columns of 0s in order to make them comparable in dimensions. Also for napping data (see Risvik *et al.*, 1994; Pages, 2005), the method can be useful for reducing individual differences.

3.9.3 Three-way Component Methods

Since sensory data are essentially three-way data tables (see Chapter 2), three-way component methods are sometimes used for quality control of sensory panels (Dahl *et al.*, 2008; Brockhoff *et al.*, 1996; Bro *et al.*, 2008). This has been demonstrated already above in Section 3.4 by the use of the simplest of these approaches, the Tucker-1 method. The other methods can, however, also sometimes be useful although they are somewhat more complex and not so often used for routine quality control.

3.9.4 Clustering of Individual Scores Matrices

The Tucker-1 method descried above is used for detecting outlying assessors*attribute combinations. In Dahl and Næs (2009), the Tucker-1 methods is used in a slightly different way, namely as a basis for cluster analysis (Chapter 16). The procedure is based on fuzzy clustering of the individual assessor matrices using the distance to a low-dimensional Tucker-1 model as the distance criterion. This means that assessors which do not fit to the joint Tucker-1 model will be determined to be outliers. More details of this method will be given in Chapter 16.

Another and closely related method is the one proposed in Dahl and Næs (2004), which is based on the Procrustes distance between the individual assessors matrices instead of the Tucker-1 model. This means that it allows for isotropic scaling and rotation differences before computation of the distances between individuals. The method is like the Tucker-1 clustering method suitable for detecting individual who at an overall level are the most different from the rest.

Both these techniques are most suitable at an early stage of the investigation for determining which assessors that are most different from the rest. In Dahl and Næs (2004) it is shown how the consensus from the panel is improved when assessors who were detected as most different are eliminated from the panel (see Figure 6.3). Related methods can be found in Qannari *et al.* (2000) and Qannari and Meyners (2001).

References

Amerine, M.A., Pangborn, R.M., Roessler, E.B. (1965). *Principles of Sensory Evaluation of Food*. New York: Academic Press.

Arnold, G.M., Williams, A.A. (1987). The use of generalised procrustes techniques in sensory analysis. In J.R. Piggott (ed.), *Statistical Procedures in Food Research*. London: Elsevier Science Publishers, 244–53.

Bro, R., Qannari, E.M, Kiers, H.A., Næs, T., Frøst, M.B. (2008). Multi-way models for sensory profiling data. *J. Chemometrics* 22, 36–45.

Brockhoff, P.B. (2003b). Statistical testing of individual differences in sensory profiling. *Food Quality and Preference* 14(5–6), 425–34.

Brockhoff P.B., Skovgaard I. (1994) Modelling individual differences between assessors in sensory evaluation, *Food Quality and Preference* 5, 215–24.

Brockoff, P, Hirst, D., Næs, T. (1996). In T. Næs, E. Risvik (eds), *Multivariate Analysis of Data in Sensory Science*. Amsterdam: Elsevier.

Brockhoff, P.B., Sommer, N.A. (2008). Accounting for scaling differences in sensory profile data. Proceedings of 10th European Symposium on Statistical Methods for the Food Industry, pp. 283–90. Louvain-La-Neuve, Belgium.

Dahl, T. and Næs, T. (2004). Outlier and groups detection in sensory panels using hierarchical cluster analysis with the Procrustes distance. *Food Quality and Preference* 15, 195–208.

Dahl, T., Tomic, O. Wold, J.P., Næs, T (2008). Some new tools for visualising multi-way sensory data. *Food Quality and Preference* 19, 103–13.

Dijksterhus, G. (1995), Assessing panel consonance, *Food Quality and Preference* 6, 7–14.

Dijksterhuis, G. (1996). Procrustes analysis in sensory research: In T. Næs, E. Risvik (eds), *Multivariate Analysis of Data in Sensory Science*, pp. 185–217. Amsterdam: Elsevier.

Gower, J.C. (1975), Generalized Procrustes analysis, *Psychometrica* 45(1), 3–24.

Gower, J.C., Dijksterhuis, G. (2004). *Procrustes Problems*. Oxford: Oxford University Press.

Hirst, D. d Næs, T. (1994). A graphical technique for assessing differences among a set of rankings. *Journal of Chemometrics* 8, 81–93.

Hunter, E.A., McEwan, J.A. (1998). Evaluation of an international ring trial for sensory profiling of hard cheese. *Food Quality and Preference* 9(5), 343–54.

Kendall, M.G. (1948). *Rank Correlation Methods*. London: Charles Griffin.

Lê, S., Pagès, J., Husson, F. (2008). Methodology for the comparison of sensory profiles provided by several panels: application to a cross-cultural study. *Food Quality and Preference* 19(2), 179–84.

Lea, P., Rødbotten, M., Næs, T. (1995). Measuring validity in sensory analysis. *Food Quality and Preference* 6(4), 321–6.

McEwan, J.A. (1999). Comparison of sensory panels. A ring trial. *Food Quality and Preference* 10(3), 161–71.

Næs T. (1990) Handling individual differences between assessors in sensory profiling, *Food Quality and Preference* 2, 187–99.

Næs, T., Solheim, R. (1991). Detection and interpretation of variation within and between assessors in sensory profiling. *Journal of Sensory Studies* 6, 159–77.

Næs, T., Hirst, D., Baardseth, P. (1994). Using cumulative ranks to detect individual differences in sensory profiling. *Journal of Sensory Studies* 9, 87–99.

Pages, J (2005). Collection and analysis of perceived product inter-distances using multiple factor analysis: Applications to the study of 10 white wines from the Loire Valley. *Food Quality and Preference* 16, 642–9.

Qannari, E.M, Wakeling, I., Courcoux, P., MacFie, H.J.H. (2000). Defining the underlying sensory dimensions. *Food Quality and Preference* 11, 151–4.

Qannari E.M., Meyners M. (2001) Identifying assessor differences in weighting the underlying sensory dimensions, *Journal of Sensory Studies* 16, 505–15.

Risvik, E., McEwan, J.A., Colwell, J.S., Rogers, R., Lyon, D.H. (1994). Projective mapping: A tool for sensory analysis and consumer research. *Food Quality and Preference* 5, 263–9.

Romano R., Brockhoff P.B., Hersleth M., Tomic O., Næs T. (2008) Correcting for different use of the scale and the need for further analysis of individual differences in sensory analysis, *Food Quality and Preference* 19(2), 197–209.

Ten Berge, J.M.F. (1977). Orthogonal Procrustes rotation for two or more matrices. *Psychometrica*, 42, 267–76.

Tomic, O., Nilsen, A., Martens, M. and Næs, T. (2007). Visualization of sensory profiling data for performance monitoring. *LWT – Food Science and Technology* 40, 262–9.

Tomic, O, Luciano, G., Nilsen, A., Hyldig, G., Lorensen, K., Næs, K. (2010a). Analysing sensory panel performance in a proficiency test using the PanelCheck software. *European Food Research and Technology* 230(3), 497.

4

Correction Methods and Other Remedies for Improving Sensory Profile Data

When controlling panel performance as discussed in Chapter 3, one will often detect assessors that are different from or less reliable than the rest. An important question is what should be done with the data from these assessors before further analysis. Here we will discuss various ways of responding to this question with focus on pre-processing of data before averaging. There are a number of aspects that need to be taken into account, both related to different use of the scale and differences in the assessors' ability to repeat their own scores for the same sample. Different methods based on standardisation and weighted averages will be discussed and compared. Towards the end of the chapter some attention will also be given to pre-processing of data before three-way analysis. The concept of external validation will also be discussed as a way of validating and determining what is the best pre-processing strategy. Some conceptual problems of external validation will also be highlighted.

The methodology presented here is based on the methods presented and discussed in Chapter 15 and Chapter 17. There is a strong relation between this chapter and Chapter 3.

4.1 Introduction

Individual differences related to misunderstandings, different use of the attributes, individual differences in noise level and different use of the scale can often be reduced by more extensive and more targeted panel training. Despite these possibilities, one will usually also like to use the data at hand, for instance for computing panel averages, even if some of the assessors are different from or less reliable than the rest.

Misunderstandings and different use of attributes are generally difficult to handle by data analysis, except of course for the obvious remedy of just eliminating the problematic observations and possibly trying to replace them by for instance some type of missing value imputation (see Chapter 17). Procrustes analysis (see Chapter 17) is another method which can sometimes be useful for the purpose (Arnold and Williams, 1987).

In the present chapter, we will give main attention of attribute-wise pre-processing methods before ANOVA or averaging over assessors. Focus will be on individual differences in reliability and in use of the scale. If differences in reliability are detected, it may be natural to weigh down the most unreliable assessors when computing the panel average. Using information about replicate variability or information from assessor comparisons is usually better for this purpose than using residuals from ANOVA since the latter uses more information about the structure of the data and may therefore lead to overoptimistic results in subsequent analyses. Differences in use of the scale are not directly related to the reliability of the assessors, but it may be important to correct for them before averaging in order to make the data comparable across assessors.

The structure of the data considered here is the same as used in Chapter 3 and depicted in Figure 2.1 (J samples, I assessors and K attributes). In some cases also external data of the same samples (J samples) will be used. The replicate structure assumed here is the standard one of random replicates discussed in Chapter 5, but the same ideas carry over also to more complex replicate structures.

4.2 Correcting for Different Use of the Scale

Scaling differences have two different aspects, differences in mean/position and in range/span (see Figure 3.1 in Chapter 3). Usually both phenomena are considered as nuisance factors with no positive impact on further analysis. It has been shown by Næs (1990) and by Romano *et al.* (2008) that much of the interaction effect can be attributed to different use of the scale. It may therefore sometimes be advantageous to pre-process the data before further analysis, being it ANOVA or multivariate analyses based on panel averages (see Chapter 6).

Mean differences are very simple to correct for by just subtracting the mean for each assessor and attribute combination. Note that when applying ANOVA methods, correction for position is done automatically if an assessor effect is incorporated in the model (Lea *et al.*, 1997).

The simplest method for reducing differences in range is to divide scores for each assessor and attribute combination by the corresponding standard deviation, i.e. standard deviation obtained across samples and replicates. This means that for each combination of attribute (k) and assessor (i) one computes the quantity

$$y_{ijkr}(new) = y_{ijkr}(old)/s(y_{ik}) \tag{4.1}$$

where the standard deviation is taken over all samples and replicates. A more sophisticated approach is to compute the standard deviation taking the replicate structure into account

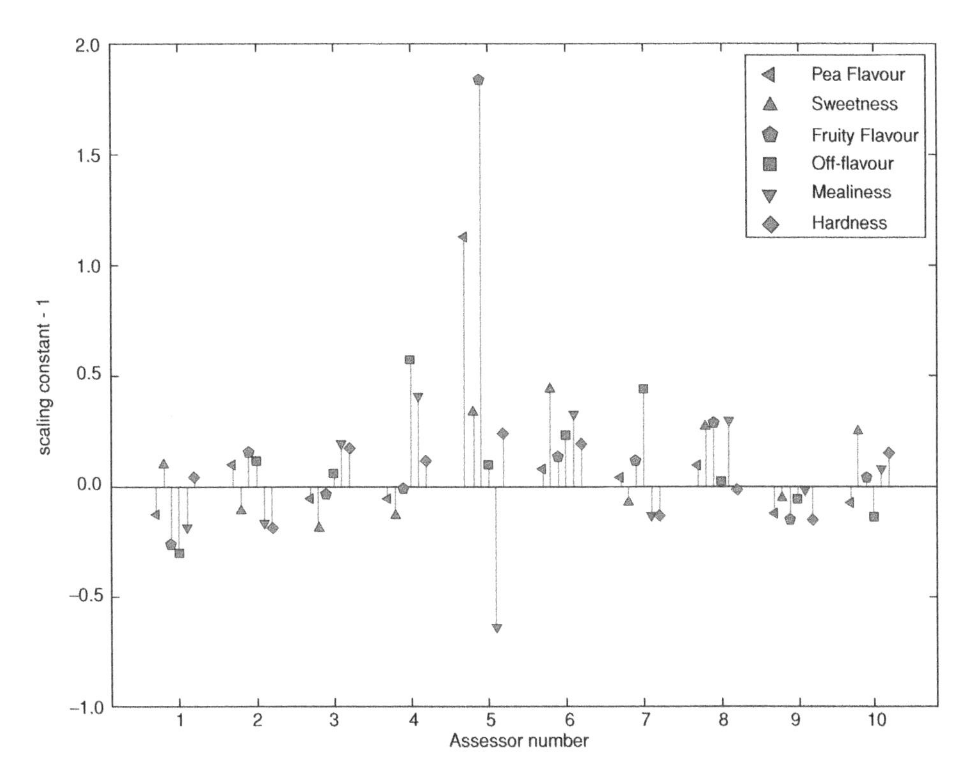

Figure 4.1 ***Scaling constants.*** *Graphical illustration of the values* c_{ik} *found by minimising the expression in equation (3.2). The data set is from descriptive analysis of peas. Positive values of* $c_{ik}-1$ *indicate the use of the scale to be smaller than the average of the panel. Negative values are interpreted the opposite way. The illustration is taken from Næs (1990). Adapted with permission from Elsevier.*

(see Chapter 13). This method is simple, intuitive and has shown to provide reasonable results. Usually, the data will be centred for each individual prior to the transform.

A somewhat better, but also slightly more complex method, is to scale the different assessors with the purpose of making them as similar as possible in a least squares (LS) sense (see Næs, 1990). The criterion to minimise for this procedure is the same as used in Equation (3.2). After estimation of the correction constants, new data are obtained by multiplying the original score values by the estimated values. Figure 4.1 shows the scaling constants for a concrete example with 6 attributes and 10 assessors. The scaling constants minus 1 are presented in order to highlight the differences between those values which are higher and those that are lower than 1. A scaling constant equalling 1 corresponds to no change. The values are sorted according to assessor. As can be seen, in some cases there are relatively large differences in scaling constants between the assessors. For assessor number 6, all values are positive which means that this assessor systematically uses a smaller part of he scale than the rest and therefore must be multiplied by a constant larger than 1 to be similar to the rest. For assessor number 9 the situation is the opposite. Assessor number 5 varies strongly in his scaling according to which attribute is used.

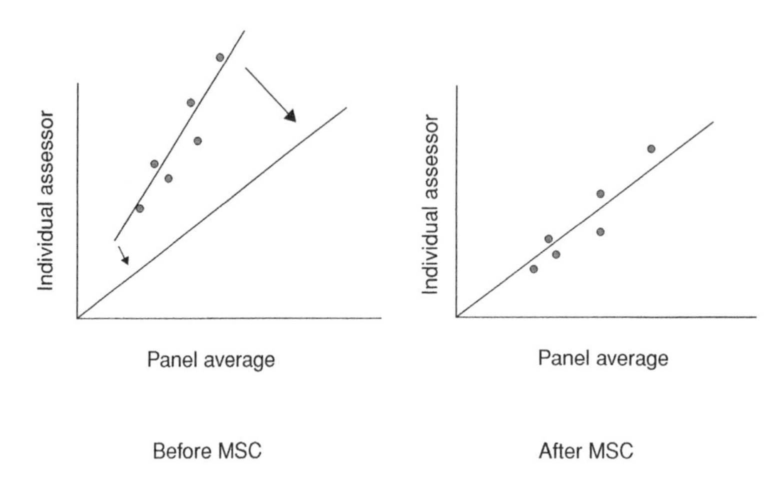

Figure 4.2 Illustration of the MSC transform. *The scores from one single assessor are plotted vs. the panel average. The straight 45 degrees line corresponds to perfect fit. In the panel to the left the assessor's scores are systematically higher than the panel average scores. The assessor also uses a larger span to distinguish the products. In the panel to the right both the level and range effect have been corrected for by subtraction and division according to formula (4.3).*

Martens (2008) has proposed to use a method inspired by the MSC transform (Geladi *et al.*, 1985) for spectroscopic data for the purpose of reducing scaling differences. The method is based on the model

$$y_{ijk} = a_{ik} + b_{ik}\bar{y}_{jk} + \varepsilon_{ijk} \tag{4.2}$$

where a_{ik} and b_{ik} are scaling constants to be determined from the data and the $\bar{y}_{jk}$ is the panel average (across assessors and replicates) for product j and attribute k. The y on the left-hand side of the equation is here averaged over replicates. The a's in this equation correspond to differences in position for the different individuals and the b's to differences in range/span. The constants are estimated for each individual (i) and attribute (k) separately using regular LS regression. After estimation, a new data set is obtained by computing

$$y_{ijkr}(new) = (y_{ijkr}(old) - \hat{a}_{ik})/\hat{b}_{ik} \tag{4.3}$$

In the same way as for the method above, the MSC approach also provides information, using the residuals from model (4.2), about individual fit and possible opposite order of the scores. Note that the MSC transform can also be used the other way round with the average on the left-hand side in the Equation (4.2) and the correction done accordingly.

Yet another method was proposed in Romano *et al.* (2008). The technique is based on fitting the Brockhoff model (Brockhoff and Skovgaard, 1994) to the data and then computing the differences between the raw data and the predicted values from the model (the residuals) thus obtaining corrected data to be used for further analysis.

Comparisons of the different methods were made in Romano *et al.* (2008) and in Tomic *et al.* (2010b). In the former of these papers, only the first two methods and the latter were

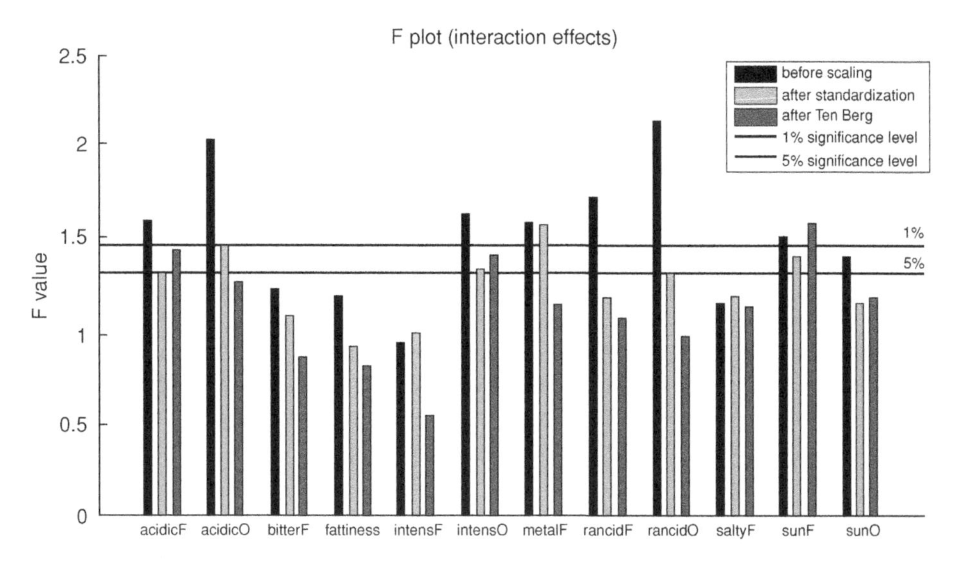

Figure 4.3 ***Significances and interactions.*** *Illustration of F-values for the interaction effect in 2-way ANOVA mixed models. Illustration taken from Romano et al. (2008). Reprinted from Food Quality and Preference, 19, Romano et al, Correcting for different use of the scale and the need for further analysis of individual difference in sensory analysis, 197-209, 2008, with permission from Elsevier.*

tested with focus on the ability to reduce the interactions in ANOVA tests. It was found using a data set from sensory differences of cheese, that all methods performed in a similar way. It seems that the choice of method is more related to personal preferences than to strong performance differences.

It can be argued that if the scaling differences play an important part of the interactions, the standard ANOVA assumptions are more realistic after correction than before correction. This is discussed further in Chapter 5, Section 5.2.2 where also an alternative modelling approach for handling scaling differences in ANOVA is presented.

It should be mentioned that after elimination of level differences for each assessor and attribute, the data set will be centred around zero. It may therefore be advantageous for interpretation purposes to add the grand average across the whole data set obtained before centring to all corrected profiles before interpreting them. This has, however, no effect for any of the analysis methods considered in this book, since mean centring is very often conducted anyway.

4.3 Computing Improved Panel Averages

Unreliable assessors, in the sense that they provide very noisy data or are very different from the rest, can reduce the validity of sensory panel averages. In such cases it may be tempting to reduce their influence by giving them less weight when computing the panel average.

4.3.1 The General Weighted Average Formula

When computing a weighted average, it will here always be assumed that the data have been scaled by one of the methods above prior to averaging. The general formula for the weighted average profile for sample j can then be written as

$$\bar{y}_{jk} = \sum_{i=1}^{I} \hat{y}_{ijk} w_{ik} / \sum_{i=1}^{I} w_{ik} \tag{4.4}$$

where the weights w are computed according to one of the criteria to be discussed below. Here the $\hat{y}$ represents raw data (here averaged over replicates) or raw data that have been smoothed before averaging as will be discussed below. Note that setting one of the assessor weights equal to 0 corresponds to eliminating that assessor from the new weighted average. Note also the resemblance between this and weighted least squares fitting based on estimated weights (iteratively reweighted regression (Weisberg, 1985).

4.3.2 Different Types of Weights

If the design of the study is used when computing the weights, one may end up with averages that are overfitted with the respect to hypothesis and tests to be considered later. For computing the weight, we therefore recommend the use either relations to the average profile or differences between replicate measurements.

A possible weight to use is the correlation coefficient (or the square or it, Chapter 11) between an individual assessor and the average profile, possibly with the actual assessor eliminated. Assessors with a weak relation to the average will thus be weighted down. Another possibility is to use 1 divided by a function of the standard deviation of the random noise. A standard deviation of the noise can be obtained by pooling standard deviations of the replicated measurements for each sample/assessor/attribute combination or by using residuals from a Tucker-1 modelling (see Chapter 14) of the raw data. The Tucker-1 method has the advantage that one does not use explicitly the information about which observations that are replicates.

4.3.3 Raw Data or Filtered Data in the Average?

When computing weighted averages according to (4.4) one must decide whether to use the original data as input or data that have been obtained by a 'filtering' procedure. Such a filtering can for instance be obtained using a Tucker-1 model with a limited number of components in it, treating the rest of the components as noise (see e.g. Tomic *et al.* (2010b)). The filtered data are then obtained directly by multiplying the joint scores with the individual loadings, leaving the noise out from further analysis. A possibility is to use for instance four components, which in most cases is higher than the number of significant components. This filtering method has performed well in applications, but no clear conclusion can be drawn regarding what is the overall best practice.

4.3.4 STATIS

STATIS (see e.g. Schlich *et al.*, 1996; Chapter 17) is another method which can be used for obtaining a weighted average. The method automatically weighs the contributions of the different matrices by using the so-called RV coefficients. This is an index measuring the degree of correspondence between data matrices and in this way can be used to put less emphasis on the individuals which are most different from the rest. STATIS ends up with a consensus matrix representing a type of average for all the assessors. This matrix can be used like regular average data matrices as described in for instance Chapter 6.

4.4 Pre-processing of Data for Three-Way Analysis

In the same way as above, there are two aspects that need to be taken into account when trying to improve three-way analysis (see Chapter 17 and Figure 2.1) by the use of pre-processing. The first is to correct for scale differences and the other is to take the reliability of the assessors into account. The first of these aspects is handled in exactly the same manner as above.

The general structure of many three-way methods is the minimisation of the criterion

$$Q = \sum_{i=1}^{I} \|\mathbf{Y}_i\mathbf{H}_i - \text{model}_i\|^2 \tag{4.5}$$

where $\| \|$ is the Frobenius norm, the summation is over assessors, $\mathbf{H}_i$ is a matrix which transforms the Y-data, and 'model' indicates the type of model or restriction assumed. The index i is added to the model in order to indicate that the model can depend on the individual i. All the Tucker models, GPA, GCA etc. (see Chapter 17) can be formulated within this framework if we allow for incorporating $\mathbf{H}_i = \mathbf{I}$ as an option. As above, the $\mathbf{Y}$ data sets are usually mean centred and scaled. Sometimes restrictions are used in the minimisation.

The simplest way of incorporating weights w_i in this criterion is to replace (4.5) by

$$Q = \sum_{i=1}^{I} w_i \|\mathbf{Y}_i\mathbf{H}_i - \text{model}_i\|^2 \tag{4.6}$$

The idea is that the assessors with the least reliability get the least weight in the sum. As before, the information content in w_i should not depend on the design of the experiment. Solutions to this criterion can for most of the methods mentioned be obtained quite easily. For methods like Tucker-1 and Tucker-2 the solution can be found by simply multiplying the $\mathbf{Y}_i$ by the square root of w_i and finding the solution as before. It is also for some of the methods possible to weigh differently for the different variables, but this is not discussed further here. For more information about scaling and how it can be solved in three-way analysis, we refer to Smilde *et al.* (2004).

References

Arnold, G.M., Williams, A.A. (1987). The use of generalised procrustes techniques in sensory analysis. In J.R. Piggott (ed.), *Statistical Procedures in Food Research*. London: Elsevier Science Publishers, 244–53.

Brockhoff P.B., Skovgaard I. (1994) Modelling individual differences between assessors in sensory evaluation, *Food Quality and Preference* 5, 215–24.

Geladi, P, MacDougall, D., Martens, H. (1985). Linearisation and scatter-correction for near-infrared reflectance spectra of meat. *Applied Spectroscopy* 39, 491–500.

Lea, P. Næs, T., Rødbotten, M. (1997). *Analysis of Variance of Sensory Data*. New York: John Wiley & Sons, Inc.

Martens, H. (2008). Personal communication regarding the use of the MSC in sensory analysis.

Næs T. (1990) Handling individual differences between assessors in sensory profiling, *Food Quality and Preference* 2, 187–99.

Romano R., Brockhoff P.B., Hersleth M., Tomic O., Næs T. (2008) Correcting for different use of the scale and the need for further analysis of individual differences in sensory analysis, *Food Quality and Preference* 19(2), 197–209.

Schlich, P. (1996). Defining and validating assessor compromises about product distances and attribute correlations. In T. Næs, E. Risvik (eds), *Multivariate Analysis of Data in Sensory Science*. Amsterdam: Elsevier.

Smilde. A., Bro, R., Geladi, P. (2004). *Multi-Way Analysis*. Chichester: John Wiley & Sons, Ltd.

Tomic, O. Forde, C. Delahunty, C. Martens, H., Næs, T. (2010b). Assessor weighting in sensory profiling using the Tucker-1 method. (in prep).

Weisberg, S. (1985). *Applied Linear Regression*. New York: John Wiley & Sons, Inc.

5

Detecting and Studying Sensory Differences and Similarities between Products

This chapter handles the use of sensory data for detecting and quantifying differences between products. It will be demonstrated how sensory analysis companied with ANOVA and PCA can be used for testing for sensory differences and for interpretation. In the first part of the chapter focus will be on the use of ANOVA in two-way models with assessor and product effects. Then, the model is extended to accommodate several experimental factors. Alternative replicate structures will be described together with how to model them. For the analysis of multivariate sensory responses, two different strategies based on combining PCA and ANOVA will be discussed: One of these methods uses PCA on the sensory data first and then relates the PCA scores to the experimental design (PC-ANOVA). The other one uses the two methods in the opposite order (ASCA).

The present chapter is based on the theory described in Chapter 13 and Chapter 14 and is strongly related to Chapter 8 which handles conjoint analysis.

5.1 Introduction

One of the most important ways of using sensory analysis is for detecting and understanding differences between food products. This is the case both for product development situations and for quality control. In the following we will describe typical problems and useful methodology for such situations. We will here focus on techniques related to sensory profiling and leave the discussion of so-called difference tests to Chapter 7 (see Chapter 2 for a description of the difference).

In order to make valid conclusions, it is important to select products carefully. For instance, if one is interested in the effect of the salt content on the sensory perception of bread, one must include products with a fully relevant range of salt level in the study. Likewise, if a sausage producer is interested in how a specific product compares to similar products in the market, he/she must include products that span the entire variation in the study. In order to achieve this, one needs both field specific as well as statistical knowledge.

Focus in this chapter will be specifically on how experimental design factors affect the sensory properties of the products. It is thus a strong resemblance between the methods discussed here and the conjoint methods treated in Chapter 8 for consumer data. The data and focus are different, but much of the same underlying methodology can be applied. Relations between sensory data on one side and other external data such as chemical and spectroscopic measurements on the other will be handled in Chapter 6.

We will first focus on analysis of models with only one sensory attribute. Afterwards, these tests will be extended and generalised to the more general situation with several sensory response variables. Main attention will be given to random replicates, but some attention will also be given to alternative replicate strategies.

5.2 Analysing Sensory Profile Data: Univariate Case

The methods treated in this chapter all concentrate on one attribute at a time. For simplicity we have therefore left the attribute index k out of the notation.

5.2.1 Regular Two-Way Mixed ANOVA

As an example we will here consider the pea study reported in Bech *et al.* (1997) and Smith *et al.* (2003). In this study 16 ($J = 16$) frozen pea samples from the Danish market were profiled in 3 replicates ($R = 3$) using 13 attributes ($K = 13$). The design of the study is given in Figure 5.1. The first problem considered is whether there are differences between the samples and for which samples the differences are the largest. The simplest approach is of course to compute the panel average for each sample and compare them with each

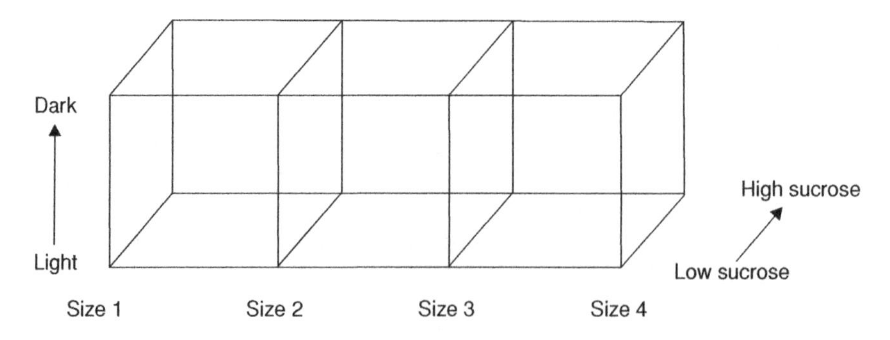

Figure 5.1 Design of the pea study used as an example.

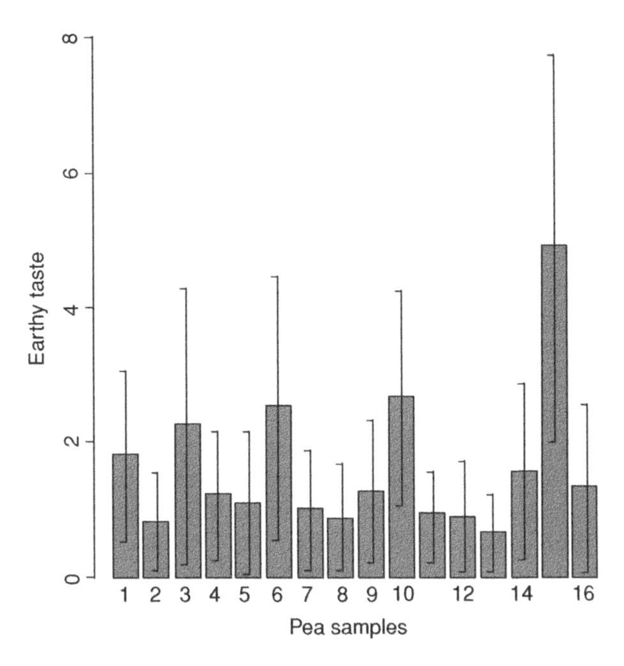

Figure 5.2 *Means and STD plot of earthy taste attribute for the pea example.*

other graphically as is done in Figure 5.2, where also the standard deviation within each sample is shown. Mean values will, however, in practice always be different and one will usually be interested in knowing whether differences are large enough to be determined as significantly different. The most common way for investigating this is to use ANOVA based on the model

$$Y_{ijr} = \mu + \alpha_i + \beta_j + \alpha\beta_{ij} + \varepsilon_{ijr} \tag{5.1}$$

where the α_i is the assessor effect, the β_j is the products effect, the $\alpha\beta_{ij}$ is the interaction and ε_{ijk} is the random replicate error. Usually, the assessor effects and therefore also the interactions are assumed random (see Chapter 13).

The most common hypothesis tested in this model is

$$\mathrm{H}_0 : \beta_1 = \beta_2 = \ldots = \beta_J = 0 \tag{5.2}$$

which corresponds to no average product differences. Note that if the assessors are considered fixed, the test for the hypotheses (5.2) can be quite misleading as was discussed in Næs and Langsrud (1998). The generally recommended test statistic for the hypothesis in (5.2) is

$$F = \frac{MS(Product)}{MS(Interaction)} \tag{5.3}$$

see e.g. Lawless and Heymann (1999), Lea *et al.* (1997) and Chapter 13.

If there really are no average differences between the products, this F statistic will be F-distributed with *J-1* and *(I-1)(J-1)* degrees of freedom and can thus easily be compared

Table 5.1 2-way ANOVA table for the attribute earthy taste for the pea example in Bech et al (1997).

Source	DF	SS	MS	F (mixed)	p-value (mixed)
Assessor	11	173.8	15.8	6.5	<0.0001
Product	15	610.7	40.7	16.9	<0.0001
Assessor*Product	165	398.4	2.4	1.9	<0.0001
Error	384	479.0	1.3		
Total	575	1661.9			

with the table of F-values. The test is, however, usually conducted by computing the *p*-values. The smaller these p-values are, the more likely is it that the hypothesis is wrong. Usually, one will say that a p-value smaller than for instance 0.05 is a clear indication of significance, but it should be stressed that there is no sharp limit between significance and none significance. The hypothesis that all the assessors centre their values at the same point, can be tested by a similar test, i.e. by dividing the MS for assessor with the interaction MS, as is also described in Chapter 13. In order to get a complete overview of differences between products it is recommended to also check for interaction effects. For the interaction effect there is no difference between the fixed and mixed effect tests, in both cases the denominator of the F-test is the MS(Error).

The main result of such an ANOVA test will by any software be presented in an ANOVA table as the one given in Table 5.1, in this case for the earthy taste attribute for the pea example. The basic idea of the ANOVA is the decomposition of the total variability of the 576 observations (16 products, 3 replicates and 12 assessors) of earthy taste into, in this case, four possible sources of variability: Product, Assessor, Interaction, Error. This decomposition is directly shown as the column of sums of squares (SS). For interpretational purposes the column of mean squares (MS = SS/DF) provides the more relevant information: On a comparable scale these values show the sizes of the various sources of variability. We see that the main product variability/differences by far is the largest contributor (MS = 40.71) in this case. The last column of the table tells us that all effects are significant.

5.2.1.1 *Post Hoc Investigations*

In addition to having an overall assessment of the effect of a factor, one will also be interested in knowing for which level of the factor that the actual differences are. This is done by computing the averages accompanied with some information about the statistical uncertainty, for instance their standard errors. Note that plots of average values are also important for assessing the practical significance of the differences found. In fact, there are several possible choices for what type of uncertainty information that can be provided:

1. Standard deviation
2. Standard error
3. Confidence interval (approximately twice the standard error)
4. LSD bars (LSD=Least Significant Difference)

As mentioned, in Figure 5.1 the choice (1) was used to visualise the raw basic variability within each sample – including potential assessor-to-assessor differences. All of the more statistical post hoc choices (2)–(4) will depend on the statistical model used, and the choice of fixed versus mixed model can make a major difference. Even though (2) and (3) seem to be the most popular ones, they actually suffer from one or more important deficiencies if the purpose of the bar plot is to highlight significant product differences and similarities: neither (2) nor (3) actually focus on product differences, they focus on the individual product averages. In such cases conclusions regarding significant differences can be misleading. For instance in the 2-way mixed ANOVA recommended here, the main differences (main effects) of assessors will enter the uncertainty of the product averages themselves, but will *not* enter in the uncertainty of product differences – hence error bars based on (2) and (3) (though extracted properly from a mixed model software) could be completely uninformative about potential significant product differences. On top of this comes the problem that several post hoc tests are made and one can easily end up with significant tests even if all the hypotheses are true. It is generally recommended to apply some sort of protection against the significance-by-chance results that appear in such cases.

There are several different approaches for doing this protection (see e.g. Scheffe, 1959). In this book we will focus on Tukey's correction as is described in Chapter 13 (see also Lea *et al.*, 1997) for a broader discussion in a sensory context): In Figure 5.3, the so-called least significant difference (LSD) lines, that is plus/minus half the 95 % LSD-value, are given using a Tukey adjustment. Using this type of plot one can compare the products and their bars pair-wise: if the LSD lines for two samples overlap, the products are not significantly different.

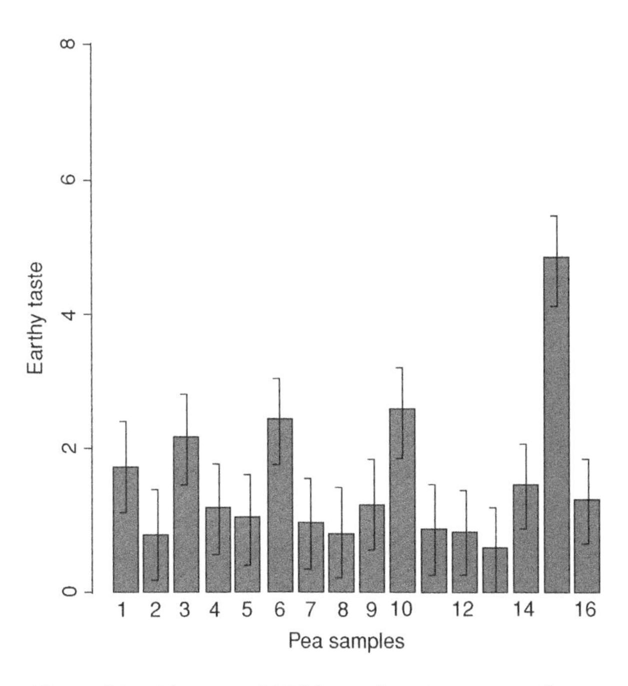

Figure 5.3 *Means and LSD bars of earthy taste attribute.*

5.2.1.2 An Alternative Hypothesis

An alternative hypothesis which was put forward by Næs and Langsrud (1998) is the combined hypothesis

$$H_0 : \beta_j + \alpha\beta_{ij} = 0 \quad \text{for all } i \text{ and } j \tag{5.4}$$

which takes both the main effects for products and interactions into account. In other words, it tests both the average effect and the individual differences at the same time. The interpretation of this hypothesis is that there is no difference between the products at all, neither at the average nor at the individual level. For the test of this hypothesis it does not matter whether the assessors are assumed random or not, which is an advantage of the approach.

The hypothesis is easily tested by a combined F-statistics

$$F = \frac{(SS(Product) + SS(Interaction)) \,/\, (DF(Product) + DF(Interaction))}{MS(Error)} \tag{5.5}$$

(see Næs and Langsrud (1998) for details).

5.2.1.3 A Nonparametric Analogue

Yet another possibility is to use a nonparametric test, like the Friedman's test for two-way ANOVA (Friedman, 1937). The advantage of such a test is that no assumption is made about normality; only the individual ranking of products within each assessor are taken into account. The test, is, however, somewhat less intuitive and is less strong if the regular assumptions are reasonable. It is also much more difficult to generalise to more complex situations.

5.2.2 An Alternative ANOVA Model

Another possibility is to use the alternative model recently developed by Brockhoff and Sommer (2008) which is based on doing a more elaborate modelling of the interaction effect. The motivation for doing this is the general empirical finding that quite often a considerable part of the interaction effect is due to different use of the scale (see Chapter 4). In the standard two-way ANOVA (5.1) such differences will together with real disagreement enter the interaction MS. If strong scaling effects are present, the assumptions behind the general ANOVA approach, where the interactions are assumed to be independently distributed, become less realistic.

The new model is based on the same idea as for the so-called *Brockhoff assessor model* put forward in a sensory context in Brockhoff and Skovgaard (1994). In the new model the scaling part of the interaction is modelled by incorporating the product effect also as a covariate in the model. The model can be written as

$$Y_{ijr} = \mu + \alpha_i + \beta_j + s_i(\bar{y}_j - \bar{y}) + d_{ij} + \varepsilon_{ijr} \tag{5.6}$$

where the α_i's correspond to the assessor effects, the β_k's to the random product effects, s_i is the random scaling effect, d_{ij} is the random (remaining) interaction term, $\bar{y}$ is the mean of all observations and $\bar{y}_j$ is the mean for product j. Model (5.6) is a version of a so-called random coefficient model, a well-known type of linear mixed model. The benefits of using this model is that it is taken into account that the interaction can be due to scaling differences

Table 5.2 *Extended ANOVA table for the attribute earthy taste – F tests and* p*-values based on the model (5.6).*

Source	DF	SS	MS	F	*p*-value
Assessor	11	173.8	15.8		
Product	15	610.7	40.7	3.1 (DF = 86.6)	0.0005
Scaling	11	144.9	13.2	8.0	<0.0001
Remaining interaction	154	253.5	1.7	1.3	0.0173
Error	384	479.0	1.3		
Total	575	1661.9			

or due to real disagreements. This approach provides an extended ANOVA table, where the interaction from the usual two-way ANOVA is further decomposed into the two sources, scaling and disagreement.

The consequence for the test of product differences is that the scaling mean square becomes the one to use in the denominator, although with denominator degrees of freedom somewhere in between the scaling DF and the remaining interaction DF, see Brockhoff and Sommer (2008) for more details. The model (5.6) is generally expected to give similar results (in terms of significances) as compared to the model (5.1) above whenever the assessors use the scale in a homogeneous way.

In Table 5.2 the extended ANOVA table for the earthy attribute is given. Note how the DF and SS values for scaling and remaining interaction (disagreement) adds up to the interaction term from Table 5.1. The F-test for product differences is now only 3.09 – still very significant but not as extreme as before. As can also be seen, a large portion of the interactions effect is actually coming from the differences in scaling. In such cases, the present approach is considered better than regular ANOVA as discussed in Chapter 5.2.1.

5.2.3 Alternative Replicate Structures

In (5.1) it is assumed that the replicates are randomised within the full experiment, which means that there is no systematic replicate effect in the setup. An example of this type of structure is when 2 single carrots from each of 4 varieties are served to each of, say, 5 assessors in a single tasting session. Hence, 10 carrots of each variety are used, and each assessor evaluates $4 \times 2 = 8$ carrots. The random replicate situation is a common situation, but other possibilities also exist, for instance.

A. The same products are served in several separate tasting sessions, possibly with several days in between.
B. The product units come in a structured way – typically in a (two-level) hierarchy

In the first of these cases (A), there is a need for an additional systematic effect in the model to accommodate the systematic session or replicate effect. A possible model to use is

$$Y_{ijr} = \mu + \alpha_i + \beta_j + \nu_r + \alpha\beta_{ij} + \alpha\nu_{ir} + \beta\nu_{jr} + \varepsilon_{ijr} \tag{5.7}$$

In most cases one will also treat the replicate effect as random since one is not interested in the actual levels of the replicates. One then ends up with a mixed model with only one fixed effect, namely the product effect β_j – the other five effects are all random.

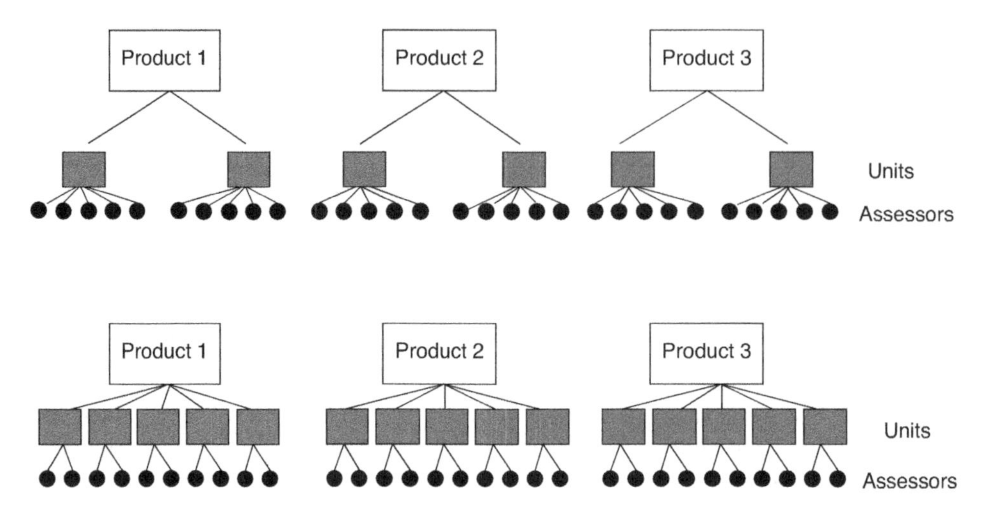

Figure 5.4 ***Different replicate structures.*** *In the upper figure is depicted the situation discussed in the text with two animals (units) from each of the three breeds. In the lower part is depicted a less frequently used situation with five animals from each breed.*

An example of case (B) is sensory analysis of pork chops where the focus is on animal breed. If we assume that there are three breeds ($J = 3$), five assessors ($I = 5$) and two animals ($R = 2$) available for each breed (see Figure 5.4.a), a natural way of setting up this experiment is to give one chop from each animal to each assessor, leading to $3 \times 2 \times 5 = 30$ chops in total. The 2 replicates (animals) are said to be nested (see Chapter 12) within each breed. This is called replicate structure 1 in Lea *et al.* (1997). Note that the number of animals within each group fits with the number of replications ($R = 2$) within each assessor and that the number of slices fits with the number of assessors ($I = 5$).

Alternatively, the two-level product/unit hierarchy could come with 5 animals within each breed, one for each assessor, and the two replicated chops from each animal could be served to the same assessor (see Figure 5.4.b). In this case the animal follows the assessor and the chops correspond to the replicates. This is called replication structure 4 in Lea *et al.* (1997) and is used less often because of higher costs.

The ANOVA model that is natural to use for replication structure (B), Figure 5.4.a is

$$Y_{ijr} = \mu + \alpha_i + \beta_j + \alpha\beta_{ij} + \delta_{jr} + e_{ijr} \tag{5.8}$$

including the additional hierarchical level δ_{jr} in the replication. This additional effect will be considered random as well.

In practice one will then have to choose between one of these ANOVA models ((5.1), (5.7) or (5.8), or maybe even one of the averaging approaches described below) when presented with a typical sensory profile data set. If one did not carry out the experiment oneself, a number of clarifying questions would need to be asked regarding the actual setting of the situation before analysis can be done.

The general point to make here is that it can be quite risky to miss important interaction/product unit effects whereas it has very little influence to include unnecessary effects in the model. Note that from a modelling perspective the three models are nested in each

Table 5.3 3-way ANOVA table for the attribute earthy taste.

Source	DF	SS	MS	F(mixed)	*p*-value
Assessor	11	173.8	15.8	6.7	<0.0001
Replicate	2	7.1	3.5	2.5	0.1151
Product	15	610.7	40.7	15.4	<0.0001
Assessor*Product	165	398.4	2.4	2.0	<0.0001
Product*Replicate	30	43.6	1.5	1.2	0.2303
Assessor*Replicate	22	25.6	1.2	1.0	0.5232
Error	330	402.7	1.2		

other: (5.1) is a sub model of (5.8) which again is a sub model of (5.7). This means that using (5.7) in situations that would really only call for (5.8) or even just (5.1) is generally OK, as the additional unnecessary effects simply are expected to show up nonsignificant in the analysis, and will in any case be quite small and affect the results of the analysis very little. It can only potentially induce a moderate loss of efficiency/power. Using (5.7) as the default analysis will therefore protect against missing important interactions, and often give similar results as the more simple models (5.8) and (5.1) in cases where one of these would have been the proper choice. Depending on the product-unit-replication-session structures the interactions may have different interpretations.

The 16 pea samples discussed above were actually evaluated in sessions with a full replication completed before beginning the next one, calling for the 3-way ANOVA expressed in model (5.7). In Table 5.3 the 3-way mixed ANOVA table is given as presented by any statistical software. As can be seen, all the effect involving replicate are nonsignificant while the others are all significant.

5.2.4 Averaging over Assessors or Replicates

It is possible to do analysis of variance also on averaged data (over either assessors or replications). When averaging over assessors the resulting model can either be written as:

$$y_{jr} = \mu + \beta_j + e_{jr} \tag{5.9}$$

if the replication situation corresponds to either model (5.1) or (5.8) or as

$$y_{jr} = \mu + \beta_j + \nu_r + e_{jr} \tag{5.10}$$

if the replication situation corresponds to model (5.7). When averaging over replicates only the model becomes:

$$y_{ij} = \mu + \alpha_i + \beta_j + e_{ij} \tag{5.11}$$

Note that the interaction is now part of the random error.

Model (5.9) is a standard one-way ANOVA as also used in Chapter 3 for analysing replicated data for a single assessor. Models (5.10) and (5.11) are both standard single-rep two-way ANOVA models (randomised block settings). It is, however, important to mention that when basing the analysis on average data, valuable information about individual differences is lost. (if not handled by one of the other methods discussed in Chapter 3).

5.2.5 Multi-Way ANOVA for Tests in Designed Experiments

In the models above, the products are modelled as individual levels of one single factor called the product effect. It is, however, not uncommon that the products investigated are constructed using some kind of experimental design. In the pea example, the 16 samples are really stemming from a $4 \times 2 \times 2$ full factorial, see Figure 5.1. The 16 products could then be decomposed into the 3 main effects, 3 two-factor interactions and one 3-factor interaction providing additional insight. The model becomes

$$y = \mu + \text{assessor} + \text{product main effects} + \text{two factor interactions} + \text{three factor interaction} + \text{noise} \quad (5.12)$$

or the possible extensions described in (5.7) and (5.8) if the replicate structure is more complex. In practice, we have the choice of considering the products as represented by just a single product factor or to utilise the decomposition into sub-factors. The advantage of the decomposition is that both the average effects of treatment factors can be investigated as well as their interactions.

If data are averaged over assessors, the ANOVA becomes a simple multi-way fixed effects analysis as described in Chapter 13. If the assessor is part of the analysis, a mixed model, with assessor and assessor interactions considered random, has to be used. Most computer software packages have today an option for analysing mixed models of this type. An example of a three-way model with one fixed effect is given in Chapter 13. A mixed model analysis of the three-way pea data can be found in Table 5.4. As can be seen, all the design factors involved are significant with the sucrose as the most important (measured by the MS's). It should be mentioned that when many tests are done at the same time, the chance of obtaining significant effects increases. This problem can be solved by a certain correction technique called Bonferroni-correction. This type of multiple comparison corrections will not be pursued further here, except when discussing post hoc testing in Chapter 13.9. In all cases, we recommend a pragmatic approach to significance testing, with less emphasis on the exact size of the p-values, as also discussed in Chapter 13.1.

For analysing fractional factorial designs, similar procedures can be used, incorporating the appropriate effects in the model (see also Ellekjær *et al.*, 1996; Johansen *et al.* 2010a). This means that if two effects are confounded, only one of them can be incorporated in the model. We refer to Chapter 13 for further discussion of this aspect.

5.2.6 Analysing Designed Experiments with Regression Analysis

If there are more than two levels of a factor that is essentially continuous, one has a choice between treating the design factor as continuous or as categorical (Chapter 15). The categorical treatment has the advantage that it is more general and can be used for all types of factors. In addition one does not need to specify a model; each factor level represents its own average value. The continuous treatment has the advantage that the estimates and tests may become more powerful since the number of model parameters is reduced. In addition it provides an estimated function that can help interpretation and also be used for prediction purposes later.

Table 5.4 *Multi-way ANOVA table for the attribute earthy taste. F-tests are based on the mixed model where all assessor main and interaction effects are random. Note the warning given in Chapter 5.2.5 when several tests are made simultaneously.*

Source	DF	SS	MS	F(mixed)	*p*-value
Size	3	188.3	62.8	27.8	<0.0001
Colour	1	34.6	34.6	16.8	0.0018
Size*Colour	3	91.0	30.3	10.4	<0.0001
Sucrose	1	79.0	79.0	36.3	<0.0001
Size*Sucrose	3	174.2	58.1	21.3	<0.0001
Colour*Sucrose	1	7.4	7.4	4.3	0.0621
Size*Colour*Sucrose	3	36.1	12.0	5.6	0.0034
Assessor	11	173.8	15.8	12.3	0.2521
Assessor*Size	33	74.6	2.3	0.7	0.8688
Assessor*Colour	11	22.8	2.1	0.8	0.6199
Assessor*Size*Colour	33	96.6	2.9	1.4	0.1945
Assessor*Sucrose	11	23.9	2.2	1.0	0.5436
Assessor*Size*Sucrose	33	90.1	2.7	1.3	0.2535
Assessor*Colour*Sucrose	11	19.0	1.7	0.8	0.6421
Four-way interaction	33	71.4	2.2	1.7	0.0085
Error	384	479.0	1.3		

With one factor, a continuous model version of model (5.1) is

$$y_{ijr} = \mu + \alpha_i + \beta x_j + \alpha\beta_{ij} + e_{ijr} \tag{5.13}$$

where x_j is the value of the continuous design variable for product j and the other effects are defined as above. If this model does not fit well to the data, one can extend by incorporating a polynomial function of x. (see Chapter 15). As was stressed in Chapter 15, one should always centre continuous x-values before using them in combination with categorical variables. How to check adequacy of the model is discussed in Chapter 15. If wanted, the interaction can also be made into a regression effect which depends on assessor.

For the pea example, the size factor was coded as 1, 2, 3 and 4, where 1 is the smallest pea size and 4 the largest one. These numbers could then be used as a regression variable x, assuming that the actual size differences between the four categories were equidistant. In Table 5.5 the results of an extension of model (5.1), where the Assessor and Assessor*Product effects are random, is shown. The sums of squares (SS) used here are based on the type I philosophy discussed briefly in Chapter 13.5. It is seen that there are indeed significant linear main and interaction effects of the size, but also that this does not explain entirely the effects of size, since the effects related to the categorical factor "size" are also still significant. The interaction effects of colour*sucrose and size*colour*sucrose were ignored here as they were considerably smaller (though significant) than the included effects.

More complex optimisation designs like for instance the central composite design or mixture designs (Cornell, 1990; Myers and Montgomery, 1995) should always be treated by regression methods. In many cases one will average the data over assessors before analysis, in this way avoiding the random effects in the model. For an example of the use of regression analysis for a mixture design we refer to Figure 5.5. In this case, data from

Table 5.5 *ANOVA table for analysis of size effects as a regression variable for the attribute earthy taste. Type I decomposition. F-tests based on the mixed model where the assessor and assessor-by-product interaction are random. In this table SizeQ denotes the covariate effect (linear trend) of the size.*

Source	DF	SS	MS	F(mixed)	*p*-value
Assessor	11	173.8	15.8	6.0	<0.0001
SizeQ	1	107.5	107.5	41.1	<0.0001
Size	2	80.8	40.4	15.5	<0.0001
Colour	1	34.6	34.6	13.3	0.0004
Sucrose	1	79.0	79.0	30.2	<0.0001
SizeQ*Colour	1	63.9	63.9	24.4	<0.0001
Size*Colour	2	27.2	13.6	5.2	0.0064
SizeQ*Sucrose	1	55.6	55.6	21.3	<0.0001
Size*Sucrose	2	118.6	59.3	22.7	<0.0001
Assessor*Product	169	441.9	2.6	2.1	<0.0001

a mixture design in three ingredients is used to estimate the regression surface using a polynomial model (see Dingstad *et al.*, 2004). As can be seen, the highest level of firmness can be found versus the lower left corner, i.e. with the highest concentration of the ingredient x_2. The lowest levels can be found with low level of x_2 and with about equal amounts of the other two ingredients.

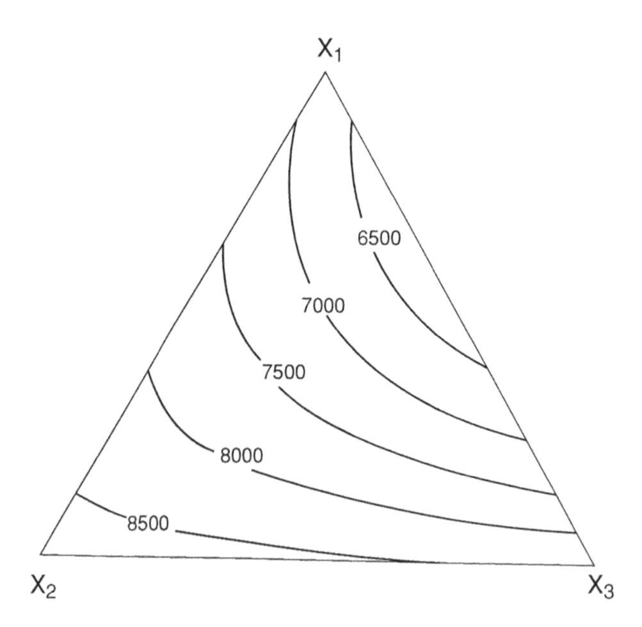

Figure 5.5 ***Contour plot for a mixture situation****. Firmness (in this case measured by an instrument) of sausages as a function of three different ingredients that were varied according to a mixture design. The illustration is taken from Dingstad et al (2004). Reprinted from Chemometrics and intelligent laboratory systems, 71, Dingstad, Westad and Næs. Three case studies illustrating the properties of ordinary and partial last squares and partial least squares regression in different mixture models. 33–45, 2004, with permission from Elsevier.*

5.3 Analysing Sensory Profile Data: Multivariate Case

5.3.1 PCA on Averages over Assessors and Replicates

A very simple, straightforward and probably the most applied approach for getting an overview of multivariate sensory data is to use Principal Components Analysis (PCA) on the product-by-attribute matrix, after having averaged out both replicates and assessors. One possibility is to interpret the scores and loadings directly without further analysis (see e.g. Baardseth *et al.*, 1992; Baardseth *et al.*, 1995), but sometimes the PCA results can also be taken as the basis for further analyses as will be discussed below.

As for the univariate approach, taking averages in this way represents loss of information about individual differences and one will therefore in many cases prefer to use one of the methods to be discussed below. An advantage of the simple PCA approach is that it can be used for all possible cases, with or without replicates and also in cases where no specific experimental design is underlying the products.

An example of a PCA for average data is given in Figure 5.6. The yoghurt data set consists of 9 strawberry yoghurt samples that are described with 19 sensory attributes. The yoghurts on the right-hand side of the PCA scores plot are those with high intensity of sweetness, fattiness, colour tone and intensity, as well as artificial odour and flavour. Yoghurts 3, 4 and 5 are the samples that have most strawberry odour and astringency. Yoghurt 2 has high intensity of whiteness and yoghurt flavour. Yoghurts 6 and 9 are the most different and yoghurts 3, 4, and 5 are the most similar. The latter are clearly characterised as being bitter, astringent and with strawberry odour. The two principal components explain 80 percent of the total variance in the data set.

5.3.2 The Use of ANOVA for Multivariate Sensory Data

In situations with several responses, it is always possible to use ANOVA for each of them separately. If the variables are highly correlated, this approach is, however, questionable since the significance tests will be strongly dependent on each other. In addition, the strategy provides no information about how the design variables are related to the multivariate structure of the data set. Therefore, it is usually better to use a multivariate ANOVA method (so-called MANOVA). In the following we will briefly mention the standard variant of MANOVA, but give main attention to a couple of modifications that are more graphically oriented and thus better suited for revealing the important structures of the data.

5.3.3 MANOVA and CVA

The standard classical MANOVA is based on assuming the same model as above (for instance (5.1) or (5.7)) for each of the variables and then allow for correlation structure of the random error terms. The assumption is that the K-dimensional vector of error terms is distributed as a multi-normal vector with expectation 0 and covariance matrix equal to Σ, i.e.

$$\mathbf{e}_{ijr} = (e_{ij1r}, e_{ij2r}, \ldots, e_{ijKr})^T \approx \mathbf{N}(\mathbf{0}, \boldsymbol{\Sigma}) \quad (5.14)$$

Under this assumption it is possible to test similar hypotheses as above, for instance whether all main effects for products are equal to zero. The test procedures that have been developed are quite complex theoretically and are not discussed further here, but most computer software at a certain level are able to do the tests. The standard MANOVA

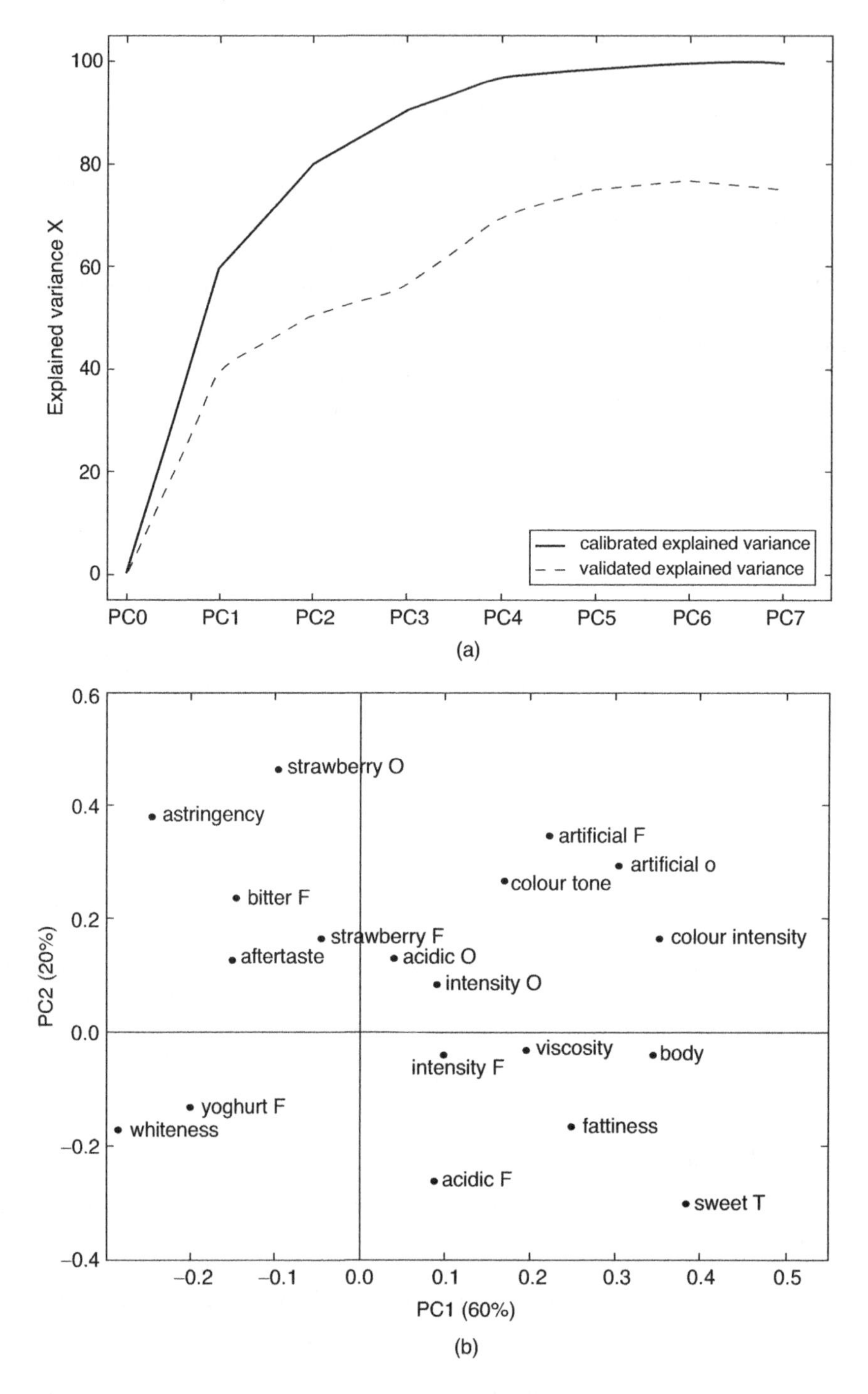

***Figure 5.6 a,b,c,d. PCA on averages over assessors and replicates**. The data set is the descriptive sensory yoghurt data as described in the text. a). Explained variances. b). Loadings.*

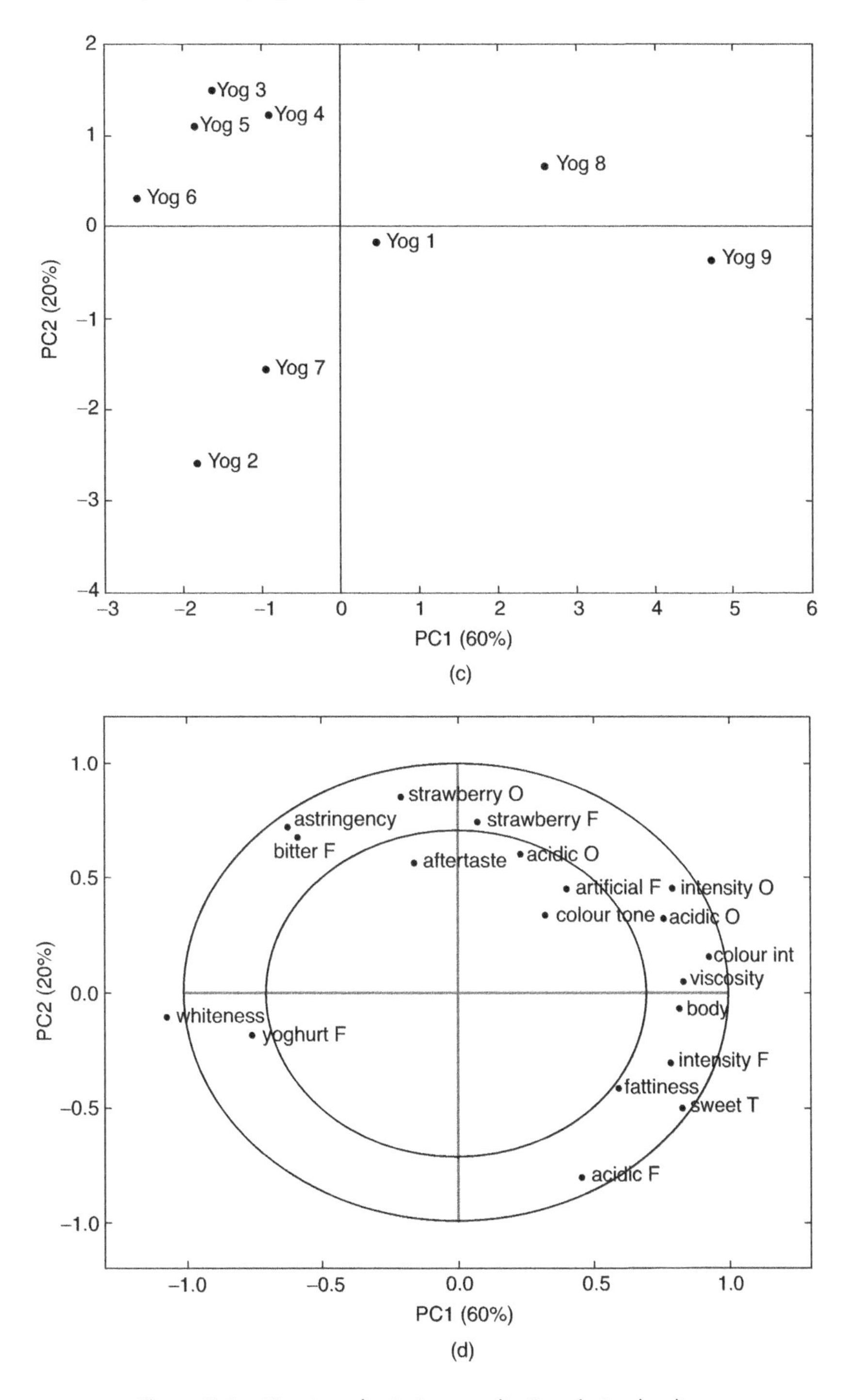

Figure 5.6 *(Continued) c). Scores. d). Correlation loadings.*

approach is only possible to use if the number of objects is larger than the number of variables in total.

A related approach which can be handled by some MANOVA software packages is Canonical Variates Analysis (CVA), which is basically a multi-component extension of Fishers' Linear Discriminant Analysis (LDA, Mardia *et al.,* 1979): Instead of finding components that maximise the variation across product averages as is done for the above PCA, one finds components that maximise the F-statistics of between product differences relative to the within product error. The within product error can if wanted be taken to be the multivariate assessor-by-product interaction hence providing a mixed version of the CVA.

An improvement of regular MANOVA particularly suited for cases with collinear response variables was developed by Langsrud (2002). This method is named 50/50 MANOVA. The method can be used for various types of hypotheses, but main attention in Langsrud's work was given to so-called type II tests, which are slightly different from those given main attention here (see Chapter 15). Langsrud (2002) also presented some examples of how the results can be presented graphically.

Incorporating alternative replicate structures are more difficult to handle within a MANOVA framework. It is in principle possible, but none of the established software packages or easily available textbooks focus on it.

5.3.4 PC-ANOVA

This is a method which is based on some of the same ideas as the 50/50 MANOVA method, but which is less formal when concerns the multivariate significance tests. On the other hand it has a stronger linked to visual inspection of the results.

The idea behind the approach is very simple and is based on two separate steps (Ellekjær *et al.*, 1996). First the PCA is used on the sensory data as they are and then the scores from the PCA are used as dependent variables in the ANOVA vs. the design factors.

The first step of the analysis is based on using PCA on the unfolded sensory data set with I^*J^*R rows and K columns (unfolded vertically as opposed to horizontal unfolding as depicted in Figure 3.8). In other words, the different assessor matrices are put under each other with the attributes representing the columns of the matrix. Note that replicates (R) are kept as a part of the analysis. The scores and loadings of the PCA are first plotted before the scores are used as dependent variables in an ANOVA model of the same type as above ((5.1) and (5.7)). The tests of significance are done the same way and reported the same way as before. It was proposed in Luciano and Næs (2009) that the scores are also plotted in various averaged ways in order to highlight the effects of the design variables. Note that this is identical to computing the effects in the univariate ANOVA models and then projecting these effects down onto the loadings as obtained by the PCA.

This method can be used for any ANOVA model, also for the different replicate structures discussed above. Unbalanced situations can be handled and multiple post hoc comparison tests can be used (Chapter 13). Combined hypotheses of the type mentioned above (see Equation (5.4) and Næs and Langsrud, 1998) can also easily be incorporated as well as situations where the product effect is composed of different experimental factors. In other words, the method shares the same properties as regular ANOVA, but provide additional information about the multivariate structure since PCA plots are available.

In Figure 5.7a and b a plot is shown of the scores and loadings of a sensory data set based on a candy product. In this situation there are $K = 9$ attributes, $J = 5$ products, $I = 10$

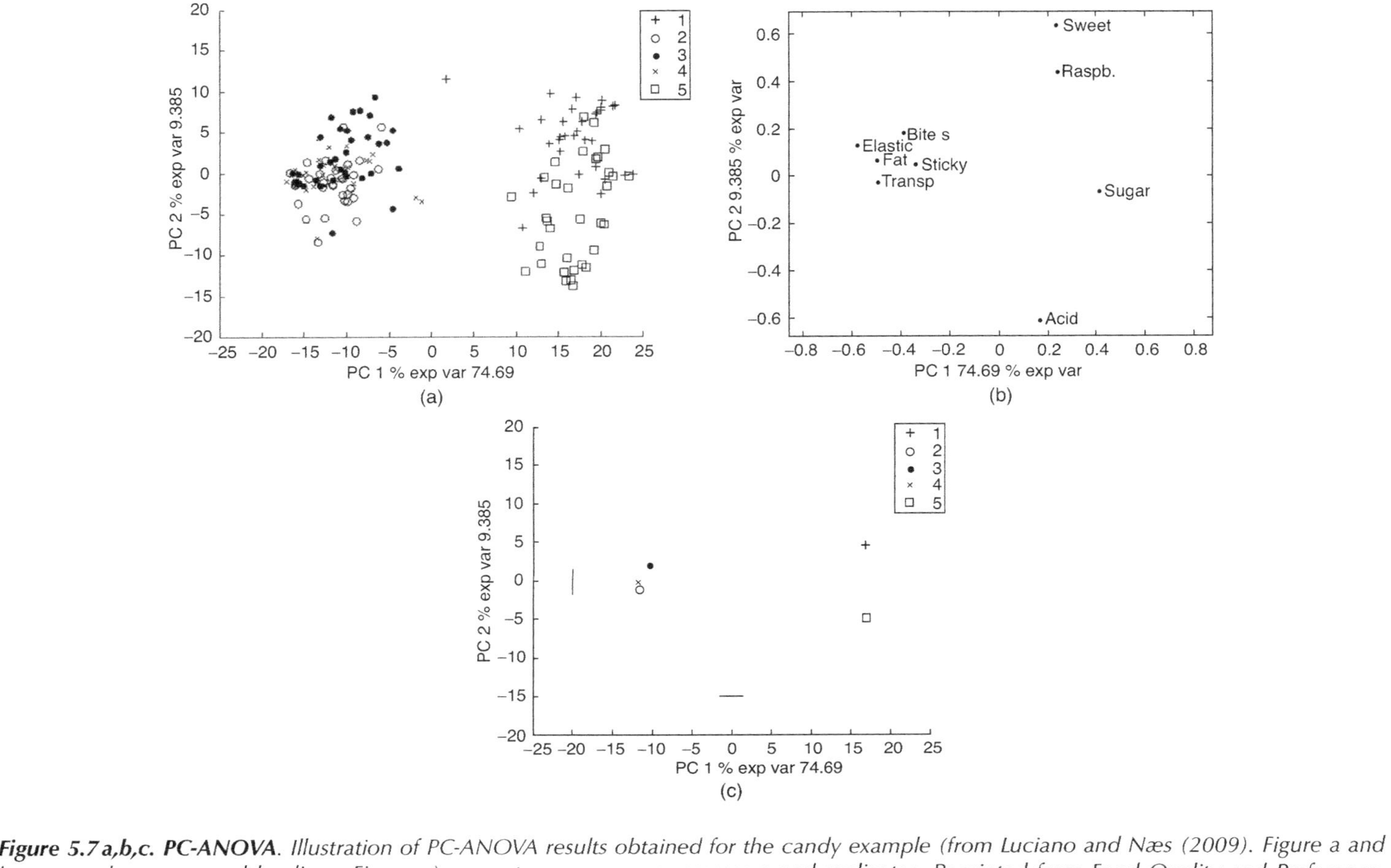

***Figure 5.7 a,b,c. PC-ANOVA**. Illustration of PC-ANOVA results obtained for the candy example (from Luciano and Næs (2009). Figure a and b present the scores and loadings. Figure c) presents averages over assessors and replicates. Reprinted from Food Quality and Preference, 20, Luciano and Næs, Interpreting sensory data by combining principle component analysis and analysis of variance, 167–175, 2009, with permission from Elsevier.*

Table 5.6 a.b *PC-ANOVA for two components. The candy example (Luciano and Næs (2009)). In the upper table are presented results for principal component 1 and in the other results for principal component 2.*

Source	DF	SS	MS	F	*p*-value
Assessor	10	224.9	22.5	1.3	0.2789
Product	4	31470.6	7867.7	444.7	<0.0001
Assessor*Product	40	707.7	17.7	1.8	0.0116
Error	110	1109.3	10.1		
Total	164	33512.5			
Assessor	10	278.3	27.8	1.1	0.4166
Product	4	1573.8	393.5	15.0	<0.0001
Assessor*Product	40	1053.3	26.3	2.2	0.0006
Error	110	1305.2	11.9		
Total	164	4210.6			

assessors and $R = 3$ replicates. The attributes were transparency, acidity, sweet taste, raspberry flavour, sugar coated texture tested with a spoon, biting strength in the mouth, hardness, elasticity in the mouth and stick to teeth in the mouth. In Figure 5.7c is given an example of scores averaged over the assessors and replicates in order to highlight product effects. The small horizontal and vertical lines centred at 0 represent the error variance of the ANOVA model for the two components and can be used for comparing the product differences with the noise level. As can be seen from Figure 5.7a, there are clear differences between products 1 and 5 on one side and the other three on the other. Along the second axis there also seems to be a differences between products 1 and 5. These differences can be interpreted the usual way using the loadings. The first axis differences are for instance related to sweet, raspberry, acid and sugar versus the rest of the attributes. As can be seen from the ANOVA table (Table 5.6), for both axes the product effect is significant. The same is true for the interactions, but the assessor effect is not significant for any of the two first factors.

5.3.5 ASCA

The ASCA (ANOVA-simultaneous component analysis) method is another important contribution to the analysis and interpreting of multivariate sensory data. The method has similarities with the PC-ANOVA since it is based on the same two basic methods, namely PCA and ANOVA in combination. The difference is basically that the ASCA uses the two operations in opposite order (see Jansen *et al.*, 2005; Luciano and Næs, 2009).

The ASCA method is based on first computing the main effects and interactions effects as they are defined by the ANOVA. Usually, one will use the standard restriction of summing all the effects to 0 (over the levels, see Chapter 13). In balanced designs, the main effects and interactions can be computed simply as differences between means as discussed in Chapter13. The effects are then collected in matrices corresponding to all the attributes. There will be one effect matrix for each effect in the model. After all these matrices have been computed, they will be analysed separately by PCA. This leads to one PCA plot for each effect in the model. For the two-way situation with interactions, there will thus be three PCA's and three different PCA plots. For the interactions, which essentially produce a three-way matrix, it is common to use either Tucker-1, which is essentially an unfolded PCA, or PARAFAC (see Jansen *et al.,* 2008), Chapter 14 and Chapter 17).

Since no information about random variation is available in the plot, it is not obvious how to make a direct assessment of significance of the differences between products or assessors along the different axes. A possible extension of ASCA would be to add some type of confidence ellipses based on for instance the bootstrap or other re-sampling schemes (see e.g. Pages and Husson, 2005).

ASCA can as PC-ANOVA easily handle imbalance and more complex error structures. The only modification that has to be done is that a restricted maximum likelihood (REML, Pawitan, 2001) estimate is needed for improved estimation of effects. REML is a method that takes the more complex error structure into account when estimating the effects. The standard LS estimates can also be used since they are unbiased, but the REML estimates are more precise. Also incorporating factorial designs in the product structure is possible with the ASCA. As long as the effects can be estimated, the method can be used. As for the PC-ANOVA, the ASCA method can also be used to analyse the joint effect of for instance the main effects of products and the interactions (see e.g. Jansen *et al.*, 2005). Since it provides a separate PCA plot for each factor separately, each of the PCA models will generally have a lower dimension than for PC-ANOVA. For more information about ASCA we refer to Jansen *et al.* (2005), Jansen *et al.* (2008), Vis *et al.* (2007) and Luciano and Næs (2009).

References

Baardseth, P., Naes, T., Mielnik, J, Skrede, G., Hølland, S., Eide, O. (1992). Dairy ingredients effects on sausage sensory properties studied by principal component analysis. *J. Food Science* 57(4), 822–8.

Baardseth, P. Næs, T., Vogt, G. (1995), Roll-in shortenings effects on Danish pastries sensory properties studied by principal component analysis. *Lebensm.-Wiss u.-Technol* 28, 72–7.

Bech, A., Hansen, M., Wienberg, L. (1997). Application of house of quality in translation of consumer needs into sensory attributes measurable by descriptive sensory analysis. *Food Quality and Preference* 8, 329–48.

Brockhoff, P.B., Sommer, N.A. (2008). Accounting for scaling differences in sensory profile data. *Proceedings of 10th European Symposium on Statistical Methods for the Food Industry*, pp. 283–90, Louvain-La-Neuve, Belgium.

Cornell, J. (1990). *Experiments with Mixture* (2nd edn). New York: John Wiley & Sons, Inc.

Dingstad, G., Westad, F., Næs, T. (2004). Three case studies illustrating the properties of ordinary least squares and partial least squares regression in different mixture models. *Chemometrics and Intelligent Laboratory Systems* 71, 33–45.

Ellekjær M.R., Ilseng M.R., Næs T. (1996). A case study of the use of experimental design and multivariate analysis in product improvement. *Food Quality and Preference* 7(1), 29–36.

Friedman, M. (1937). The use of ranks to avoid the assumption of normality implicit in the analysis of variance. *Journal of the American Statistical Association* 32(200): 675–701.

Jansen, J., Hoefsloot, J. Van Der Greef, M., Timmerman, E., Westerhuis, J., Smilde, A.K. (2005). ASCA: analysis of multivariate data obtained from an experimental design. *Journal of Chemometrics* 19(9), 469–81.

Jansen, J. Bro, R., Huub C. *et al.* (2008). PARAFASCA: ASCA combined with PARAFAC for the analysis of metabolic fingerprinting data. *Journal of Chemometrics* 22, 114–21.

Johansen, S., Næs, T., Øyaas, J., Hersleth, M. (2010a). Acceptance of calorie-reduced yoghurt: Effects of sensory characteristics and product information. *Food Quality and Preference* 21, 13–21.

Langsrud, Ø. (2002) 50–50 multivariate analysis of variance for collinear responses. *Journal of the Royal Statistical Society: Series D (The Statistician)* 51, 305–17.

Lawless, H.T., Heymann, H. (1999). *Sensory Evaluation of Food. Principles and Practices*. New York: Chapman & Hall.

Lea, P. Næs, T., Rødbotten, M. (1997). *Analysis of Variance of Sensory Data*. New York: John Wiley & Sons, Inc.

Luciano, G., Næs, T. (2009). Interpreting sensory data by combining principal component analysis and analysis of variance. *Food Quality and Preference* 20, 167–75.

Mardia, K.V., Kent, J.T., Bibby, J.M. (1979). *Multivariate Analysis*. London: Academic Press.

Myers, R.M., Montgomery, D.C. (1995). *Response Surface Methodology. Process and Product Optimisation Using Designed Experiments*. New York: John Wiley & Sons, Inc.

Næs, T., Langsrud, Ø. (1998). Fixed or random assessors in sensory profiling? *Food Quality and Preference* 9(3), 145–52.

Pages, J., Husson, F. (2005). Multiple factors analysis with confidence ellipses: a methodology to study the relationships between sensory and instrumental data. *J. Chemometrics* 19, 138–44.

Pawitan, Y. (2001). *In All Likelihood: Statistical Modelling and Inference Using Likelihood*. Oxford: Oxford University Press.

Scheffe, H. (1959). *The Analysis of Variance*. New York: John Wiley & Sons, Inc.

Smith A, Cullis B, Brockhoff P., Thompson R. (2003). Multiplicative mixed models for the analysis of sensory evaluation data. *Food Quality and Preference* 14(5–6), 387–95.

Vis, D.J., Westerhuis, J.A., Smilde, A.K., Van Der Greef, J. (2007). Statistical validation of megavariate effects in ASCA. *BioMed Central (BMC) BioInformatics* 8, 322.

6

Relating Sensory Data to Other Measurements

Sensory data are often informative enough in themselves for drawing conclusions and making decisions, but in some cases one may also be interested in finding relations between sensory data and other types of measurements. There are several reasons for this, the most important being prediction and improved understanding. In this chapter, the focus will be on the use of the multivariate regression methods in situations where one of the data sets is based on descriptive sensory analysis. Some focus will also be given to methods that relate three-way and multi-block sensory data structures to external two-way data sets. A number of examples will be given where all these aspects are illustrated.

The present chapter is closely related to Chapter 5 and based on methodology in Chapters 14, 15 and 17.

6.1 Introduction

Descriptive sensory analysis is a very important analysis technique in itself for understanding food quality, but sometimes one is also interested in relating sensory data to other types of measurements for improved interpretation, for prediction purposes or simply for validating already obtained results. Important examples of situations where relations to other data are important, are:

- *Improved interpretation.* In order to obtain improved insight it may be useful to relate the sensory data to either the chemical composition of the samples (see e.g. Helgesen and Næs, 1995) or to the experimental design used to produce them (see e.g. Dahl and

Statistics for Sensory and Consumer Science Tormod Næs, Per B. Brockhoff and Oliver Tomic

Næs, 2006). This is an important aspect also when focus is on optimising specific sensory attributes (see Chapter 5, Section 5.2.6).

- *External validation.* This situation is closely related to the former and has at least two different aspects. First of all external validation is important for obtaining improved confidence in the data and the results obtained. Secondly, external validation may be important for evaluating and comparing various pre-processing procedures (Romano *et al.*, 2008; Næs, 1990; Dahl and Næs, 2004) or simply for finding the best consensus for the panel. Comparisons with the design of the experiment or the chemical composition of the samples may provide information about which approach is most suitable for a given situation. Typically the pre-processing with the best predictive relation to the external data will be considered the best. We refer to Chapter 4 for a discussion of possible pitfalls when using external validation.
- *The 'replacement' of sensory panel data by quick chemical or spectroscopic measurements.* Sensory analysis can sometimes be rather time-consuming and it may be difficult to have a stable, reliable and well-calibrated sensory panel available at all times in a production situation. Chemical methods and in particular online spectroscopic techniques are known to be very reliable and informative in a production environment, and can be used also for nondestructive measurements online. If one is able to establish a reliable prediction equation for the sensory data based on the spectroscopic readings, the prediction equation may be applied in industrial quality control with substantial time and cost reduction (Martens and Martens, 1986; Kjølstad *et al.*, 1990).
- *Preference mapping.* This is the name used for the important area of relating sensory data to consumer acceptance data (see e.g. McEwan, 1996). Typically one has collected the consumer acceptances for a number of samples that one has previously analysed by sensory analysis. Finding relations between the two sets of data is suitable for identifying the most important drivers of liking and the most liked products. This type of methodology is handled in Chapter 9.

The data sets used in this chapter are a sensory data set as depicted in Figure 2.1 and a data set containing other types of measurements for the same samples. The analysis methods described in this chapter require that the sample dimension (vertical dimension) must be the same for both data sets. In most of the cases treated here, the sensory data will be averaged across assessors such that finding the relation between the sensory and external variables can be done by regular multivariate regression (Figure 6.1). We will also discuss briefly a number of different ways of computing this type of averages. This aspect is, however, primarily discussed in Chapter 4 in connection with how to improve sensory panel data.

The present chapter is strongly related to Chapter 5 which discusses analysis of variance of sensory data based on factorial designs. The main difference between the two chapters is that the focus here is on regression methods where the external data are considered to be continuous and in some cases also highly collinear. This requires a different focus and provides different tools for interpretation although the theoretical basis is the same.

6.2 Estimating Relations between Consensus Profiles and External Data

Multivariate regression analysis (Chapter 15) is the appropriate methodology to apply for finding relations between complex data sets. It is, however, not always obvious which

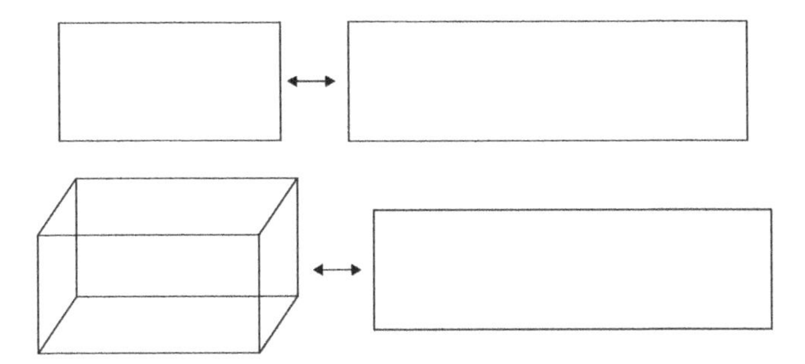

Figure 6.1 ***Illustration of data sets used.*** *The upper panel depicts a situation where the two two-dimensional matrices are related to each other; for instance sensory average data and spectroscopic data (columns) for a number of products (rows). In the lower panel, a three-way sensory data table is related directly to a regular two-way data matrix. The vertical dimension, i.e. the product dimension, is again the same in both cases.*

regression method to use or which way the regression relation is to be developed; should the sensory data be predicted from the external data or should they be used as predictors? In the first three examples mentioned in the previous section, the most natural thing to do is to use the sensory data as response data (**Y**) and the chemical or design data as predictor data (**X**). There are, however, also situations in which for instance the human nose is very capable of detecting a substance that is difficult to measure by chemistry. In such cases, the sensory data may be the natural data set to be used as predictor variables (**X**). For the preference mapping situation there is already a strong tradition of considering regression equations both ways (Chapter 9).

When relating Y-data to X-data by the use of regression methods, there is always a need for a model and an estimation methodology. A number of different alternatives exist (Næs *et al.*, 2002; Martens and Martens, 2001), but here we will focus on linear models and the estimation methods principal component regression (PCR) and partial least squares (PLS) regression as described in Chapter 15. Experience has shown that the linear models are useful for almost all purposes in this area. The linear regression model for relating two data sets can be written as

$$\mathbf{Y} = \mathbf{XB} + \mathbf{E} \tag{6.1}$$

where the regression coefficients **B** are to be estimated from the data.

Although regression methods are the most important and the main focus of this chapter, it should be mentioned that other methods can sometimes also be useful for the problem areas listed above. For instance, a number of methods exist that treat two data sets simultaneously without introducing any predictive direction. An important method in this class is canonical correlation analysis (CCA, see e.g. Mardia *et al.*, 1979), but for the typical applications treated here this method is unsuitable because it cannot handle situations with few samples and many collinear variables, which is typical in sensory analysis. Another possibility which is more useful here is the Tucker-1 method (Chapters 14 and 17) which belongs to a class of methods often referred to as multi-block methods. This is a technique which tries to approximate both data sets (or several) by using data compression through a common scores model. This implies that Tucker-1 tries to find some common directions, components or latent variables that can be used to predict both **X** and **Y** in the best possible way. The method is described in some detail in Chapter 13. We refer to Westerhuis *et al.* (1998) and Kohler *et al.* (2008) for further discussion of this method.

6.2.1 Finding the Best Consensus Sensory Profiles

The simplest way of establishing a panel consensus to be used in Equation (6.1) is by the use of regular averaging of all individual profiles for each sample separately. A number of similar alternatives exist based on giving the assessors a weight according to their reliability (Chapter 4). Another method is the STATIS method (see e.g. Schlich, 1996; and Chapter 17) which weighs the individual profile matrices using the so-called RV coefficient. Procrustes rotation (see Chapter 17 and Dijksterhuis, 1996) is another useful method which is based on fitting a matrix of consensus profiles to the original sensory profiles as well as possibly allowing for isotropic scaling and rotation of the individual profile matrices. The methods below can be used for any of these alternatives.

6.2.2 Linear Regression Methodology

PCR (see e.g. Næs and Martens, 1988; Gunst and Mason, 1977, 1979; Joliffe, 1982) is a useful method for finding relations between one y-variable and several X-variables (see Chapter 14). With multiple Y-variables, one simply regresses all of them onto the same PCA space obtained by the X-data. This method is useful since it solves the collinearity problem and provides several plotting and diagnostic possibilities. Another alternative is to use the PLS-2 regression method, which seeks latent variables based on both **X** and **Y**. The method provides plots similar to the PCR plots. Which of the methods to use is in many cases a matter of taste, but often one will prefer PLS for this purpose since it seldom provides inferior predictive ability and at the same time may give similar prediction results using a smaller number of components. For prediction purposes it is also possible to use the PLS for each of the Y-variables separately, but this approach is less simple to use for an overall interpretation of all the results.

The result of both methods is first of all a set of regression coefficients, the estimated **B** matrix in Equation (6.1), in addition to scores plot and loadings plot and a number of diagnostic tools for detecting outliers. The scores are usually plotted in two- or three-dimensional scatter plots and are useful for detecting the relations between the objects in a low-dimensional representation. The loadings are useful for understanding the relation between the scores and the original variables, thus providing an important tool for understanding the relation between the objects. The regression coefficients can also be interpreted, but this is usually more difficult in situations with collinear data and are therefore mostly used for predictive purposes, i.e. for predicting **Y** from **X**. As soon as **B** is estimated, the predictions of **y** for a future sample can be obtained by plugging the new **x** into the equation and obtaining predicted **y** by

$$\hat{\mathbf{y}}^{\mathbf{t}} = \mathbf{x}^{\mathbf{t}}\hat{\mathrm{B}} \tag{6.2}$$

The quality of the relation obtained should generally be validated by using either cross-validation or prediction testing on an independent data set. We refer to Chapter 15 for more information about this subject.

If there are multiple sets of external data, one may put them adjacent to each other and use regular PLS or PCR on the combined matrix. A simple alternative is to use multi-block PLS (Westerhuis *et al.*, 1998) which is essentially a regular PLS on the combined matrix which provides additional results which are useful for interpreting how the different blocks are involved in the equation. Another possibility is the LS-PLS method (Jorgensen *et al.*, 2004). This method has the advantage that different types of data can be used (both design matrices and collinear data blocks) and that different dimensionality of the blocks can be allowed. The method is invariant to individual weighting of the blocks. The method also provides direct information about incremental contributions of the different data blocks which may be important for interpretation and practical use of the results.

6.2.3 Symmetric Multi-Block Methodology

The Tucker-1 method is based on computing the PCA for the merged data matrix

$$\mathbf{Z} = (\mathbf{X}, \mathbf{Y}) \tag{6.3}$$

a horizontal concatenation of matrices **X** and **Y**. It is therefore often referred to as multi-block PCA (see e.g. Hanafi and Kiers, 2006). Tucker-1 used in this way corresponds to finding the common scores matrix **T** that best represents the joint variability in the matrix **Z** (see Chapters 3 and 17). If the two blocks have very different variance, it may be natural to do some sort of scaling before the analysis.

The results from this method will as for the regression methods be presented in scatter plots of the common scores **T** and two individual loadings plots. They can be presented together since interpretation of the joint information is the main aspect. The fit of the model can here be determined by looking at the residuals from both blocks and comparing the variability with the total variance.

6.2.4 Examples of Use

6.2.4.1 Predicting Sensory Data by Chemical Spectroscopy

Predicting sensory properties by the use of for instance spectroscopy is an important application area in an industrial context. One is primarily interested in the sensory properties, but they may be difficult to obtain as frequently as wanted. Therefore it is useful to investigate the possibility of predicting them by fast and reliable online or at-line spectroscopy.

There are a number of publications focusing on this problem. Martens and Martens (1986) investigated a large data set consisting of sensory properties and NIR (near infrared, see e.g. Williams and Norris (2001)) measurements of pea samples of different variety and degree of maturity. The purpose of the study was to evaluate the potential of NIR spectroscopy to determine the sensory properties of the peas. The PLS regression method was used to estimate the regression coefficients. The results were very good in terms of explained variance and indicated strongly that it is possible to predict the most important sensory properties by this type of spectroscopic method.

Other examples with different types of food products and also different types of spectroscopy can be found in Ellekjær *et al.* (1994), Hildrum *et al.* (1995), Kjølstad *et al.* (1990) and Wold *et al.* (2006).

6.2.4.2 *Understanding Sensory Profiles in Terms of Chemical Measurements*

Helgesen and Næs (1995) reported a study of fermented lamb sausages by the use of sensory and chemical analysis. The different brands, 14 in total, were characterised and similarities and differences were identified and interpreted by sensory profiling. For understanding more about which chemical components that were responsible for the detected sensory effects, a chemical profile was also established for the sausages. The relation between the two data sets was obtained by the use of PLS-2 regression with the sensory data as responses (**Y**) and the chemical data used as **X**. The relations obtained were reasonably good providing confidence in both data sets (explained variance in sensory data from the chemistry was equal to 49 % using two components only). Significance of the two first components was confirmed by cross-validation. The results were interpreted by investigating scores plot and the loadings plot for both **X** and **Y** (Figure 6.2).

The most striking tendencies are that water activity (Aw) and water content (H_2O) were negatively related to the sensory attributes bitterness and hardness and positively to juiciness and acidic odour and flavour. Protein and fat content were related to the same sensory variables in the opposite way. This indicates that high water activity, low protein and salt content gave sausages with high juiciness, low hardness and bitterness and vice versa. The second component was related to fat content, spiciness and sweet flavour. The second component indicates that in this data set high fat content, low water content and low pH were related to the spicy sausages with low level of smoke flavour.

6.2.4.3 *Optimising Product Properties*

Assume for instance that a consumer test has indicated a region of optimal sensory properties and none of the tested objects or products has exactly these properties. If a model can be established between sensory properties and the ingredients, it may be possible to use this model to identify the amount of the different ingredients that give the desired sensory properties. Such a model is as usual typically obtained by using some type of experimental design (mixture design, central composite, Sivertsen *et al.*, 2007) and regression fitting. Optimisation of a single sensory property is generally a simple task using a contour plot as indicated in Chapter 11. With a few properties in focus one can do the same for each individual property, but with a high number of properties, multivariate optimisation is required. This is generally a more difficult task, but a simplified way of doing this is proposed in Sivertsen *et al.* (2007) based on contour plots of various derived quality criteria.

6.2.4.4 *External Validation of Sensory Data*

Although sensory data are often good and reliable data there may sometimes be of interest to validate the results with respect to for instance chemical measurements. One possible reason may be to convince a customer or a newcomer in the area about the benefits and the quality of sensory panel data. More generally this type of external validation is useful for judging the validity and quality of the data at hand. As above, one will define the sensory data as **Y** and the external data as **X** and estimate the relation using for instance the PLS regression method. Good predictive relations will be considered positive for the confidence of the sensory data.

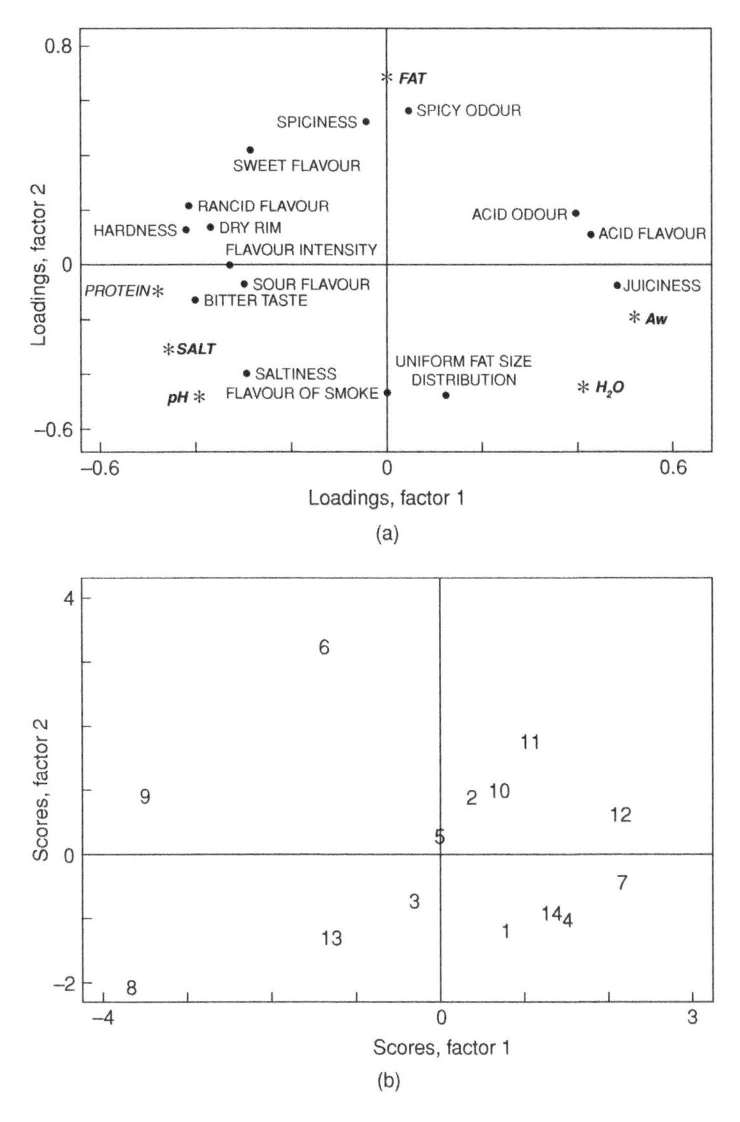

Figure 6.2 PLS loadings and scores plot for external validation. *PLS a) loadings and b) scores plots for sensory data of dry fermented lamb sausages (Helgesen and Næs (1995)). The data considered here come from the comparison of chemical (**X**) and average sensory (**Y**) data. Reprinted from Food Quality and Preference, 20, Helgesen and Næs, Selection of dry fermented lamb sausages for consumer testing, 109–120, 1995, with permission from Elsevier.*

An example of this can be found in Dahl and Næs (2004) where the sensory properties of green peas were again compared to spectroscopic measurements. Prediction results using principal component regression (PCR) are presented in Figure 6.3 for different subsets of assessors. Clustering indicated that assessor 6 was very different from the rest and that assessors 1, 2, 4, 7, 9 and 10 were the most reliable. The dendrogram for this clustering

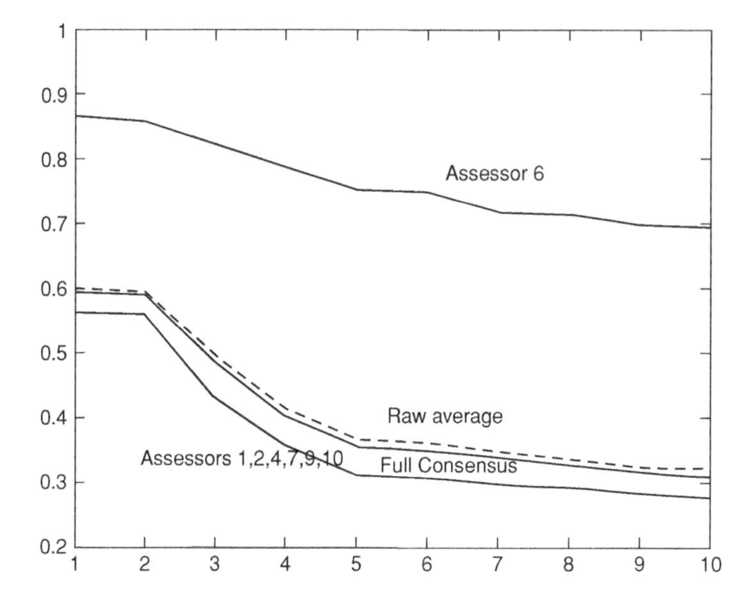

Figure 6.3 Prediction of sensory data by NIR spectroscopic measurements of peas. *The figure shows average squared prediction residuals as a function of the number of principal components used (PCR, see Chapter 15, see also Dahl and Næs (2004)). The different curves correspond to prediction of averages based on different combinations of sensory assessors. The upper curve is based on assessor 6 alone and the curve at the bottom (the best) is for a subset of assessors. The raw average and the consensus from the generalised procrustes analysis (GPA) lie in between. Reprinted from Food Quality and Preference, 15, Dahl and Naes, Outlier and group detection in sensory panels using hierarchial cluster analysis with the Procrustes distance, 195–208, 2004, with permission from Elsevier.*

can be found as an example in Chapter 15 (Figure 15.3). The prediction ability curves in Figure 6.3 confirm that the consensus profile based on the reliable assessors provides a better relation to the spectroscopic data than the consensus based on all assessors.

6.3 Estimating Relations between Individual Sensory Profiles and External Data

6.3.1 Using Sensory Measurements as Response Data

Also in this case, the regression direction can go in both directions. If the sensory data are treated as **Y**, the most common way of modelling the data is by the use of some multi-block or multi-way component analysis of the sensory data accompanied with subsequent or simultaneous regression of the low-dimensional consensus representation of the sensory data onto the external data.

In more mathematical terminology, this is the same as modelling the sensory data by either PARAFAC, GCA or one of the Tucker methods (see Chapter 17, Tucker, 1964; Smilde *et al.*, 2004) and then regressing the low-dimensional consensus representation (**T**) of the samples onto the external data (**X**). For the Tucker-2 method, which is a very natural three-way method to use for sensory data, a solution has been developed which does both steps in one single process. This means that the multi-block modelling is influenced also by the external data. The model for this particular situation is the following

$$\mathbf{Y_i} = \mathbf{XBW_iP^T} \tag{6.4}$$

where the $\mathbf{W_i}$ are the individual differences matrices, the $\mathbf{P}$ is the common loadings matrix while the external regression relation $\mathbf{XB}$ has now replaced the common scores $\mathbf{T}$ in the original Tucker-2 model (see Van der Kloot and Kroonenberg, 1985; Kroonenberg and De Leeuw, 1980).

For the other methods, the two steps are done separately. For Tucker-1 this means that first a PCA of the unfolded sensory matrix is done and then the common scores $\mathbf{T}$ are used as Y-data in a regression equation onto the external data $\mathbf{X}$. The same holds for GCA and also for the PARAFAC.

6.3.2 Examples: Using Sensory Measurements as Response Data

6.3.2.1 Predicting the Average and Understanding Individual Differences

The Tucker-2 method was used in Næs and Kowalski (1989) for relating sensory Y-data to NIR data for green peas of different maturity and variety (see also Martens and Martens, 1986). The predictive ability of the sensory data was compared for several methods. The results obtained using model (6.4) gave good predictive ability. The regression coefficients $\mathbf{B}$, the common scores and also the individual matrices $\mathbf{W_i}$ were determined. The latter were used to investigate the structure of the individual differences. The individual differences were quite large in this case related to whether they focused on the common component 1 or 2 or both. Note that the method can also be used to predict individual profiles by multiplying $\mathbf{x}$ with $\mathbf{B}$ before $\mathbf{W_i}$ and $\mathbf{P}$. How to interpret $\mathbf{W_i}$ matrices is discussed further in Chapter 17.

6.3.2.2 External Validation for Determining the Best Data Compression

The focus of the study in Dahl and Næs (2006) was to consider the relation between the Tucker-1 method and the GCA (Carroll, 1968). It was proved that the two methods could be linked through the introduction of a general class of methods depending on a parameter α. The two methods were obtained by two different settings of α. A procedure was also proposed for determining which low-dimensional representation, GCA, Tucker-1 or something in between, that had the best relation to the external data. The results were obtained by simply computing the first two latent variables for a grid of α values, and then for each case investigating the relation to the external data $\mathbf{X}$. In this case the experimental design was used as external data. The quality criterion used for deciding which α value to use was the size of the F-value from regular ANOVA. A compromise between the two extremes (GCA and Tucker-1) was found to give the best results and was therefore considered the best of the possibilities.

6.3.3 Using Sensory Measurements as Predictor Data

One possible and simple approach when Y-data are used as predictor data (see Chapter 6.2 for arguments) is to use multi-block PLS regression (see e.g. Westerhuis *et al.*, 1998). This is essentially a regular PLS regression of the unfolded data matrix followed with interpretation tools for investigating the individual differences. Another alternative is to use the LS-PLS (Jørgensen *et al.*, 2004) approach mentioned above, but this approach is most useful for situations with only a small number of blocks (here assessors). A third possibility is the three-way PLS (Bro, 1996) which is essentially a combination of PARAFAC and regression. The common sample scores are used in the regression but they are determined in such a way as the relation to the external data is as good as possible.

References

Bro, R. (1996). Multiway calibration. Multilinear PLS. *Journal of Chemometrics* 10, 47–61.

Carroll, J.D. (1968). Generalisation of canonical analysis to three or more sets of variables. *Proceedings of the 76th convention of the American Psychological Association*, Vol. 3. 227–8.

Dahl, T., Næs, T. (2004). Outlier and groups detection in sensory panels using hierarchical cluster analysis with the Procrustes distance. *Food Quality and Preference* 15, 195–208.

Dahl, T., Næs, T. (2006). A bridge between Tucker-1 and Carroll's generalised canonical analysis. *Computational Statistics and Data Analysis* 50(11), 3086–98.

Dijksterhuis, G. (1996). Procrustes analysis in sensory research: In Næs T. and Risvik E. (eds). *Multivariate Analysis of Data in Sensory Science*, pp. 185–217. Amsterdam: Elsevier.

Ellekjær, M.R., Isaksson, T., Solheim, R. (1994) Assessment of sensory quality of meat sausages by the use of near infrared spectroscopy. *Journal of Food Science* 59(3), 456–64.

Gunst, R.F., Mason, R.L. (1977). Biased estimation in regression. An evaluation using mean squared error. *Journal of American Statistical Association* 72, 616–28.

Gunst, R.F., Mason, R.L. (1979). Some considerations in the evaluation of alternate prediction equations. *Technometrics* 21, 55–63.

Hanafi, M,. Kiers, H. (2006). Analysis of K sets of data, with differential emphasis on agreement between and within sets. *Computational Statistics and Data Analysis* 51, 1491–1508.

Helgesen, H., Næs, T. (1995). Selection of dry fermented lamb sausages for consumer testing. *Food Quality and Preference* 6, 109–20.

Hildrum, K.I., Isaksson, T., Næs, T., Nilsen, B.N., Lea, P. (1995) Near infrared reflectance spectroscopy in the prediction of sensory properties of beef. *Journal of Near Infrared Spectroscopy* 3(2), 81–7.

Joliffe, I.T. (1982). *Principal Components Analysis*. New York: Springer Verlag.

Jørgensen, K., Segtnan, V., Thyholt, K., Næs, T. (2004). A comparison of methods for analysing regression models with both spectral and designed variables. *J. Chemometrics* 18, 451–64.

Kjølstad, L., Isaksson, T., Rosenfeld, H.J. (1990) Prediction of sensory quality by near infrared reflectance analysis of frozen and freeze dried green peas (Pisum sativum). *Journal of the Science of Food and Agriculture* 51(2), 247–60.

Kohler A, Hanafi M, Bertrand D, *et al.* (2008) Interpreting several types of measurements in bioscience. In: P. Lasch, J. Kneipp (eds), *Modern Concepts in Biomedical Vibrational Spectroscopy*, New York: John Wiley & Sons, Inc.

Kroonenberg, P., DeLeeuw, J. (1980). Principal components analysis of three-mode data by means of alternating least squares algorithms, *Psychometrika* 45, 69–97.

McEwan, J.A. (1996). Preference mapping for product optimization. In T. Næs & E. Risvik (eds.), *Multivariate Analysis of Data in Sensory Science,* Vol. 16, *Data Handling in Science and Technology*, pp. 71–102, Amsterdam: Elsevier Science B.V.

Mardia, K.V., Kent, J.T., Bibby, J.M. (1979). *Multivariate Analysis*, London: Academic Press.

Martens, M., Martens, H (1986). Near-infrared reflectance determinations of sensory quality of peas. *Applied Spectroscopy* 40, 303–10.

Martens, H., Martens, M. (2001). *Multivariate Analysis of Quality: An Introduction*. Chichester: John Wiley & Sons, Ltd.

Næs T. (1990) Handling individual differences between assessors in sensory profiling, *Food Quality and Preference* 2, 187–99.

Næs, T., Martens, H. (1988). Principal components regression in NIR analysis. *J. Chemometrics* 2, 155–67.

Næs, T., Kowalski, B. (1989). Predicting sensory profiles from external instrumental measurements. *Food Quality and Preference* (4/5), 135–47.

Næs, T., Iskasson, T., Fearn, T., Davis, T. (2002). *A User-Friendly Guide to Multivariate Calibration and Classification*. Chichester: NIR Publications.

Romano R., Brockhoff P.B., Hersleth M., Tomic O., Næs T. (2008) Correcting for different use of the scale and the need for further analysis of individual differences in sensory analysis, *Food Quality and Preference* 19(2), 197–209.

Schlich, P. (1996). Defining and validating assessor compromises about product distances and attribute correlations. In T. Næs, E. Risvik (eds), *Multivariate Analysis of Data in Sensory Science*. Amsterdam: Elsevier.

Sivertsen, E. Bjerke, F., Almøy, T., Segtnan, V., Næs, T. (2007). Multivariate optimisation by visual inspection. *Chemometrics and Intelligent Laboratory Systems* 85, 110–18.

Smilde, A., Bro, R., Geladi, P. (2004). *Multi-Way Analysis*. Chichester: John Wiley & Sons, Ltd.

Tucker, L.R. (1964). The extension of factor analysis to three-dimensional matrices. In N. Frederiksen, H. Gulliksen (eds), *Contributions to Mathematical Psychology*. New York: Holt, Rinehart & Winston, 110–82.

Van Der Kloot, Q.A., Kroonenberg, P.M. (1985), External analysis with three-mode principal components models. *Psychometrika* 50(4), 479–94.

Westerhuis, J.A., Kourti, T., MacGregor, J.F. (1998). Analysis of multiblock and hierarchical PCA and PLS models. *Journal of Chemometrics* 12, 301–21.

Williams, P., Norris, K. (2001). *Near-Infrared Technology in the Agricultural and Food Industries*. Minnesota: American Association of Cereal Chemists.

Wold J.P., Veberg A, Nilsen A.A. (2006). Influence of storage time and color of light upon photooxidation in cheese. A study based on sensory analysis and fluorescence spectroscopy. *Int. Dairy Journal* 16: 1218–26

7

Discrimination and Similarity Testing

This chapter handles primarily what is usually known as sensory discrimination testing with major focus on the four basic sensory discrimination test protocols: Duo-Trio, Triangle, 2-AFC and 3-AFC. This includes the basic binomial based hypothesis testing and power/sample size considerations. A thorough treatment of how to also use these methods for similarity testing is given. The use of confidence bands is recommended and exemplified. Also included is a treatment of how to work with replicated difference tests. Throughout, as well a Thurstonian approach as a simple binomial analysis approach is taken in parallel. Instead of providing statistical tables, which is common in most text books the open source R-package sensR (Christensen and Brockhoff, 2009b) is used throughout.

The chapter is related to Chapter 11.

7.1 Introduction

The sensory discrimination methods are characterised by being relatively simple tasks for the assessor. This is why these methods are widely used for as well consumer studies as for trained sensory panels. The most basic and mostly used sensory difference tests are the following six: Duo-Trio, Triangle, 2-AFC, 3-AFC (AFC = Alternative Forced Choice), A-not A and same-different. In the Duo-Trio test the panelist (which in this chapter then also could be a consumer) is presented with three samples of two types, e.g. AAB. One of the As is then identified to the panelist as a control and he/she is asked to select which of the remaining two samples that match the control. Also in the Triangle test three samples are presented including both sample types, e.g. AAB. Now the panelist is asked to select the odd sample. In the 2-AFC test a pair of two different samples is presented to the panelist, but now the question is on a specific sensory characteristic, and he/she is asked to select the strongest (or the weakest) sample. Similarly in the 3-AFC, the panelist is asked to select

Statistics for Sensory and Consumer Science Tormod Næs, Per B. Brockhoff and Oliver Tomic

the strongest (or the weakest) sample among three samples presented. These four methods share the characteristic that each panelist is presented with two different samples. This is why these methods sometimes from a psychological perspective are said to be 'without response bias', see e.g. O'Mahony (1992). This is not the case for the A-not A and same-different tests where each or some panelist is presented with only one of the two samples. In the A-not A method panelists are familiarised with both A and not-A samples before the test. Then in the test a panelist is presented with only one sample and is asked if the sample is 'A' or 'not A'. Finally in the same-different test each panelist is presented with one of the possible pairs of samples, AA, AB (or BA) or BB, and is asked if the sample pair is the 'same' or 'different'. For all test protocols the panelist is forced to make a choice/give an answer.

7.2 Analysis of Data from Basic Sensory Discrimination Tests

A number of these test protocols simply produce what could be termed one-sample binomial data: we observe x 'successes' in N independent trials, see Chapter 11. Statistical computations, be it hypothesis tests, power and sample size calculation and/or confidence bands for binomial data situations are often presented by the use of the normal distribution, since for N large enough (relative to the binomial p) the normal distribution approximates the binomial distribution adequately. This tradition, employed in most introductory textbooks, is to a large extent a reminiscence from the pre-computer time. Alternatively, all computations can be based directly on the binomial distribution leading to 'exact' statistical results, but with the downside, that it becomes less straightforward to provide simple computational formulae. Today, practitioners would either use dedicated statistical tables from scientific papers, books or standards, see e.g. Schlich (1993) and ISO 4120 (see reference list), or a dedicated statistical software. In the examples here we have used the open source R-package sensR (Brockhoff and Christensen, 2010; Christensen and Brockhoff, 2009b).

Since there is a basic guessing chance of getting the answer correct, $^1/_2$ for the duo-trio and 2-AFC, 1/3 for the triangle and 3-AFC, the observed proportion of correct answers does not directly estimate the proportion of individuals in the (relevant) population that would detect the product difference with the chosen test protocol. But there is an easy transformation between the two proportions.

In Figure 7.1 the three overall levels of how to approach this type of product testing is illustrated. The levels are here denoted '0', '1' and '2'. Level 1 analysis is what we actually directly observe, the proportion of correct answers $\hat{p}_c = x/N$, which is a direct estimate of the underlying proportion p_c. This is related to the proportion of individuals in the (relevant) population that would detect the product difference, here denoted by p_d, according to the formulae:

$$\text{Duo-trio, 2-AFC}: \; p_c = \frac{1}{2} + p_d\frac{1}{2} \quad \text{triangle, 3-AFC}: \; p_c = \frac{1}{3} + p_d\frac{2}{3} \tag{7.1}$$

Level 2 analysis focuses on the transformation the other way around:

$$\text{Duo-trio, 2-AFC}: \; p_d = \frac{p_c - 1/2}{1/2} \quad \text{triangle, 3-AFC}: \; p_d = \frac{p_c - 1/3}{2/3} \tag{7.2}$$

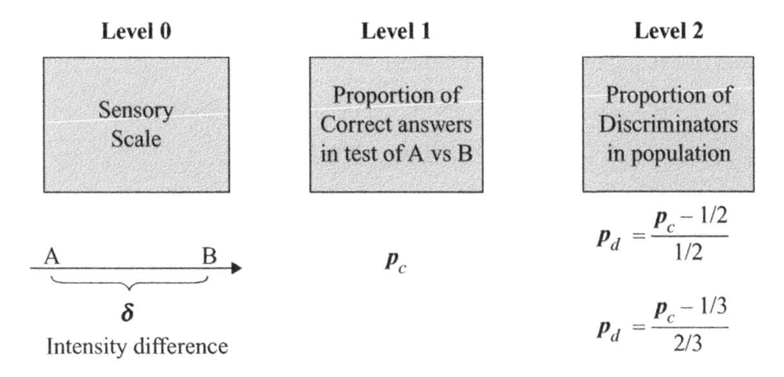

Figure 7.1 The three overall levels of sensory discrimination/similarity testing.

That is, in Level 2 analysis the results are expressed as the proportion of individuals in the (relevant) population that would detect the product difference with the chosen test protocol.

The weakness of the Level 1 and 2 approaches, working directly on the count scale, is that it is test protocol dependent: the number of expected correct answers for the same products depends heavily on which test that is carried out. This has been pointed out by several authors, see e.g. Ennis (1993a), and amounts to the lack of a common framework for comparing the underlying sensory differences across different testing paradigms/protocols. It does not mean that the Level 1 and 2 statistical analyses are wrong, but rather that they are limited in the sense that it is not possible to compare results for different test protocols with each other. The Thurstonian approach of transforming the number of correct answers into an estimate, called d-prime (d'), of the underlying (relative) sensory difference, is the solution of this deficiency of the count data statistical approach. This can be done for any of the tests and is what we denote Level 0 analysis in Figure 7.1.

7.3 Examples of Basic Discrimination Testing

7.3.1 Level 1 Analysis

Assume that we have 10 out of 15 correct responses in each of the four discrimination test settings covered here. The Level 1 analysis simply tells uniformly for each of the four situations that the estimated proportion of correct answers is

$$\hat{p}_c = 0.667$$

with a 90 % confidence band of [0.423, 0.858]. This is the so-called exact binomial confidence interval, which e.g. means that 0.423 is the smallest possible value of p_c which is accepted in a level 5 % one-tailed hypothesis test with the given data.

For the duo-trio and 2-AFC situations the hypothesis test of no difference can be formulated as

$$\begin{aligned} H_0 &: p_c = 1/2 \\ H_A &: p_c > 1/2 \end{aligned} \tag{7.3}$$

For the actual observation the p-value for this test is equal to 0.1509, obtained by using the exact binomial test (Chapter 11) computed as

$$\begin{aligned} p - \text{value} &= \text{P(Observing 10 or larger "at random")} \\ &= \text{P}(x \geq 10),\, x \sim \text{bin}(15,1/2) \\ &= 0.1509 \end{aligned} \tag{7.4}$$

Similarly, for the triangle and 3-AFC tests, the hypothesis of no effect can be formulated as

$$\begin{aligned} H_0 &: p_c = 1/3 \\ H_A &: p_c > 1/3 \end{aligned} \tag{7.5}$$

with a p-value for this data equal to

$$\begin{aligned} p - \text{value} &= \text{P}(x \geq 10),\, x \sim \text{bin}(15,1/3) \\ &= 0.0083 \end{aligned} \tag{7.6}$$

Note how the conclusion for the duo-trio and the 2-AFC is that there is no difference between the products while there for the triangle and 3-AFC test is a clear significant effect.

In general, p-values could be read off in tables (Schlich, 1993), in sensory textbooks or they can be computed by standard or dedicated software.

7.3.2 Level 2 Analysis

In the level 2 analysis one uses the simple transformations given above for both the estimate and the lower and upper confidence limits (setting the lower limit to zero in case the transformation turns negative). For the actual data, the duo-trio and the 2-AFC tests, give an estimate equal to 33.3 % and the confidence interval (0 %, 72 %) and the triangle and the 3-AFC tests give an estimate equal to 50 % with a confidence interval equal to (13 %, 79 %).

The exact intervals described here are generally better than the classical normal based approximate interval given by:

$$\hat{p}_c \pm z_{1-\alpha/2}\sqrt{\hat{p}_c(1-\hat{p}_c)/N} \tag{7.7}$$

where the z is the upper $\alpha/2$ percentile of the standard normal distribution. For large samples they will give more or less the same results, but for small samples the exact interval is more correct. Another practical advantage of the exact intervals is that they are computable even in the extreme cases of observing either 0 or 100 % correct answers in these test – situations where the normal based approach fails to provide an answer – as the answer is *not* in these cases that we know for sure that $p = 0$ or $p = 1$!

7.3.3 Level 0 Analysis: Thurstonian Modeling

The level 0 analysis is the so-called Thurstonian computation of the underlying sensory difference d'. Below, we discuss in more detail the principles of the Thurstonian approach. Here, just think of it as some attempt of quantifying the degree of difference between the two products that were tested. If the proportion of correct answers is really big – close to 1 – we would interpret this as showing us that the two products have a large difference, whereas

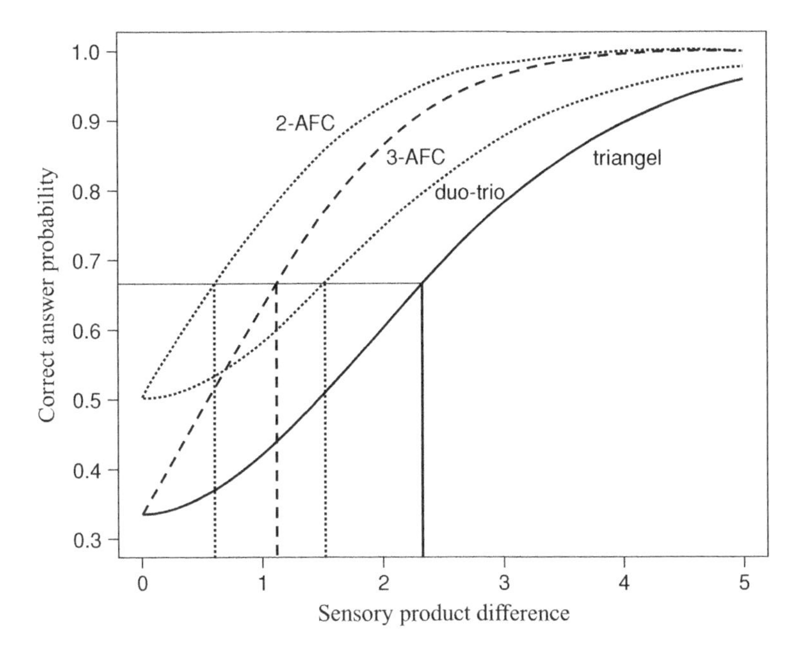

Figure 7.2 *The four psychometric functions.*

a small proportion of correct answers – close to the guessing probability – we interpret it as a small difference. However, it is clear that the same proportion of correct answers in different test protocols cannot generally correspond to the same sensory difference: for instance 5 out of 10 correct answers in a triangle test indicates some sensory difference whereas the same result for the duo-trio corresponds exactly to the guessing level, that is, no difference. Less obvious but still fairly reasonable, the triangle and the 3-AFC tests are not expected to provide the same number of correct answers on the same products: And similarly the duo-trio and the 2-AFC tests are expected to provide different answers to the same products. The Thurstonian approach provides an explicit mathematical model between the proportion of correct answers and the underlying sensory difference for each of the four test protocols. In Figure 7.2 these four functions are depicted together with an indication of the 10/15 outcome for the four protocols as in Table 7.1. To estimate a sensory difference, the d', say, for a triangle outcome of 10/15, the triangle function in Figure 7.2 is used in an inverse way: find 2/3 on the Y-axis and read off the point on the X-axis. The results, that is, the estimated sensory difference between the products for each of the four protocols are given in Table 7.1 together with their 90 % confidence limits. For the two situations with no significant product difference, the lower confidence limit of the sensory difference is zero. The triangle test estimates the highest sensory difference.

Whereas the exact confidence interval has been used regularly for the Level 1 analysis of such data it is common to provide a normal based approximate interval for the d' in the level 0 analysis, which would look like:

$$d' \pm z_{1-\alpha/2} SE_{d'} \tag{7.8}$$

Table 7.1 *Analysis for the results 10 correct out of 15 tests for each of the four test protocols. The 90% confidence intervals are given (exact binomial intervals for level 1 and 2, likelihood intervals for level 0).*

	Level 1 Analysis			Level 2 Analysis			Level 0 Analysis			
n = 15 x = 10	P_c	Low	upp	P_d	low	upp	d′	low	upp	p-value
Duo-trio	2/3	0.423	0.858	0.333	0	0.717	1.52	0	2.66	0.1509
2-AFC	2/3	0.423	0.858	0.333	0	0.717	0.61	0	1.40	0.1509
Triangle	2/3	0.423	0.858	0.500	0.134	0.788	2.32	1.23	3.43	0.0085
3-AFC	2/3	0.423	0.858	0.500	0.134	0.788	1.12	0.41	1.85	0.0085

where the standard error $SE_{d'}$ is found by standard statistical maximum likelihood methods, see e.g. Pawitan (2001). This approximate interval for d' shares deficiencies with the approximate binomial interval. For some reason no one seems to have done the obvious here: transforming the exact binomial interval into an interval for d' by transforming the interval endpoints by the same nonlinear function which is used to compute the d'. In the examples here we have provided something even better than this: we give the so-called likelihood intervals, which are found in a similar way to the exact intervals but using the likelihood function (Pawitan, 2001) as criterion rather than the binomial probabilities. The likelihood intervals can just as the exact binomial intervals always be computed and they are in fact just a little bit better than the exact intervals in the simplest of cases, see Boyles (2008). Further, one more important feature is the generality of the method – it can be straightforwardly adapted to many different situations and data types – something less obvious for the exact intervals. As for the exact binomial intervals we need a computer to do the job for us and we have used the sensR package in R (Brockhoff and Christensen, 2010).

In Table 7.2 the analysis of another set of four outcomes, one for each protocol, is given. With responses ranging from 10 to 17 correct answers out of 20, the four protocols provide basically the same estimate of sensory difference d' in each case: 1.42–1.47. In Figure 7.2 it would more or less correspond to following/extending the duo-trio vertical marker, since in Table 7.1 the duo-trio d' is around the same level, showing how the four different protocols

Table 7.2 *Analysis for the results 10 correct out of 15 tests for each of the four test protocols. The 90% confidence intervals are given (exact binomial intervals for level 1 and 2, likelihood intervals for level 0).*

		Level 1 Analysis			Level 2 Analysis			Level 0 Analysis			
n = 20	DATA x	P_c	Low	upp	P_d	low	upp	d′	low	upp	*p*-value
Duo-trio	13	0.650	0.442	0.823	0.300	0	0.645	1.42	0	2.40	0.1316
2-AFC	17	0.850	0.656	0.958	0.700	0.313	0.916	1.47	0.71	2.31	0.0013
Triangle	10	0.500	0.302	0.698	0.250	0	0.547	1.47	0	2.38	0.0919
3-AFC	15	0.750	0.544	0.896	0.625	0.313	0.844	1.43	0.8	2.11	0.0002

are expected to give quite different number of correct answers to tests of the same two products. The product difference is only significant for the 2-AFC and the 3-AFC, not for the duo-trio and triangle. The 3-AFC provides the clearest significance statement (smallest p-value) and the most precise estimate of the actual sensory difference (the most narrow confidence band for d'). This illustrates that the 3-AFC is the most powerful method among these four.

7.4 Power Calculations in Discrimination Tests

In Table 7.2 10 out of 20 correct answers in a triangle test experiment did not show a significant difference between the products. First of all now, it could be tempting in this situation to calculate the power of detecting the observed difference $p_c = 0.5$ with $N = 20$. However, such 'post hoc power analysis' is generally not advisable, see e.g. Hoenig and Heisey (2001). The proper statistical tool to use at this point is the confidence interval: the interval really shows explicitly the quality of the information in the data: a narrow interval shows that we are quite certain about the statement we have made, whereas a wide interval shows nicely that we do not have much information such that the 'accept result' is simply due to lack of knowledge.

Instead let us as an example find the power of the triangle test with $N = 20$ (using the test level 0.05) for three different scenarios:

$$\begin{aligned}&\text{Small difference (25\% effect) corresponding to } p_d = 0.25 \Leftrightarrow p_c = 0.5000\\&\text{Medium difference (37.5\% effect) corresponding to } p_d = 0.375 \Leftrightarrow p_c = 0.5833\\&\text{Large difference (50\% effect) corresponding to } p_d = 0.50 \Leftrightarrow p_c = 0.6667\end{aligned} \tag{7.9}$$

These are the scenario levels proposed in Schlich (1993), but clearly the definition of what is 'small' or 'large' can depend heavily on the situation at hand. The critical value of the level 0.05 hypothesis test with $N = 20$ is equal to $x_c = 11$. This means that observing less than 11 correct answers would (wrongly) accept the hypothesis of no difference. The power of the large difference scenario (50% effect), can then be computed as

$$\begin{aligned}Power &= P(x \geq 11), x \sim bin(20{,}0.6667)\\&= 0.908\end{aligned} \tag{7.10}$$

This means that the chance of detecting a large difference, as defined here, with the triangle test with $N = 20$ is 90.8 %. This is usually considered as a reasonable power for a relevant scenario. For the small difference scenario the power becomes

$$\begin{aligned}Power &= P(x \geq 11), x \sim bin(20{,}0.5000)\\&= 0.412\end{aligned} \tag{7.11}$$

And for the medium difference scenario the power becomes

$$\begin{aligned}Power &= P(x \geq 11), x \sim bin(20{,}0.5833)\\&= 0.704\end{aligned} \tag{7.12}$$

Table 7.3 Sample sizes needed to obtain a 90% power using a level 0.01 test.

Test protocol	d′ = 1 ("Small")	d′ = 1.5 ("Medium")	d′ = 2.5 ("Large")
Duotrio	478	121	52
2-AFC	47	22	14
Triangle	425	105	46
3-AFC	35	17	11

The experiment is not good for finding medium differences and completely unsuited for detecting small differences.

Using the sensR software one can easily compute power for any constellation of N, p and significance level α and also compute the N needed for obtaining a given power. As an example; in order to obtain a power of 90 % for a small (25 %) difference in a triangle experiment when the significance level is 0.01, is found to be equal to $N = 115$, a rather large number in this context.

When working with power considerations in the experimental planning phase, defining sensory differences like above within the Level 1 and 2 analyses approaches has again the clear limitation that the approach is test protocol dependent. As already pointed out a 50 % effect for the 3-AFC test would not correspond to a 50 % effect for the triangle test, so the approach only works within a situation where we already decided the protocol to use. The solution is to work with the Level 0 (Thurstonian) approach, which of course requires tables or software that includes the link to this level. This is included in the sensR package, which we can use to (easily) provide the sample sizes in Table 7.3 needed to obtain a 90 % power using a level 0.01 test for the four different protocols. The superior power of the 3-AFC becomes clear here, so whenever possible this should be chosen instead of the triangle test.

7.5 Thurstonian Modelling: What Is It Really?

The focus in the Thurstonian approach is on quantifying/estimating the underlying sensory difference d' between the two products that are compared in the difference test. This is done by setting up mathematical/psycho-physical models for the cognitive decision processes that are used by assessors in each sensory test protocol, see e.g. Ennis (1993a). This is done by first of all thinking of each individual observation in the test situation as a realisation of a random (but unobservable) underlying sensory variable with some distribution stemming from perceptual and sample replication variability. In Figure 7.3 this is illustrated for a three-sample protocol, such as the triangle and 3-AFC, the test consisting of two A-products and one B-product. However, the two distributions illustrate that within each product type we allow for some variability and the fact that the two distributions overlap illustrates that we are in the confusable stimulus area, which means that most of time A-products are perceived less intense than B-products but sometimes a B product falls to the left of one or both of the A-products.

For the triangle test, the usual model for how the cognitive decision process is taking place is that the most deviating product would be the answer – sometimes called a tau-strategy.

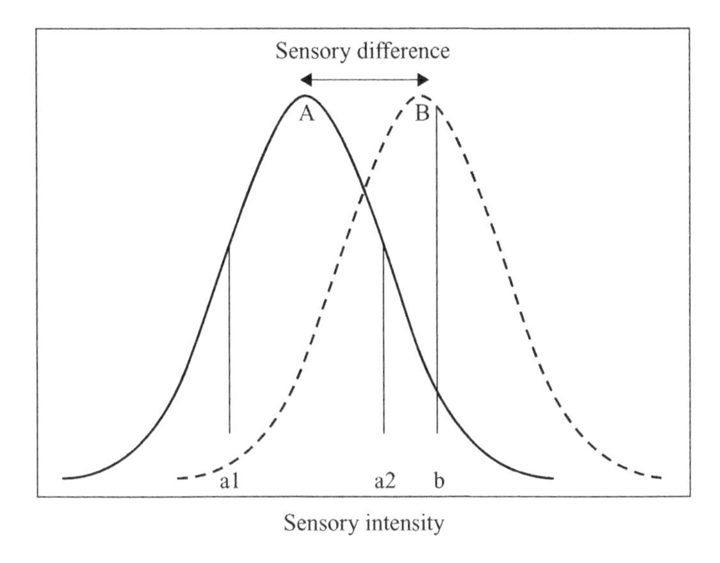

***Figure 7.3 Thurstonian modelling**. The basic idea of Thurstonian modelling of a 3-sample test like the triangle and/or the 3-AFC.*

More specifically, this means that the assessor identifies the two products that are closest to each other in terms of sensory perception and points out the third as the odd sample. For the 3-AFC test, the usual model is that assessors use a so-called beta-strategy where the product with the largest intensity would be the answer. In the hypothesised example in Figure 7.3 the 3-AFC test would provide the correct answer 'b' whereas the triangle test would wrongly claim 'a1' to be the answer. Using basic probability calculus on 3 realisations from two different normal distributions, differing by exactly the true underlying sensory difference d', one can deduce the probability of getting the answer correct for each strategy. This function is called the psychometric function and relates the observed number of correct answers to the underlying sensory difference d'. The four psychometric functions are shown in Figure 7.2. We do not give the formal expressions here (Brockhoff and Christensen, 2010; Ennis, 1993a), but all of them have origins going much further back in time. Specifically the 2-AFC relates directly back to Thurstone (1927). Within the field known as signal detection theory, see e.g. Green and Swets (1966) or MacMillan and Creelman (2005), methods of this kind were further developed, originally with special emphasis on detecting weak visual or auditory signals. Further developments of such methods and their use within food testing and sensory science have developed over the last couple of decades with the numerous contributions of D. Ennis as a cornerstone (Ennis, 2003, 1993b; Ennis *et al.*, 1988, 1998).

7.6 Similarity versus Difference Testing

All the hypothesis tests carried out until now in this chapter are *difference* test, where the implicitly assumed purpose of the experiment is to find evidence of a product difference. Often these sensory protocols are used with the opposite purpose: to find evidence of a sufficient degree of similarity between products. For instance, in the ISO 4120 standard two specifically different approaches are pointed out for these two different purposes. In fact, the approach recommended for similarity testing using the triangle test in ISO 4120 corresponds to the following approach that can be used for any test protocol:

1. Specify the wanted degree of similarity
2. Claim similarity with confidence $1 - \beta$ if the $100(1 - \beta)\%$ upper confidence limit is within this specification

This approach is superior to what is usually denoted the 'power approach' of similarity testing as described in for instance Chapter 11, p. 255 in Bi (2006). The approach is based on claiming similarity when the usual discrimination null hypothesis is not rejected (but doing so only for setups chosen with reasonable power levels). One problem of this approach is that with a large sample size, small differences within the specified degree of similarity would be detected as significant, that is as nonsimilar.

More recently the concept of formal equivalence testing has been adopted for this purpose, since this is really what fits the situation of similarity testing, see e.g. Bi (2005). This corresponds to interchanging the roles of the null and the alternative hypotheses compared to the difference tests above combined with using the specified level of similarity. The null and the alternative hypotheses for the similarity (equivalence) test of a specified $p_d \leq 0.25$ similarity level are: (using here 1/3 as the guessing probability)

$$\begin{aligned} H_0 &: p_d \geq 0.25 \Leftrightarrow p_c \geq 0.50 \\ H_A &: p_d < 0.25 \Leftrightarrow p_c < 0.50 \end{aligned} \tag{7.13}$$

With e.g. 55 correct answers in 150 tests, the exact p-value for this equivalence test is easily found as a binomial probability:

$$\begin{aligned} p - value &= P(x \leq 55), x \sim bin(150, 0.5) \\ &= 0.000684 \end{aligned} \tag{7.14}$$

Note how this computation is exactly the same as when we found the exact upper confidence limits above. This means that the rule set up in the ISO 4120 standard is exactly the one-tailed equivalence test, even though this is not stated in the standard. The β's in point 2 above (and in Table A.2 of the standard) are the α's for the equivalence test. In other words, by providing the 90 %-confidence intervals as in Tables 7.1 and 7.2, we can directly perform the level $\alpha = 0.05$ equivalence test by simple inspection: is the upper confidence limit below the specification level. And since, confidence limits are given on any of the three levels of analysis, this can be done on any level including the Level 0 Thurstonian d' analysis.

For the proper planning of a similarity test, power and sample size considerations should be approached with this in mind instead of just using the discrimination test power/sample size as given above. In the ISO 4120 standard only the latter is recommended. In Bi (2006) the more suitable approach is described. In sensR the exact version of the proper similarity test power and sample size computations are available in simple functions.

The idea of using confidence intervals for the unification of discrimination and similarity testing was put forward by MacRae (1995) and Carr (1995), but has only to some extent been adopted by the sensory community. One of the reasons is that there has been and still is some controversy about how to perform 2-tailed equivalence tests in various settings, see e.g. Ennis *et al.* (2008) and Ennis and Ennis (2010a) (and three additional discussion contributions: Castura (2010), Bi (2010) and Ennis and Ennis (2010b). This controversy is a version of an ongoing and rather fundamental discussion between 'Fisherians' and 'Neymanians' in the statistical literature about this issue, see e.g. Berger and Hsu (1996) and Pearlman and Wu (1999). It is beyond the scope of this book to discuss this controversy further.

7.7 Replications: What to Do?

In the treatment above it was implicitly assumed that no replications were present. More precisely, the number of observations N corresponds to either the number of consumers/panelists in the experiment (each carrying out one and only one test) or (less typical) to the number of tests carried out by a single person. Often, it is tempting and easy to have the individuals in a study perform more than a single test each.

7.7.1 The Naive Approach and Various Approaches that Handle Individual Differences

The setting is now that N assessors/consumers each carried out K tests and we record the number of correct answers for each: $x_1, \ldots, x_N$. The fundamental additional data analytical issue is that there may be individual differences between assessors in their ability to detect differences. Such differences is a basic concept in all sensory data and we saw in Chapter 5, how a statistical solution for this is to include the effect of the individuals as a random effect in the models. In Chapter 13 this is treated in further detail. However, the combination of binomial data with the inclusion of individual random effects leads to models of higher complexity, less well known by nonstatisticians and sometimes less supported by standard software than the linear mixed models within the normal distribution framework treated in most other chapters of this book. Some of the models which have been considered are generalised linear mixed models, beta-binomial models and binomial mixture models. In Ennis and Bi (1998) the standard beta-binomial model is recommended for these situations and in Brockhoff (2003), Meyners (2007a) and Bi (2006) the *corrected* beta-binomial is recommended. We believe the latter is the more proper approach, but it remains to be thoroughly investigated how big a difference it really makes whether using one or the other approach. For 2-tailed situations, e.g. a simple paired consumer preference test, where the consumer is asked to point out whether he/she prefers product A or B, this debate vanishes.

Simpler, model free tests were also suggested by Meyners (2007b), which basically amounts to the identification of the situation as a frequency table and applying a standard chi-square test (see Chapter 11) approach comparing observed with expected tables though with the use of simulation based (exact) tests instead of using the chi-square distribution. However, Meyners (2007b) only treats 2-tailed situations leaving only a comment at the end that similar tests could be carried out for the 1-tailed situations in focus here. And no confidence limits are provided with this approach.

First of all, for the basic difference hypothesis test, e.g. for the triangle and/or the 3-AFC tests with the hypothesis of $p_c = 1/3$ for all assessors, it has been argued (Kunert and Meyners, 1999) that the 'naïve' binomial test, completely ignoring the replication issue, is still a valid test. This means that if this test becomes significant it is giving the valid conclusion, that the products are different. However, the true power of this test cannot be evaluated without taking the heterogeneity (individual differences) into account. In Brockhoff (2003) the power of the naïve test was investigated using different possible models for the heterogeneity. It was shown that the loss of power compared to the homogeneous scenario actually was surprisingly low. An important point also made in Brockhoff (2003) was that much more powerful tests for the hypothesis of no difference exist, when heterogeneity is present, and such tests, based on likelihood theory could be deduced from the contents of the paper.

7.7.2 The Beta-Binomial Approaches

To approach individual differences in this context it is natural to extend the basic binomial model to include individual heterogeneity as a parameter in the model. Letting the individual probabilities (be it observed p_{ci} or chance corrected p_{di}) follow a beta-distribution makes good sense in this case since a beta-distribution gives values in the 'probability' interval (0,1). This means that some individuals will have a lower p_{ci} and some will have a higher p_{ci} than others. For each individual, the number of correct answers out of all the K replications still follows a binomial distribution with a p_{ci} from that particular individual. The reason for the popularity of this approach for the directly observed proportions p_{ci} is due to the fact that the induced distribution of the overall number of correct answers can be explicitly expressed (the 'beta-binomial' distribution). Hence standard statistical theory and methods can easily be implemented for this model and it has been suggested and applied in many different areas outside the sensory field, see e.g. Skellam (1948).

The problem of this Level 1 approach is that the model in fact allows for individual p_{ci}'s to be smaller than the guessing probability, meaning that such an individual in the long term performs worse than just guessing. This cannot happen, at least not under the commonly applied Thurstonian model where the perceptual variation for the two products is assumed the same. A solution can be obtained by the Level 2 approach, where the beta-distribution is assumed on the p_{di}s instead, so an extreme of zero in the beta-model corresponds to the guessing level for p_{ci}. Unfortunately, the deduced expression for the distribution of the total number of correct answers is no longer the simple beta-binomial distribution, but a somewhat more complicated distribution called the corrected beta-binomial distribution. This distribution was first deduced by Morrison (1978) but further introduced, treated and recommended in the sensory context by Brockhoff (2003). In Bi (2006) a somewhat simpler version of the probability density function was given where the integral computation is substituted by a finite sum, now in an exact way. The sensR package includes both the standard beta-binomial as the corrected version. The individual heterogeneity/variability can also be expressed/modelled at Level 0 – an approach introduced in Christensen and Brockhoff (2008). Figure 7.4 illustrates how the individual variability can be expressed at each of the three levels leading to three different statistical models, two of which are the most commonly used (Level 1 and 2).

7.7.3 An Example Based on the Naive Approach

Consider the following example of replicated triangle test data: $N = 15$ assessors each completed $K = 12$ triangle tests with the following 15 individual numbers of correct

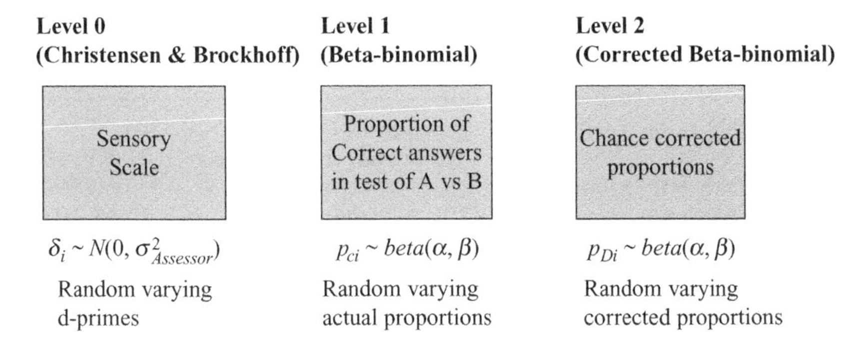

Figure 7.4 *Models for replicated difference tests in light of the three overall analysis levels.*

answers: 2, 2, 3, 3, 4, 4, 4, 4, 4, 4, 4, 5, 6, 10, 11. This means that in 70 out of 180 triangle tests the answer was correct. Let us first consider a naïve analysis of the data, not taking the individual differences into account. The estimates and the 90 % confidence intervals (found as described above in Tables 7.1 and 7.2) are:

$$\widehat{p}_c = 0.389(0.328,0.452), \quad \widehat{p}_d = 0.083(0,0.375), \quad d' = 0.798(0,1.191).$$

The exact one-tailed p-value for the test of no difference is 0.0678. The conclusion would then be that there is no significant product difference.

For comparison, the approximate standard error of $\widehat{p}_c$ would be, using formula (7.7):

$$SE(\widehat{p}_c) = \sqrt{(7/18)(11/18)/180} = 0.0363$$

and hence the standard error for $\widehat{p}_d$ is:

$$SE(\widehat{p}_d) = 1.5SE(\widehat{p}_c) = 0.0545$$

Using formula (7.7), the 90 % confidence intervals for these two parameters turn out to be very close to the given exact ones. The standard error for d' (as e.g. given directly by sensR) is $SE(d') = 0.275$ giving an approximate 90 % confidence interval, cf. Equation (7.8), of d' equal to (0.346, 1.25), which in fact is quite far from the given likelihood based interval.

Looking at the individual triangle outcomes, it seems not quite satisfactory to conclude that there is no product difference. For two of the assessors there is a clear indication with 10 resp. 11 correct answers out of 12 that the products definitely do differ, since an individual based analysis would give very small p-values for these two outcomes. Another flaw of the naïve analysis is that the uncertainties of the estimated population parameters will be underestimated due to the fact that observations are not independent – 12×15 observations is not the same as 180 observations.

7.7.4 The Example Handled by the Beta-Binomial Model

In Table 7.4 is given a complete maximum likelihood based analysis for both the standard beta-binomial as well as the corrected beta-binomial using the sensR package (Christensen

Table 7.4 Results of replicated difference tests for example data.

	Standard Beta-binomial	Corrected Beta-binomial
	$\hat{p}_c = 0.3949 \Leftrightarrow \hat{p}_D = 0.0924$	$\hat{p}_D = 0.0978 \Leftrightarrow \hat{p}_c = 0.3985$
	$SE(\hat{p}_c) = 0.0526$	$SE(\hat{p}_c) = {}^2/_3\, SE(\hat{p}_D) = {}^2/_3\, 0.0660 = 0.0440$
	Homogeneity Test : $\chi_1^2 = 6.26(0.0123)$	Homogeneity Test : $\chi_1^2 = 13.06(0.0003)$
	Joint difference test : $\chi_2^2 = 8.70(0.0129)$	Joint difference test : $\chi_2^2 = 15.49(0.0004)$
d'	0.843	0.869
SE(d′)	0.382	0.312

and Brockhoff, 2009b). A maximum likelihood based analysis provides the optimal set of parameter estimates by maximising the likelihood function, approximate standard errors by differentiating the log-likelihood function and the option of doing hypothesis tests by comparing the size of the optimised likelihood function for different models/hypotheses (Pawitan, 2001). The likelihood function expresses how likely parameter values are given the observed data and using the imposed statistical (probability) model. For further use and discussion of likelihood methods for sensory discrimination tests see Christensen and Brockhoff (2009a) and Brockhoff and Christensen (2010).

The standard beta-binomial analysis directly provides results on the observed level (level 1). The corrected beta-binomial analysis provides results on the chance-corrected level (Level 2). They can each easily be transformed to the other level and for both of them also to the Thurstonian level (Level 0). First note that the estimate of d' itself changes a little bit (as compared to the naïve approach) for both of the methods – mostly for the corrected analysis. Also note that in both analyses the standard error for $\hat{p}_c$ and hence also d' is larger than in the naïve analysis. The standard errors given by the corrected beta-binomial model are clearly smaller than the ones given by the standard beta-binomial in this case. Two hypothesis tests are given in each analysis: they are both likelihood ratio tests, which means that they are results of an optimisation of the nonlinear likelihood function in each model and no closed form expressions can be given.

The results are clear here. If we acknowledge that heterogeneity is an important aspect, then the results in Table 7.4 show that the products are indeed different. The corrected beta-binomial model gives considerably smaller p-values that the standard beta-binomial and a substantially different conclusion as compared to the one coming from the naïve analysis.

A truly Thurstonian version of a random individual effect model would correspond to having individual d'-primes varying randomly on the underlying sensory scale. Such a model has not been reported in the sensometrics literature, but was discussed by Christensen and Brockhoff (2008).

The planning of replicated difference/similarity tests is even more a challenge in light of the different possibilities for data analysis. The results of Brockhoff (2003) can be used to achieve a minimum level of power for the naïve test. In Meyners and Brockhoff (2003) strategies are given on how to design replicated difference test to achieve a certain level of standard errors in the subsequent replicated data analysis.

7.8 Designed Experiments, Extended Analysis and Other Test Protocols

In Brockhoff and Christensen (2010) it was shown how the Thurstonian Level 0 statistical analysis as performed above can be identified as doing a so-called generalised linear model analysis (see Chapter 15). This is a well known class of models in statistics (McCullagh and Nelder, 1989) that makes it possible to carry out standard linear model analysis (ANOVA, linear regression, ANCOVA) also for nonnormal responses, like for binomial responses discussed here. With this in place, and the implementation in sensR, it is possible to combine design of experiments with discrimination/similarity testing and combine standard statistical modeling/analysis with Thurstonian modeling. The psychometric functions then play the role as so-called link functions leading to logistic regression type analyses but including the Thurstonian interpretation.

The data from the A-not A test mentioned as one of the basic, but 'biased', discrimination procedures, is also in Brockhoff and Christensen (2010) identified as a version of a generalised linear model. This is also handled in the sensR package. The data from the same-different procedure does not fall in this class, but maximum likelihood based analysis is implemented in sensR (see Christensen and Brockhoff (2009a) for details).

Similar ideas of using (extensions of) generalised linear models for the proper analysis of designed consumer discrimination test data including consumer background information and order and carry over effects is presented in Christensen *et al.* (2010). In this case A-not A tests with additional evaluations were analysed. This leads to ordinal data (see also Chapter 17).

References

Berger, R.L. and Hsu J.C. (1996) Bioequivalence trials, intersection-union tests and equivalence confidence sets. *Statistical Science* 11(4), 283–302.

Bi, J. 2005). Similarity testing in sensory and consumer research, *Food Quality and Preference* 139–49.

Bi, J. (2006). *Sensory Discrimination Tests and Measurements: Statistical Principles, Procedures and Tables*. Oxford: Blackwell Publishing.

Bi, J. (2010). Comments on D.M. Ennis' presentation on equivalence testing, *Food Quality and Preference* 21(3), 259–260.

Boyles, R.A. (2008): The role of likelihood in interval estimation. *The American Statistician* 62(1), 22–6.

Brockhoff, P.B. (2003). The statistical power of replications in discrimination tests. *Food Quality and Preference* 14(5–6), 405–17.

Brockhoff, P.B., Christensen, R.H.B. (2010). Thurstonian models for sensory discrimination tests as generalized linear models. *Food Quality and Preference* 21(3), 330–338.

Carr, T. (1995). Confidence intervals in the analysis of sensory discrimination tests: The integration of similarity and difference testing. *4th Agrostat*, 22–31.

Castura, J. (2010). Equivalence testing: A brief review, *Food Quality and Preference* 21(3), 257–258.

Christensen R.H.B. and Brockhoff P.B. (2008). Thurstonian models for replicated difference tests. Unpublished, presented at 9th Sensometrics Meeting, Brock University, Canada, August 2008.

Christensen, R.H.B., Brockhoff, P.B. (2009a). Estimation and inference in the same-different test. *Food Quality and Preference* 20(7), 514–524.

Christensen, R. H. B., Brockhoff, P.B. (2009b). sensR: An R-package for Thurstonian modelling of discrete sensory data. R-package version 1.1.0. (www.cran.r-project.org/package=sensR/).

Christensen, R.H.B., Brockhoff, P.B. and Cleaver, G. (2010). Statistical and Thurstonian models for the A-not A protocol with and without sureness. Manuscript for *Food Quality and Preference*.
Ennis, D.M. (1993a). The power of sensory discrimination methods. *Journal of Sensory Studies* 8, 353–70.
Ennis, D.M. (1993b). A single multidimensional model for discrimination, identification, and preferential choice. *Acta Psychologica* 84, 17–27.
Ennis, D.M. (2003). Foundations of sensory science. In H.R. Moskowitz, A.M. Munoz and M.C. Gacula (eds), *Viewpoints and Controversies in Sensory Science and Consumer Product Testing*. Trumbull, CT: Food & Nutrition Press.
Ennis, D.M., Bi, J. (1998). The beta-binomial model: Accounting for inter-trial variation in replicated difference and preference tests. *Journal of Sensory Studies* 13, 389–412.
Ennis, D. M., Palen, J., Mullen, K. (1988). A multidimensional stochastic theory of similarity. *Journal of Mathematical Psychology* 32(4), 449–65.
Ennis, D.M., Bi, J., Meyners, M. (2008): Response and discussion of Bi (2007), *Food Quality and Preference* 344–48.
Ennis, D.M. and Ennis, J.M. (2010a). Equivalence hypothesis testing, *Food Quality and Preference* 21(3), 253–256.
Ennis, D.M. and Ennis, J.M. (2010b). Equivalence hypothesis testing: Reply to Bi, *Food Quality and Preference* 21(3), 261.
Ennis, J.M., Ennis, D. M., Yip, D., O'Mahony, M. (1998). Thurstonian models for variants of the method of tetrads. *British Journal of Mathematical and Statistical Psychology* 51, 205–15.
Green, D.M., Swets, J.A. (1966). *Signal Detection Theory and Psychophysics*. New York: John Wiley & Sons, Inc.
Hoenig, J.M., Heisey, D.M. (2001), The abuse of power: the pervasive fallacy of power calculations for data analysis, *American Statistician* 55(1), 19–24.
Kunert, J., Meyners, M. (1999). On the triangle test with replications, *Food Quality and Preference* 10(6), 477–82.
McCullagh, P., Nelder, J. (1989). *Generalized Linear Models* (2nd end). Chapman & Hall/CRC.
MacMillan, N.A., Creelman, C.D. (2005). *Detection Theory, A User's Guide* (2nd edn). New York: Lawrence Elbaum Associates.
MacRae, A.W. (1995). Confidence intervals for the triangle test can give assurance that products are similar. *Food Quality and Preference* 6(2), 61–7.
Meyners, M. (2007a). Least equivalent allowable differences in equivalence testing. *Food Quality and Preference*, 541–7.
Meyners, M. (2007b). Proper and improper use and interpretation of Beta-binomial models in the analysis of replicated difference and preference tests, *Food Quality and Preference* 18, 741–50.
Meyners, M., Brockhoff, P.B. (2003). The design of replicated difference tests. *Journal of Sensory Studies*, 18(4), 291–324.
Morrison, D.G. (1978). Probability model for forced binary choices. *American Statistician* 32(1), 23–5.
O'Mahony, M. (1992). Understanding discrimination tests: A user friendly treatment of response bias, rating and ranking r-index tests and their relationship to signal detection. *Journal of Sensory Studies* 7, 1–47.
Pawitan, Y. (2001). *In All Likelihood: Statistical Modelling and Inference Using Likelihood*. Oxford: Oxford University Press.
Pearlman, M.D., Wu, L. (1999): The emperor's new tests. *Statistical Science* 14(4), 355–69.
Schlich, P. (1993). Risk tables for discrimination tests. *Food Quality and Preference* 4, 141–51.
Skellam, J.G. (1948). A probability distribution derived from the binomial distribution by regarding the probability of success as variable between the sets of trials. *J. Royal Stat. Soc. Ser. B*, 10, 257–261.
Thurstone, L. L. (1927). A law of comparative judgment. *Psychological Review* 34, 273–86.

8

Investigating Important Factors Influencing Food Acceptance and Choice (Conjoint Analysis)

In consumer studies of food, one of the main challenges is to identify the important factors for consumer acceptance and choice. This chapter is devoted to methodology suitable for investigating this type of problems. Methods based on consumer acceptance of the different products will be given the strongest emphasis, but a sub-chapter is also devoted to choice based methods. Main focus will be on situations where the products are based on designed experiments in a number of factors. The sub-chapter covering experimental design will both contain information about how to establish important sets of products as well as different ways of presenting the products to consumers. Factorial and fractional factorial designs are the most important techniques in this context. Different analysis methods will be discussed; both separate analysis of data for each consumer as well as joint analysis for all consumers simultaneously. Advantages and disadvantages of the methods will be discussed. The theory will be illustrated by examples.

The methodology of the present chapter is based on standard design and ANOVA techniques presented in Chapters 12 and 13. The chapter is closely related to Chapter 5 which treats similar problems in sensory analysis and Chapter 9, which is about preference mapping.

8.1 Introduction

As for sensory data, one of the most important challenges in consumer science is to identify which product factors or attributes that contribute the most to the consumers' acceptance. There exist several ways of doing this (Gustafsson *et al.*, 2003), but the simplest way is

Statistics for Sensory and Consumer Science Tormod Næs, Per B. Brockhoff and Oliver Tomic

to ask the consumer directly. These so-called self-explicated tests may give good results and are not necessarily inferior (Sattler and Hensel-Börner, 2003) to other techniques, but they also have some important drawbacks; it is difficult to estimate interactions between the factors and the method requires a mental process which is clearly not typical for a real buying situation. Sometimes the results are inconclusive and all factors may come out as equally important. These tests will not be pursued further here.

An important alternative is to use some type of factorial design (see Chapter 12) and construct 'products' by combining different levels of factors/attributes and then ask the consumer to asses the different combinations either by giving an acceptance score for each 'product' (rating based) or by choosing one of them (choice based). The idea is to let the consumers provide information about what is important without asking them directly and without requiring a mental processing which is too different from a real buying situation. The effects of the different factors and their interactions are then estimated by either ANOVA or regression types of methods (see Chapters 13 and 15). This type of methodology is in many applications referred to as conjoint analysis (Green and Rao, 1971; Green and Srinivasan, 1978, 1990; Gustafsson *et al.*, 2003; Bagozzi, 1994).

A third possibility is to just take some products as they are, either selected at random from a larger population of products or produced according to certain pre-specified criteria, measure the important product characteristics by for instance sensory profiling and then ask the consumers to express their acceptance for each product. The main challenge here is to identify which sensory attributes are the most important for consumer liking. We refer to Chapter 9 for a full discussion of this particular situation.

In the present chapter focus will be on the second alternative above, namely situations based on designed experiments in the factors or attributes that one is interested in. Focus will be on rating based methods, but rank based and choice based methods will also be discussed briefly. No consideration will be given to other and more advanced and specialised conjoint procedures such as for instance the two-factor at a time procedure described in Green and Srinavasan (1978, p. 107; see also Gustafsson *et al.*, 2003 for an overview of other possibilities).

In addition to the main focus of testing factor effects, we have also incorporated a sub-chapter about some simple analysis techniques that should be used prior to the main analysis of the factor effects and also a sub-chapter about how to relate the preference results to external consumer attributes such as demographic variables and attitudes.

8.1.1 Possible Application Areas

In the literature one can find many situations where this type of methodology is important. Frequently occurring examples are within product development, in the study of context effects, in market segmentation studies and investigations of consumer trends and general consumer attitudes (see e.g. Gustafsson *et al.*, 2003). The factors can in practice be of very different nature; some typical examples are different ways of presenting the product (packaging, labelling etc.), aspects of user-friendliness (cooking, opening of package) and information about health benefits (fat level, sugar level etc.). In most cases the factors considered are presented to the consumer using written information or information provided by illustration or graphs as discussed in for instance Moskowitz and Silcher (2006) and Gustafsson *et al.* (2003), but in some cases also real products may be used. An example of nonstandard use of the technique can be found in Hersleth *et al.* (2003) which considers the effect of different contexts (food and social environment) on consumers' preference for wine.

As can be seen, the focus is very different in all these examples, but the general structure and methodology for design and analysis are basically the same, although there are many particular cases that need a special treatment (see for instance Gustafsson *et al.* (2003) for a recent overview). Our presentation of methodologies below will therefore be held at a general level.

8.1.2 The Structure of the Data Sets

The data that will be considered and analysed in the present chapter can be split in three separate data sets or data matrices. One of the data sets contains information about the design of the experiment, both for the samples used in the study and about how the samples are presented to the consumers. The second data set contains information about the consumers' preference/acceptance values for each of the products/profiles in the experiment. In most of the cases considered here, this data set will contain information about the degree of consumer liking for each of the consumer and product combinations. For the choice based studies considered towards the end of the chapter, this data set will contain information about which product that is selected by each of he consumers within each of the choice sets presented. The third data set contains information about the consumers themselves, typically demographic information (gender, age etc.) and/or information about habits and attitudes. For the rating based situation the three data sets and their relation considered can be visualised as in Figure 8.1.

The primary interest lies in finding the relationship between the two first data sets, i.e. the design of the study and the consumer liking scores. This is usually done by ANOVA and regression based methods ending up with estimated coefficients for the different design variables and tests of their significance. The coefficients for the design variables are within the area of conjoint analysis sometimes referred to as part worths and the estimated preference values for each of the factor combinations are called utilities (Green and Srinivasan, 1978).

A second objective is to identify and understand differences between individuals in their response to the different stimuli. This can be done by so-called segmentation, but other

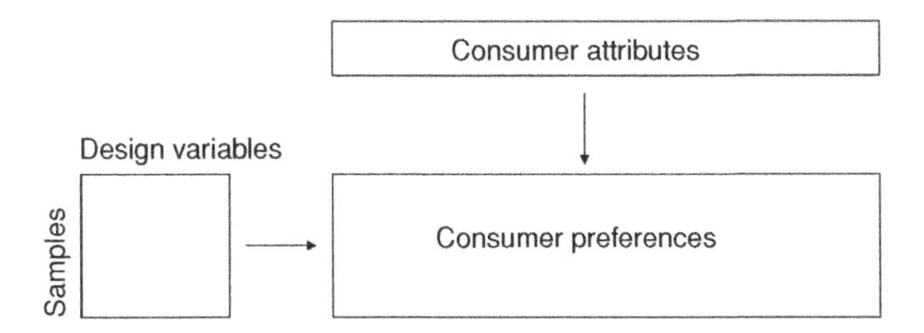

Figure 8.1 ***The data sets used and their relation.*** *The design matrix defining how the products are generated, the consumer acceptance data for the products and the additional consumer attributes such as demographic variables, attitudes and habits.*

methods are also available as will be discussed below. For this purpose, the third data set of consumer attributes or characteristics becomes useful.

A final interest which will be touched upon briefly at the end of the chapter is market simulation, i.e. to get an idea about the market potential for a certain product as compared to the rest of the products in the study.

The number of consumers used for this type of studies will usually be higher than 100 (typically between 100 and 150), but important applications have also been published with a lower number. Experience has shown that relevant differences can be detected even with a lower number of consumers (see e.g. Hersleth *et al.* (2003) where only 60 consumers were used to detect important differences). The general rule is that the more consumers that are used, the better become the results.

The number of samples or product combinations that an assessor can handle is always limited, but usually the number used is in the range of 8 or higher (in Green and Srinivasan (1978) is indicated that 30 is an upper limit), depending on the type of product attributes tested. If tasting real products is a part of the study it is important to keep the number low in order to obtain reliable data (see Chapter 9 for further discussion of this). The number of variables investigated can vary strongly, from the lower limit 2 to more than 10 in some large studies based on visual or verbal information. It is important to remember that the more variables that are used, the more samples are needed to obtain reliable information.

8.1.3 Modelling of the Factor Effects

A mathematical model relating a number of factors ($factor_1$, $factor_2 \ldots$, $factor_K$) to a consumer response variable y (acceptance, liking or purchase intent) can be written as

$$y = f(factor_1, factor_2, \ldots, factor_K) + \varepsilon \tag{8.1}$$

where ε is the random error term not accounted for by any of the factors. The model represented by the function f can take on a number of possible functional forms depending on information available and what type of factors that are involved. For choice based experiments, the y must be interpreted as a probability of choice (see Chapter 14).

In this chapter main emphasis will be given to methodology where all factors are considered categorical (nominal), without taking advantage of a possible ordinal or metric structure (see introduction). For this type of methods no functional form is assumed for the relation, each level is allowed to be estimated without any constraint related to other factors or other levels of the factor itself. The advantage of this is that little prior knowledge is required and few assumptions are made. Note that this is the only possibility when the underlying factor is actually categorical, representing for instance different colours, different labels, different contexts etc. Experimental factors that are treated as continuous variables will be discussed briefly in Chapter 8.6. The main advantage of using a regression method is that more stable estimates are obtained if the model is adequate.

The methodology discussed in Chapter 9 where focus is on sensory attributes of the products instead of designed variables is based on the same general model as in (8.1). The explanatory variables for these studies can, however, sometimes be very many and also highly collinear which calls for a quite different methodology. Although the two situations and methodologies have much in common and can be presented within a general theoretical framework (see Chapter 15), the presentation in two different chapters is here chosen due to the different nature of the applications and because of the different statistical tools available.

8.2 Preliminary Analysis of Consumer Data Sets (Raw Data Overview)

Before analysing the factor effects by the use of ANOVA methods, a preliminary analysis of the data is recommended in order to get an overview and also for detecting possible abnormalities. Typically, one may be interested in the following:

- What is the average liking and its variation/distribution for each sample in the data set and which samples are the most liked?
- What is the distribution of attitude and demographic variables over the consumers in the study? This may be of interested in itself, but also for checking representativity with respect to a given population of interest.

The most simple way of responding to the first of these questions is to simply compute average scores for each sample and present the results graphically in bar plots together with the standard deviations. An alternative is to use Box plots, which present the median, percentiles and the inter-quartile range of the observations in one single plot (see Chapter 11). The latter technique has the advantage that in addition to providing structural information, it also provides information about possible extreme observations (outliers) in the data set. More detailed information can also be obtained using histograms (see Chapter 11), for instance for each of the samples separately. Note that the same information can be made available for sub-groups of consumers defined by one or several of the consumer variables available, for instance gender or age group.

The means for the different samples/products can also be compared by using an ANOVA method if one is interested in more explicit statements about significance. The most natural model to use for this type of data is the two-way ANOVA model with samples considered a fixed effect and consumers as random effects (see Chapter 13). Since the interaction in this case (because of lack of replicates) is confounded with the random error, the model can be analysed the same way as a fixed effects model.

A related problem is to compare two a priori segments, for instance the two genders, with respect to their preferences for one of the samples. In this case the model to use is the simple one-way ANOVA model with segments representing the single 'way' in the data table. The reason for this change of model from two-way to one-way ANOVA is that the consumers in the different groups in this case are different and a systematic consumer effect is then not necessary.

The data set used for illustration here is taken from a study of apple juice with different sugar and acid levels. The study is based on a 3*2 design with 2 acid levels and 3 sugar levels, all of them based on simple manipulation of the same juice concentrate. The study is repeated in Norway and Spain with 125 consumers in each country (see Rødbotten *et al.* (2009) for more details). The means with their standard deviations and the Box plot for all the six samples are presented in Figure 8.2 and Figure 8.3 (for Norwegian consumers). In these plots the first letter in the code refers to the sugar level (L-Low, M-medium, H-High) and the second letter refers to the acid level (L-Low, H-high). Already at this

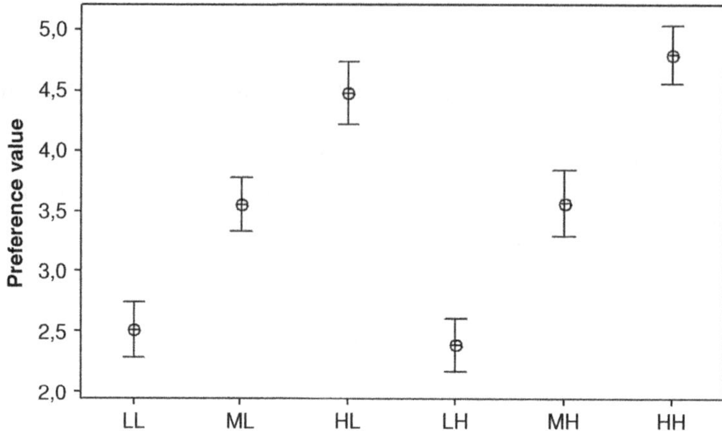

Figure 8.2 Plot of means and confidence intervals. *The plot shows the mean values and the confidence intervals for each of 6 different apple juice samples generated according to the design described in the text (see Rødbotten* et al. *(2009)). The first letter in the code refers to the sugar level (L-Low, M-medium, H-High) and the second letter refers to the acid level (L-Low, H-high). Adapted from Rødbotten et al. (2009) Food Quality and Preference.*

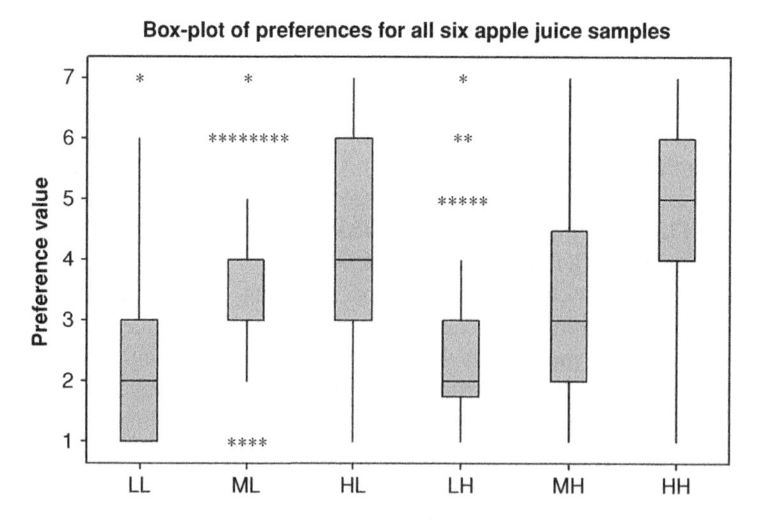

Figure 8.3 Box plots of acceptance values. *The box plot for the same apple juice data as used in Figure 8.2. (see Rødbotten* et al. *(2009)). The first letter in the code refers to the sugar level (L-Low, M-medium, H-High) and the second letter refers to the acid level (L-Low, H-high). Adapted from Rødbotten et al. (2009) Food Quality and Preference.*

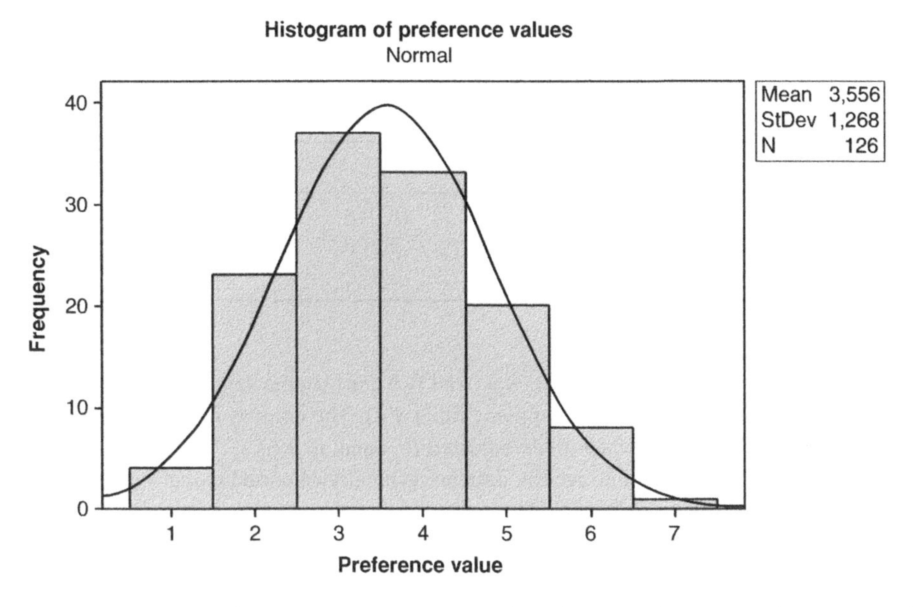

Figure 8.4 ***Histogram for consumer acceptance.*** *The values are liking values for sample with medium sugar and low acid. (see Rødbotten* et al. *(2009)).*

point a number of important aspects can be noticed. The most striking tendency is that the preference increases with sugar content both for the samples with a high and a low acid content. It is also clear that the disagreement among the assessors is quite high for all samples. The histogram in Figure 8.4 for one of the samples shows that the distribution is reasonably normal for the acceptance for this particular sample. A simple two-way ANOVA made for the Norwegian consumer panel with random consumer effect shows that both the samples and the consumer effects are significant (Table 8.1). The R^2 is as high as 49 %, while the adjusted R^2 is equal to 39 %. Interpreting R^2's for models with consumer effects is, however, questionable as discussed in Næs *et al.* (2010). We refer to Section 8.4.3 below for further discussion of this point.

Table 8.1 ***ANOVA for comparing all 6 juice samples using the Norwegian consumer panel.*** *The table has one row for each effect and one column for all the relevant quantities. The most important column is the p-value column to the right. This tells us directly the importance of the various factors. In this table and in tables to follow, the 0.000 means less than 0.001.*

Source	DF	SS	MS	F	*p*-value
Consumer	125	380,286	3,042	1,86	0,000
Sample	5	607,587	121,517	74,36	0,000
Error	625	1021,413	1,634		
Total	755	2009,286			

Table 8.2 ANOVA for comparing liking for juice sample 2 (ML-medium sugar and low acid content) in Norway and Spain. *The table has one row for each effect and one column for all the relevant quantities. The most important result is the p-value to the right. This tells us directly the importance of country for describing variability.*

Source	DF	SS	MS	F	*p*-value
Country	1	48,016	48,016	21,75	0,000
Error	250	551,968	2,208		
Total	251	599,984			

As an illustration of the use of a one-way ANOVA for his data set, we did a comparison of the two countries for one of the samples (Table 8.2). The country difference is clear. In this case the R^2 is as low as 8 % with an adjusted R^2 equal to 7.6 %.

For the second bullet point above the data are typically obtained using a questionnaire with a number of questions related to demographics (age, gender etc.) attitudes and habits. The data are typically categorical, but if the number of categories is not too small (>5), it may in some cases be possible to treat them as continuous variables. Often the number of consumers that fall in the different categories are counted and presented in tables (see Rødbotten *et al.*, 2009; Helgesen *et al.*, 1998).

Also here, one may be interested in splitting the results according to some of the consumer variables. For instance one may be interested in investigating differences in attitude for the two genders. In this case one ends up with data tables with two or more 'ways'. In some of these cases one may need more advanced statistical methodology like generalised linear models for testing the significance (see Chapter 15), but in some simple situations it is enough to use a homogeneity test based on the chi-square distribution (Chapter 11). A simple example where two different consumer groups are investigated for their differences in a particular attitude variable is given in Table 8.3 (measured on a scale between 2 and 7). As can be seen, in this case, the two genders are not significantly different.

8.3 Experimental Designs for Rating Based Consumer Studies

The first aspect to consider is which factors or attributes to concentrate on. The number of factors of interest may in some cases be quite large and it is important to make some choices in order to avoid problems with either too many experimental runs or too complex

Table 8.3 Two consumer groups vs. an attitude variable for ham data. *The response values are collected and counted in groups marked from 2 to 7. The horizontal percentages are given in parentheses. The p-value for the chi-square test is 0,884.*

	2	3	4	5	6	7	All
Group 1	3 (7,3)	8 (19.5)	18 (43.9)	7 (17.1)	3 (7.3)	2 (4.9)	41 (100)
Group 2	3 (7.5)	4 (10.0)	18 (45.0)	9 (22.5)	4 (10.0)	2 (5.0)	40 (100)

analysis of the data. With context factors or sensory testing involved in the study, it is very important to keep the number of factors and factor combinations low. In studies with only graphical or verbal information given to the consumer, the number can be higher (see e.g. Moskowitz and Silcher, 2006).

The next step is to determine the levels of the factor to be studied. The choice may be quite obvious when considering for instance a number of colours to use on a label, but in cases with continuous factors, the choice is usually less obvious. In all cases, the levels considered should be relevant for the practical problem considered. A large span may be of interest in order to understand consumer acceptance of extreme differences, but such levels may be of less interest from a practical point of view. It is useful to make this type of decision in close collaboration between the researcher who is responsible for the subject matter of the project and the statistician who is responsible for the design and analysis of the data.

When these preliminary choices have been made, there are two important aspects to take into account for the design of the study: The first has to do with which combinations of the attributes to incorporate and the second one has to do with how to present the actual combinations (objects/samples/profiles) to the consumers. Here we will discuss the two aspects separately, but also point out how they are related.

When selecting an experimental plan it is very important to use an established statistical design plan based on varying all factors simultaneously. It has been shown both theoretically and in practice that these plans are much more efficient and provide additional information as compared to other plans based on varying one of the factors at a time (Box *et al.*, 1978; see also Chapter 12).

8.3.1 Selecting Attribute Combinations: Design of Experiments

Experimental design is a term used for a series of experimental procedures that have been developed for providing as much information as possible for the least possible cost (see Chapter 12). The simplest and most well known among these methods is the full factorial design. These are experimental plans which use all possible combinations of the factors involved and which can be used both for estimating average effects (so-called main effects) as well as various types of interactions. As the number of factors and/or the number of levels increases, the number of combinations will, however, become very large. If for instance there are 6 factors in the study and each of them has 3 levels, the full experiment will consist of 3^6 combinations, which is too many for all practical purposes.

A possible way to solve this problem is to use so-called fractional factorial designs. These are designs which systematically select a fraction ($^1/_2$ fraction $^1/_4$ fraction etc.) of the full design in such a way that the most important information is still present in the data. The main problem is that it is not longer possible to distinguish between for instance main effects and two-factor interactions on one side and higher order interactions on the other. The latter type of effects can in many cases be considered moderate or even negligible compared to the main effects and two-factor interactions, so this is usually considered a sound practice even though one may lose some information.

The theory of fractional factorial design is simplest to use for situations with only two levels for each factor. In this case, the methodology is so simple that the whole design can be set up by hand and without using a computer software. For more complex situations

***Table 8.4* Design of the sausage study in *Helgesen* et al. (1998).** *Four products were combined with two factors with two levels each.*

Product	Sausage	Fat level	Price level
1	A	Low	Low
2	A	Low	High
3	A	High	Low
4	A	High	High
5	B	Low	Low
6	B	Low	High
7	B	High	Low
8	B	High	High
9	C	Low	Low
10	C	Low	High
11	C	High	Low
12	C	High	High
13	D	Low	Low
14	D	Low	High
15	D	High	Low
16	D	High	High

it is often useful to use the principle of optimal design (for instance D-optimal designs, Myers and Montgomery, 1995). There also exist a number of suggestions and tables in the literature that propose possible designs to use, usually based on the principle of orthogonal arrays (Myers and Montgomery, 1995). The theory about these designs is more complex and will not be discussed further in this book. Here main attention will be given to full factorial designs and factions of experiments with only 2-level factors.

An example of the use of a full factorial design can be found in Helgesen *et al.* (1998) who studied the effect of both product properties and information on consumer liking. Four different sausage products were tested in combination with 2 levels of information about price and 2 levels of information about fat level. A full factorial design with $4*2*2 = 16$ different combinations was used. As can be seen from Table 8.4 all the four samples are presented in combination with both levels of price and both levels of fat. This means that when concerns fat level, the true fat level is not always used. It is important in such cases that the information does not contradict what the consumer perceives when tasting the product.

An example of a fractional factorial design can be found in Hersleth *et al.* (2003) (see also Table 8.5). The situation here is based on 3 product factors related to different manipulations of a white wine crossed with 2 context factors related to food and the social setting in which the wine is consumed. A full factorial design in the five factors would lead to $2^5 = 32$ different combinations, which was considered too many. It was therefore decided to use a half fraction with only 16 combinations. The design used is a so-called resolution V design which means that all main effects and two-factors interactions can be estimated without confounding (see Chapter 12). The crosses in the table indicate the combinations that were used in the study. The data were analysed with a joint ANOVA as will be discussed further below.

Table 8.5 ***Example of a fractional factorial design***. *The design is the one used in the wine study in Hersleth* et al. *(2003). The design is a Res IV design (see Chapter 12). The – and + signs represent "without" and "with" respectively. The crosses represent the 16 combinations used in the design.*

	Product factors			Context factors			
Sample	Mal. Ferm.	Oak	Sugar	Lab without food	Reception without food	Lab with food	Reception with food
1	–	–	–		X	X	
2	+	–	–	X			X
3	–	+	–	X			X
4	+	+	–		X	X	
5	–	–	+	X			X
6	+	–	+		X	X	
7	–	+	+		X	X	
8	+	+	+	X			X

The number of levels is not only important when concerns limiting the size of the experiment. It may also, as indicated in the conjoint literature, have an effect on the span of the estimated effects. A possible explanation is that consumers use a larger span for the attributes with several levels in order to better distinguish between them. For interpretation purposes, this aspect should be kept in mind if the design has factors with different numbers of levels.

8.3.2 Presenting the Different Products to the Consumers

When the factor combinations (products) have been determined, one has to decide in which way to present them to the consumers. There are essentially two different possibilities in this case:

- All factor combinations are presented to all consumers.
- Different factor combinations are presented to different consumer groups. This is usually done in a systematic way using for instance an incomplete block design (Montgomery, 2005; Gacula *et al.*, 2009).

In both cases, the order of the profiles for each consumer should be randomised (see Chapter 12) in order to avoid systematic serving order effects in the analysis of the experiments. Randomisation should be different for each consumer or for different groups of consumers. MacFie *et al.* (1989) presents a number of useful strategies (see also Wakeling and MacFie, 1995).

The former of these alternatives is the simplest both to organise and to analyse by ANOVA techniques. The main advantage of the latter is that it allows for a smaller set of products to each of the consumers. The main problem with this strategy is that one may end up with a design with a confounding between factor effects and assessor effects that is difficult to interpret. If the consumer effect is confounded with factors (for instance higher order interactions) that are also confounded with the estimated effects (for instance main effects), the estimated effects become consumer dependent in a way that is difficult to interpret. Are the observed consumer differences in the effects due to real changes in the estimated effects themselves or related to other non-estimated effects that are confounded with them? This may create problems for both the individual analyses and for the joint approach described in Chapter 8.4. Unless one is familiar with the area of incomplete block designs, it is therefore recommended to use the first strategy. If the number of products is too large, one is therefore recommended to either reduce the number of factors or the number of factor levels.

If it can be assumed that the consumers can be collected in a few homogeneous segments, the second strategy above is easier to justify. The reason for this is that after segmentation, there may be enough different combinations of products within each segment to make unique estimation of all effects possible. This approach, however, requires that the sample set is large enough to guarantee that each segment contains enough combinations to allow for estimation of all relevant effects within each segment.

In some cases it may be difficult to randomise all factor combinations properly. For instance, in the wine study published by Hersleth *et al.* (2003) a full randomisation was not possible. Therefore a simpler strategy was chosen in which several combinations of wine attributes were tested within the same context, i.e. during the same evening when the actual context (combination of food and social setting) was set up. Without this simplification the experiment would simply have become impossible to conduct. This type of structure may lead to so-called split-plot models, where the random error is composed of two terms, a so-called whole-plot error and a sub-plot error (see Chapters 13 and 14). For the situation described in Hersleth *et al.* (2003), the different wines are the whole plot factors and the different contexts the sub-plot factors. In most consumer studies, however, with rather large random noise, the contribution from the whole plot error to the total error will in most cases be small or almost negligible. In such cases (see e.g. Letsinger *et al.*, 1996; Næs *et al.*, 2007), the data may be analysed without taking the split-plot error structure into account (as was done in Hersleth *et al.*, 2003).

8.4 Analysis of Categorical Effect Variables

8.4.1 Different Modelling Approaches

This section is about estimation of factor effects in factorial designs. We will describe a number of approaches with different properties, but with the same focus of understanding what are the most important drivers of product acceptance. The first two methods will be based on ANOVA, the third will be based on segmentation accompanied with ANOVA and the last will be based on multivariate analysis of the full profile of acceptance values.

In most cases one will analyse consumer conjoint data as they are without further pre-processing. This means that possible differences in the use of the scale (range and level differences, see also Chapter 3) will be part of the analysis itself. If wanted, however, these effects can be eliminated by mean subtraction and by dividing the individual scores by their standard deviations. Note that for the ANOVA based methods below, mean differences are handled automatically by the consumer effect in the model. See Chapter 9 for further discussion on the topic.

It will here be assumed that there are no missing values, but in most cases ANOVA can be run as usual even when some of the values are left out of the data set. If, however, there are factor combinations with no observation and this factor combination is involved in an interaction effect, one must be careful. In such cases, the interpretation of the results may become complex. For more details on incomplete designs we refer to Chapter 13 and for methods that can be used to estimate missing values we refer to Chapter 17.

In many applications of conjoint analysis, focus is on main effects only. In this book, however, we will present the methodology also with focus on interactions (see also Green, 1973). In our experience, interactions are quite frequent.

Below we will use the indices i, j and k for indicating factor effects. Note that this is different from the way they are used for sensory panel data in for instance Chapter 3.

8.4.2 Individual ANOVA Models for Each Consumer

This analysis is based on computing all factor effects separately for each consumer by the use of a simple ANOVA model. With three factors, A, B and C, a model with main effects and two-factor interactions can be written as

$$y_{ijk} = \mu + \alpha_i + \beta_j + \gamma_k + \alpha\beta_{ij} + \alpha\gamma_{ik} + \beta\gamma_{jk} + \varepsilon_{ijk} \quad (8.2)$$

where α corresponds to factor A, β to factor B and γ to factor C. If a reduced design is used, it may happen that only main effects can be computed. Note that if a reduced design is used, it must be assumed that all higher order and non-estimable terms are equal to 0. If they are important for the data and not estimated, they may influence the significance tests in a complex way.

In order to get an overall interpretation of the results, the individual models will usually need some type of 'meta-analysis' after estimation, i.e. an analysis of the similarity and differences between the coefficients for the different consumers. One possibility is for each consumer to rank the parameter estimates and compare the ranking between the consumers afterwards, either by simple voting (for the most liked etc.) or more advanced techniques like the Kendall's coefficient of concordance (Kendall, 1948). Another possibility is to use some type of cluster analysis to look for similarities as will be discussed in Chapter 10. Note that this approach allows for comparing consumers for only one or a few of the factors. One can for instance compare assessors only for factor A and forget about the rest if they are of less interest.

The main problem with this approach is that the number of samples is usually not very much higher than the number of parameters in the model. This leads to imprecise regression estimates and low power of significance tests. The few degrees of freedom for random error estimation also lead to weak model validity checks. The main advantages of the approach are the simplicity with respect to the simple ANOVA model used and the fact that individual differences are explicitly estimated.

It is important to note the similarity of this approach and the external preference mapping method discussed in Chapter 9. In both cases, the individual consumer preferences are treated separately using linear models, the only difference is the nature of the external data.

8.4.3 Joint ANOVA for All Consumers

In the joint ANOVA approach all data are used in one single model with all the systematic design factors incorporated. This means all design factors plus their possible interactions in addition to systematic consumer effects. Since many consumers are used for estimation of the factor effects in this approach, they will be estimated more precisely than in the individual ANOVA models approach above. On the other hand, the fixed effects will have another interpretation, namely as average effects over the population of consumers considered. The individual differences can be studied by the use of variance components, by residuals or by interaction plots of the random consumer effects as will be discussed below.

A typical ANOVA model of this type is the following based on a full factorial design in 3 factors, A, B and C, at two levels each. This leads to 8 different profiles tested by all consumers. The model with main effects and all two factor interactions for such a data set can be written as

$$\begin{aligned} y_{ijkn} = {} & \mu + \alpha_i + \beta_j + \gamma_k + \alpha\beta_{ij} + \alpha\gamma_{ik} + \beta\gamma_{jk} \\ & + Cons_n + (Cons^*\alpha)_{in} + (Cons^*\beta)_{jn} + (Cons^*\gamma)_{kn} \\ & + (\alpha\beta^*Cons)_{ijn} + (\alpha\gamma^*Cons)_{ikn} + (\beta\gamma^*Cons)_{jkn} + \varepsilon_{ijkn} \end{aligned} \tag{8.3}$$

where α, β and γ are the main effects for the corresponding design factors, the combinations of the letters represent the interactions and Cons is the consumer effect. The design factor effects (α, β, γ) are considered fixed while the consumer effect and its interactions with the design factors are considered random. Note that all the effects that are represented at the fixed effects level are also represented at an individual level. This is natural since population effects will always have an individual basis. In some cases, however, the interactions between consumer on one side and interactions between the design factors on the other are omitted since they are believed to be small as compared to the random error. A three factor interaction between the effects A B and C is also possible, but usually such factors play a minor role. In this particular model, if a three-way interaction is incorporated both at a population and individual level, the model is saturated and the random error becomes equal to 0.

The general assumption for the random effects is that all random effects are independent and that variance is constant over the different levels of the random effect. For instance $Var(Cons_n)$ is independent on n .In some cases, the assumption of independence may, however, be questionable if there are structures in the data related to how the different consumers appreciate the different effects. For instance if a consumer appreciates a certain level of a factor, this may be correlated to how the same consumer appreciates a certain level of another factor. This type of dependence is possible to handle using very advanced ANOVA methods (see e.g. Hand and Crowder (1996)), but these methods are beyond the scope of the present book.

As was noted above, this model is essentially a model of the average effect of A, B and C and their interactions over the whole group of consumers and the individual differences are only present in the consumer effects. In order to study the individual differences by this model, one needs to consider the structure of the random interactions between consumers and the fixed effects or the residuals e. The sizes of the variance components for the different effects provide overall information about where the size of the individual differences. Another possibility is to look at the residuals after fitting of only the fixed factor effects. These residuals contain information of the individual acceptance values and may represent an important source of information for studying individual differences. It is

Table 8.6 ANOVA for the ham data.

Source	DF	SS	MS	F	*p*-value
Consumer	80	878,438	10,980	1,31	0,064
Product	3	91,807	30,602	3,83	0,010
Information	1	6,520	6,520	3,34	0,071
Product*Information	3	10,387	3,462	2,20	0,089
Consumer*Product	240	1918,068	7,992	5,08	0,000
Consumer*Information	80	156,105	1,951	1,24	0,109
Error	240	377,488	1,573		
Total	647	3438,813			

$R^2 = 89{,}02\%$ $R^2(\text{adj}) = 70{,}41\%$

Source	Variance component
Consumer	0,32626
Product*Consumer	3,20954
Information*Consumer	0,09461
Error	1,57287

also possible to use cluster analysis on the residuals. In this way one can obtain information about which assessors that are similar to one another in their acceptance pattern.

One must also here be aware of the same problem as emphasised above, namely that if a reduced design is used, it must be assumed that all higher order and non-estimable terms are equal to 0. If they are important for the generation of the data and not estimated, they may influence the results in a complex way in the same way as described above.

The data set used for illustration in Table 8.6 and Figures 8.5, 8.6 and 8.7 comes from a study where four different hams are compared together with information about 2 different countries of origins (see Næs *et al.*, 2010). In other words, the design was a 4*2 factorial design. All 81 consumers tested all 8 combinations The model is exactly the same as used in (8.3), but with only two systematic factors in it. From Table 8.6 it is clear that the product effect is significant at 5 % level while the information effect is not. This means that the products were perceived as different while the two information levels seem to have a less important effect on the consumers liking. The main effects for the product factor are plotted in Figure 8.5. On the individual differences level, one can see that there are strong differences regarding the individual assessment of the products with a *p*-value less than 0.001 (see also Figure 8.6 for a visual illustration of the individual differences). There seems to be only a moderate additive consumer effect and very small interactions between consumer and information. The R^2 is as high as 89 % and adjusted R^2 is equal to 70 %. As discussed in Næs *et al.* (2010), there is a general tendency that the R^2 become quite large when the individual differences are incorporated in the model. In order to validate the fixed part of the model, it may then also be useful to look at R^2 for the fixed effects model only. The R^2 values are here reduced to as low values as 3.2 and 2.1 %. This clearly shows that one must be very cautious when interpreting R^2 for mixed models of this type. Both types of R^2 are relevant in their own way, but the latter values are more relevant for measuring the importance of the effect of the conjoint variables in the population.

The residuals for five of the consumers after fitting of the fixed effect model are plotted in Figure 8.7. This plot is obtained by averaging the residuals over the two information

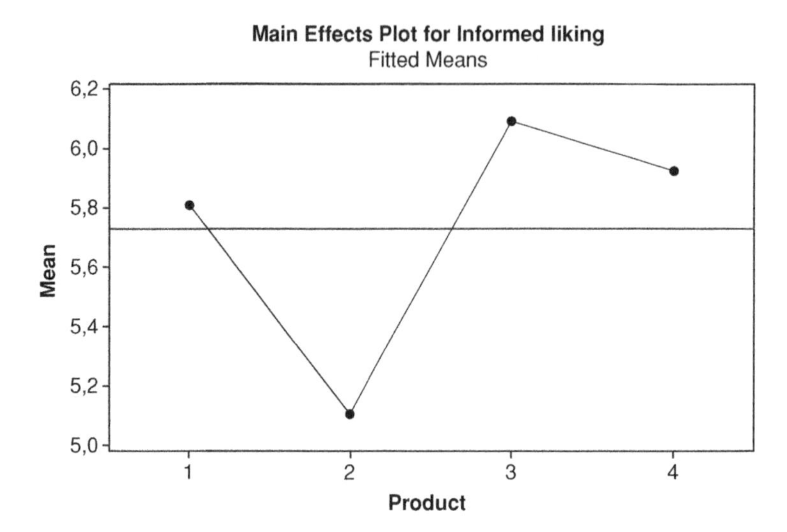

Figure 8.5 Main effects plot for the ham study. *The average effects for the four different products are plotted (standard error = 0.23). (see Næs* et al. *(2010). Reprinted from Food Quality and Preference, Næs* et al. *Alternative methods for combining dsign variables and consumer preferences with information about attitude and demographics in conjoint analysis, 2010, with permission from Elsevier.*

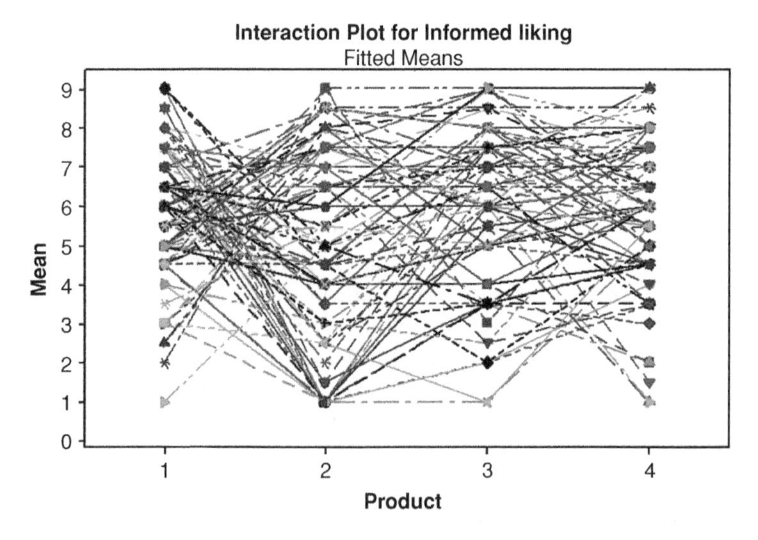

Figure 8.6 Interaction plot for the ham study. *The plot shows the interaction curves for the interactions between the consumers and products. (see Næs* et al. *(2010). Reprinted from Food Quality and Preference, Næs* et al. *Alternative methods for combining dsign variables and consumer preferences with information about attitude and demographics in conjoint analysis, 2010, with permission from Elsevier.*

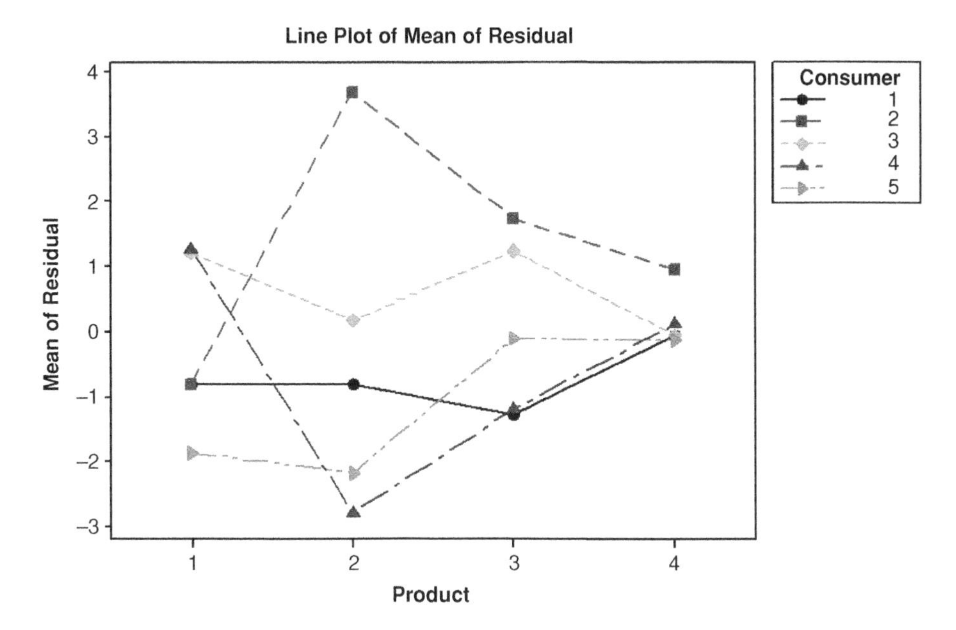

Figure 8.7 ***Individual differences in conjoint.*** *The plot presents residuals for five different consumers. The residuals are averaged over the two information levels and presented for each of the four products. (see Næs* et al. *(2010). Reprinted from Food Quality and Preference, Næs* et al. *Alternative methods for combining dsign variables and consumer preferences with information about attitude and demographics in conjoint analysis, 2010, with permission from Elsevier.*

levels, highlighting the product effects. As can be seen, the plotted consumers are very different in their assessment of product differences. It is also possible to do a full PCA of the residuals matrix (or use a cluster analysis) in order to obtain further information about individual differences, but this will not be pursued here.

The ham example is based on a full factorial setup. When using this type of modelling for fractional factorials one must be aware that in the case of confounding between the factors, only a subset of the effects can be used. If several factors are confounded with each other, only one of them can be used in the model. An example of such a situation was treated in Johansen *et al.* (2010a). In that paper, 4 factors were analysed using only 8 experiments. This leads to a resolution IV design with confounding between some of the 2-factor interactions. In this case, one has to select only one of the factors that are confounded to be used in the model. For instance, if DE is confounded with AB, only one of them is allowed in the ANOVA model. In the fractional design used in Hersleth *et al.* (2003), however, there was no confounding between two-factor interactions and all of them could be incorporated in the model.

The discussion and examples presented here refer to data obtained by asking people about their preference. It should be mentioned, however, that the same modelling is also used for experimental auctions where real behaviour is monitored (see e.g. Lange *et al.*, 2002; Combris *et al.*, 2009).

8.4.4 Segmentation and Estimation of Effects within Segments

Segmentation can in consumer science be done by using consumer attributes directly (a priori segmentation) and by using their acceptance or preference pattern (a posteriori segmentation). The idea behind the latter is that the consumers can sometimes be split in subgroups or segments with similar acceptance pattern within each group. Identifying these segments can be done in various ways as will be discussed in more detail in Chapter 10. In this chapter we will confine ourselves to mentioning that the segmentation can be based on different types of input data and different statistical methods. Examples are:

- Original preference data in combination with some type of cluster analysis (either visually based using principal components analysis (PCA) or a more algorithmic approach).
- Using the same approach for factor effects computed for each individual (see above). Note that the effects in a linear model are linear functions of the raw data and if the model is saturated (the same number of parameters and observations) they represent essentially the same information as the raw data.
- Segmentation based on residuals from least squares (LS) fitting of the consumer response values to linear functions of the design. This approach involves a joint segmentation and estimation of effects within each of the segments. Advantages of this approach will be discussed in Chapter 10.

In these cases, additional information about consumers, for instance demographics and attitudes, is best analysed after the segments have been identified. If incorporated in the segmentation directly as part of the modelling as will be done for the ANOVA below, these variables may become very important for the clustering and therefore influence the segmentation more than one is interested in when a posteriori segmentation is wanted.

8.4.5 Using PCA on the Acceptance Data Directly

Another possible analysis technique is to use PCA on the profiles of acceptance values directly. This can be done in two different ways, one based on the data matrix organised with products as rows and consumers as columns and one based on the transposed matrix. As can be noted, the former is quite similar to the internal preference mapping to be discussed in Chapter 9. For the latter method, it is our experience that it is sometimes quite useful to centre the data (the rows) for each consumer prior to the PCA, leading to a double centred matrix for the PCA analysis. If not done, the first component may be strongly related to different use of the scale.

This former method provides a score plot representing how the different products are related to each other and a loading plot which indicates the differences between the different consumers. The scores can then be related to the design of the study in order to understand the plot better. The main drawback with this approach is that no explicit information about significant differences between the products will be available. The main advantage lies in the graphical displays of the results based on scores plots and loadings plots. As for the internal preference mapping, it is natural to seek further understanding of the information in the scores plots as will be discussed in Section 8.5.

8.4.6 Using Factor Effects to Form Utilities

When the different parameters have been estimated, either globally (averages) or for each segment separately, they can be used to estimate the utilities for each of the profiles. The utilities are defined as the sum of the estimates and represent estimates of the average acceptance levels for each of products or profiles. This means that the different factor estimates are simply added together using the same model as used for estimation of the effects. For model (8.3), the estimates of the utilities (in the population) for the different combinations (profiles) can be written as

$$\hat{y}_{ijk} = \hat{\mu} + \hat{\alpha}_i + \hat{\beta}_j + \hat{\gamma}_k + \alpha\hat{\beta}_{ij} + \alpha\hat{\gamma}_{ik} + \beta\hat{\gamma}_{jk} \tag{8.4}$$

where the 'hat' on top indicates estimated effects. These utility values can for instance be used to identify the best possible combinations, i.e. those combinations with the highest value. Note that these values are estimates based on a model as opposed to the raw average value for the same combinations. They will generally not be the same, but in most cases quite similar. In Section 8.10 we will discuss how the utilities can be used for market share simulation. Note that utilities can in a similar way also be computed for individuals using model (8.2).

8.5 Incorporating Additional Information about Consumers

As stated above, one will in many cases have additional information available about the consumers, either related to demographics or to attitudes and habits. This type of information can be highly valuable for a deeper understanding of the preference pattern.

There exist a number of different ways of incorporating additional information in this type of factor studies. Some of the most important methods are based on segmentation accompanied with subsequent analysis of the relations between the segments and the additional variables. These methods will in this book be discussed in Chapter 10. In the present chapter we will confine ourselves to techniques that can be used without having to cluster the data in advance. For this class of methods, it is natural to distinguish between so-called a priori and a posteriori use of the additional consumer attributes or characteristics, related to whether one applies it in the primary analysis or after the factor effects have been estimated.

8.5.1 The Use of ANOVA with Direct Inclusion of Consumer Variables

Let us for simplicity use the same example as above with three design factors and assume that the additional information about gender of the consumers is available. The most natural way of incorporating this effect is to add a gender effect in the ANOVA model and let the consumers be incorporated as nested (see Chapters 12 and 13) under gender:

$$\begin{aligned} y_{ijknl} = {} & \mu + \alpha_i + \beta_j + \gamma_k + \alpha\beta_{ij} + \alpha\gamma_{ik} + \beta\gamma_{jk} + \phi_l + (\alpha\phi)_{il} \\ & + (\beta\phi)_{jl} + (\gamma\phi)_{kl} + \alpha\beta\phi_{ijl} + \alpha\gamma\phi_{ikl} + \beta\gamma\phi_{jkl} + Cons(\phi)_{nl} \\ & + (\alpha^{*}Cons(\phi))_{inl} + (\beta^{*}Cons(\phi))_{jnl} + (\gamma^{*}Cons(\phi))_{knl} + (\alpha\beta^{*}Cons(\phi))_{ijnl} \\ & + (\alpha\gamma^{*}Cons(\phi))_{iknl} + (\beta\gamma^{*}Cons(\phi))_{jknl} + \varepsilon_{ijknl} \end{aligned} \tag{8.5}$$

Here ϕ represents the gender effect and *Cons*(ϕ) represents the consumer effect within gender (see Chapter 13). Note that gender is also assumed to have an interaction with the

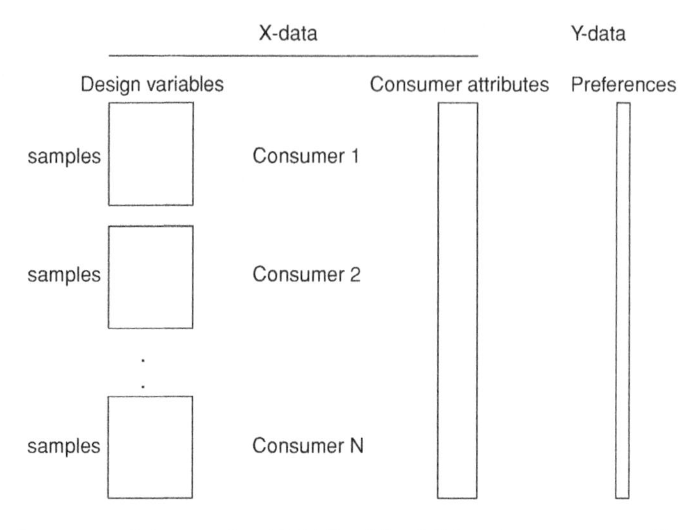

Figure 8.8 Structure of data set for ANOVA with additional consumer attributes. *(see* Næs et al. *(2010).*

design factors which means that men are allowed to respond differently than women not only in an additive way, but in a more complex way (see Figure 8.8). Usually, one will mainly look at the interaction between consumer attributes (here ϕ) and the conjoint factors in order to obtain the desired information about how they are related. Note that also here the same interactions that are present at the population level are incorporated at the individual level.

Table 8.7 *The inclusion of gender for the same data as in Table 8.6 (ham data).*

Source	DF	SS	MS	F	*p*-value
Gender	1	9,666	9,666	0,88	0,351
Product	3	92,322	30,774	3,82	0,011
Information	1	6,444	6,444	3,29	0,073
Consumer(Gender)	79	868,772	10,997	1,30	0,070
Gender*Product	3	8,421	2,807	0,35	0,790
Gender*Information	1	1,407	1,407	0,72	0,399
Product*Information	3	10,387	3,462	2,20	0,089
Product*Consumer(Gender)	237	1909,647	8,058	5,12	0,000
Information*Consumer(Gender)	79	154,697	1,958	1,24	0,107
Error	240	377,488	1,573		
Total	647	3438,813			

$R^2 = 89{,}02\%$ R^2 (adj) $= 70{,}41\%$

Source	Variance component
Consumer (Gender)	0,31928
Product*Consumer (Gender)	3,24236
Information*Consumer (Gender)	0,09633
Error	1,57287

The data used for illustration here are the same as used in the example above, namely the study of the different hams. As can be seen from Table 8.7, the gender does not seem to have any effect in this case, neither as a separate effect or in the interactions with the other variables. The other effects in the table as well as the variance components are comparable to above. An example with gender as a significant effect can be found in Helgesen *et al.* (1998).

This type of study is easiest to use for categorical consumer variables like gender, age group etc. Continuous consumer variables are also possible to use (ANCOVA, Weisberg, 1985), but this is generally not to be recommended unless one is very familiar with ANOVA methods. First of all it may be difficult to justify linearity of the covariates on the response and lack of linearity may have a problematic effect on the analysis and interpretation of the other effects. Secondly, mixing continuous and categorical variables is more difficult both from a technical and interpretational point of view, in particular when there are interactions. If continuous variables are attempted they should in all cases be centred prior to the ANCOVA.

If there is a large number of consumer variables (continuous and/or categorical) obtained for instance using a survey study, a possible way of incorporating all these variables within this framework is to use a type of cluster analysis first and then use the group membership as the categorical variable in the model (8.5). A possibility is to use a regular cluster method, another is visual inspection of a PCA plot. The advantage of the PCA method is that it also gives an automatic and immediate interpretation of the segments by using the loadings plot.

The primary advantage of the ANOVA approach is that it gives confidence intervals and significance tests for the various effects including the additional information variables used in the model and consequently an immediate and reliable assessment and ranking of the importance of the design factors and their relation to the additional consumer attributes.

The results from an ANOVA can be presented in many ways, both in tables like in Table 8.7, but also plots can be useful (see Helgesen *et al.* (1998), Figure 8.5 and Chapter 11). These types of plots should also be accompanied by indications of the standard error (see above, Section 8.4.3).

8.5.2 Direct Regression of Acceptance Data onto Consumer Attributes

For this method the consumer acceptance data are used as response variables **Y** and the additional consumer attributes as explanatory variables **X**. The consumers are organised as rows in the data sets. The analysis is then done by using PLS or PCR (see Chapter 15) directly on the two data sets. This analysis focuses on understanding the preference pattern among the samples directly as functions of the additional consumer data. This technique is like the technique above, a method which uses the consumer data in an a priori way. Note that no information is used about the attributes/design of the samples in this first step of the analysis.

The results from this approach are best visualised graphically using scores plot and loadings plots for both the X- and the Y-data. In this case, the different samples represent the Y-loadings and the additional consumer variables the X-loadings. The scores represent the consumers. If the number of samples is not more than moderately large, the loadings can be interpreted directly and without further calculations by comparing the plot with knowledge about the design. Another possibility is to relate the Y-loadings (one at a time) to the external design data by using regular ANOVA for each of the factors separately.

One simply uses the loading value as a response variable in an ANOVA model with all the relevant effects as explanatory variables (see Chapter 9 for such an example).

As was also discussed above, if the consumer acceptance data (and possibly also the questionnaire) data have a strong component of different use of the scale, it may be wise to subtract the mean from each assessor prior to PLS/PCR modelling. If this type of mean centring is not done, the first component may have a very strong relation to use of the scale. Since PCR/PLS is used on centred data, the data sets used for analysis will then essentially be double centred.

Note that if there are several groups of external data, the data set can be split accordingly and the relations between all of them can be established using a path diagram (see Chapter 17). In other words, this approach can easily be extended to a path modelling approach using the same organisation of the data (see also Olsen *et al.*, 2010).

In some cases it may be natural to highlight certain aspects of the responses instead of using the responses as they are. For instance, instead of using acceptance for all products as response variables, one can use averages or differences to highlight certain effects. For the ham example above, one could for instance use the differences between the two information levels for all the samples, as proposed and done in Næs *et al.* (2010).

Possible advantages of the present approach are the visual aspects related to the scores and loadings plots and also the fact that all the consumer variables can easily be involved directly in the analysis without any further categorisation (if linearity is reasonable) as is recommended for some of the other methods. A possible drawback is that if relations between external data and acceptance are weak, which they often are, the results may be inconclusive.

8.5.3 Relating Individual Models to External Data

The last method to be discussed here is based on using the coefficients from the individual models obtained in Section 8.4.2 above. This method thus uses the consumer attribute data a posteriori. It is worth mentioning that the individual differences terms are now absorbed into the individual differences among the regression coefficients. Instead of incorporating the consumer attributes as extra variables with interactions, the individual differences are here directly related to the size of the regression coefficients.

Relating the individual effects to consumer attributes will typically be done for each of the design effects separately by using a regression model of the effect vs. the consumer attributes. The method can also be combined with clustering based on the coefficient vector. The clusters can then be related to the external data as discussed in for instance Wedel and Steenkamp (1989, 1991). Again clustering can be done on the whole vector of effects or on only a subset.

The main advantage of this method is the direct establishment of the relations between the individual effects and the consumer attributes. This makes the approach conceptually appealing. The main problem is, however, the stability of the regression coefficients since only a relatively small number of samples are tested for each consumer. This means that significance tests etc. are weak. When used for cluster analysis or other multivariate analysis, the estimated effects have, however, less random noise than the original observation and can be used without problems.

8.6 Modelling of Factors as Continuous Variables

If the underlying processes for the factors are continuous, it is possible to use regression analysis for estimating the effects. In order to represent a real difference from the ANOVA methodology described above, there must be at least three levels for at least one of the factors in the model. Otherwise, the results will essentially be the same as when the factors are treated as categorical. Note that if a variable is categorical, representing for instance different colours, the variable is impossible to use in a regression study even when the colours are given a number. In such cases, combined (ANCOVA, Weisberg, 1985) models are needed (see Chapter 15).

The most important assumption behind a regression approach is that an adequate regression model can be found. In practice one will usually use either a linear or a polynomial model in this type of data fitting. For regression analysis, essentially the same three approaches as for the categorical variables can be used, namely individual models for each consumer, joint modelling of all consumers and segmentation followed by separate modelling within the segments.

8.6.1 Individual Regression Model for Each Consumer Separately

Using individual models is sometimes more natural here than for categorical data since regression models have a smaller number of parameters to be estimated. A simple situation is the one with two factors x_l and x_2 related to the response y by a linear model

$$y = \beta_0 + \beta_1 x_1 + \beta_2 x_2 + \varepsilon \tag{8.6}$$

The parameter β_0 is here the so-called intercept and the other two β's are the linear regression coefficients for the two factors x_1 and x_2. Note that the model has only three parameters to be fitted which is smaller than for an ANOVA model. With K levels for one of the factors and J for the other, an ANOVA model requires a minimum of $1 + (K - 1) + (J - 1)$ parameters. With K and J equal to 3, this will lead to 5 parameters to be fitted as opposed to 3 for the regression model.

If the model does not fit the data well, the model can easily be extended to contain also nonlinear effects. Usually a polynomial extension is used in such cases, often called an ideal point model since it can be used to estimate optimal acceptance level within the experimental region. Using a polynomial model, however, increases the number of parameters. For two input variables, the number of parameters is 6 if both nonlinear terms and interactions/products are incorporated.

The model shape and adequacy can be tested by the use of regular regression diagnostics, for instance residual plots as described in Chapter 15.

8.6.2 Joint Analysis of All Consumers

The same structure as in model (8.6) can be used here, but now it is necessary to add the consumer effect in the same way as was done for model (8.3). Using n as index for the consumers and j as index for product number (J^*N observations in total), a possible model is the following:

$$y_{jn} = \beta_0 + \beta_1 x_{j1} + \beta_2 x_{j2} + Cons_n + x_{j1}\beta_{1n} + x_{j2}\beta_{2n} + \varepsilon_{jn} \quad (8.7)$$

Usually, the x's are the same for each individual, but the model also allows them to be different. As can be seen, the individual differences are now taken care of by the use of the categorical and random additive consumer effect 'Cons' and individual regression coefficients for the two design variables β_{1n} and β_{2n}. Both the additive Cons and the two individual regression coefficients are assumed to be random. The model (8.7) thus assumes that in addition to a linear population structure, there are individual and possibly different regression lines for each single consumer. Again, the best is to mean-centre the variables before analysis. Analysing this type of models requires so-called ANCOVA methodology. As usual, the methods in Chapter 15 can be used to test adequacy of the model shape.

8.6.3 Segmentation

Again the same model as in (8.6) is used, either a linear or polynomial. How to do segmentation is based on the ideas and principles discussed in Chapter 10.

8.7 Reliability/Validity Testing for Rating Based Methods

As soon as the factor effects are computed, the model should be validated. As also discussed in Chapter 15, for validating a model properly it is simply not enough to just compare the fitted model values with the raw data since the data used for fitting the model will always fit better than new data. This may lead to overoptimistic results regarding the usefulness of the model. The best is of course to compare a model with real behaviour of consumers, but this is often very difficult in practice and one usually needs to rely on data based on consumer ratings.

8.7.1 Prediction Testing

For the joint modelling approach the absolutely best strategy is to select some new data using the same factors (and possibly by using new consumers) and then to compare the estimated values obtained by the model with the true average values given by the consumers. Alternatively, one can run a new data analysis of the same type for the new consumers and compare the factor effects directly. Note that this type of validation gives a validation not only of the model shape and the assumptions made, but also of the actual estimated values of the model effects.

For comparing model values with measured values in a new data set, it may be useful to compute the correlation between the measured and predicted values and to compare this with the correlation coefficient obtained in the estimation. Alternatively, one can compute the root mean square error of prediction (RMSEP) of the predicted vs. the measured values (Martens and Næs, 1989; Chapter 15). It may also be useful to look at the residuals plotted as a function of the predicted value or as a function of one or several of the experimental factors in order to reveal possible systematic tendencies. Note that the correlation coefficient and RMSEP computed this way can also be used to choose among different models tested.

If new consumers are not available, an alternative is to use the same consumers and present them, possibly after some time, with either the same profiles as used before, a subset of the profiles or another fraction of the full factorial design than the one used in the conjoint study itself. The latter may possibly be the best since then no aspect of memory will be an issue. Note that this type of validation is also a joint validation of the model shape and the true parameter estimates.

For the individual modelling approach, validation is more difficult, but the latter approach of presenting the same profiles again later, another fraction or a subset may also here be used as a validation strategy.

For the segmentation approach more or less the same strategies can be used. When using the same consumers in the validation it is already known which segment he/she belongs to and comparison goes as above within the actual segment he/she is in. When data from new consumers are used, a first allocation of consumer to the best fitting segment must be done. This can be done in a simple way by just comparing the squared residual values obtained by fitting the consumers to the different predicted values within a group. The consumer is then allocated to the group to which he/she has the smallest squared residual value. We refer to Chapter 15 for further discussion of this type of methodology which is often referred to a discriminant analysis.

8.7.2 Cross-Validation

If no new data are available, it is possible to do much of the same as described above by using so-called cross-validation (see Chapter 15), i.e. by splitting the data in two or several subsets and then test on samples that have not been part of the estimation. The drawback with this method is that it only validates the model shape since the estimates are different in each segment. Another drawback is that the number of samples in each segment is smaller than the full set and the validation results are not totally representative for the data set used for fitting. Also in this case plotting the residuals may be an important additional tool for detecting possible discrepancies between model and data (see Chapter 15 for how to do this).

Note that in the conjoint literature, the different validation strategies are given different names. According to Green and Srinivasan (1978) the type of validation just described is called reliability testing of the model. When a model is used to test real behaviour, it is sometimes called external validation (Green and Srinivasan, 1978). Since we have chosen a general approach in this chapter, we decided to use terminology proposed and used in the statistical literature.

8.8 Rank Based Methodology

Rank based studies present all products at the same time to each of the consumers and the consumer are asked to rank them according to either acceptance or purchase intent. This means that it is most suitable for situations with a limited number of possibilities. Note that it is impossible to use in context studies.

The data for each consumer are given as ranks, i.e. numbers between min = 1 and max = 'the total number of products'. This type of data is useful, but when concerns analysis, the possibilities are more limited than for rating data. The standard Friedman's ANOVA test can be used for checking the effect of a single effect, but methodology for more complex analyses, for instance for fractional factorial designs with many factors, are not easily available. The reason for the additional complication is the strong dependence among the numbers since for each consumer the sum of the ranks is fixed. This creates problems for standard ANOVA models which essentially assume independence among the residuals.

An important general method which can be used is the method of optimal scaling (see Chapter 17 and Overalls). This is a method which transforms the rank data into regular interval scale data for each consumer. The idea is based on first defining which model to use and then transforming the data in such a way that they fit to the model as well as possible. The underlying assumption is that the rank data represent a simplification of the real underlying data generation process which is continuous. The transformation used is flexible, except that the new data points satisfy the same ranking as the original ranks themselves. The model and data transform are fitted at the same time using the ALS strategy (see e.g. Kendall, 1948; Young, 1981; Green, 1973). The method is attractive, but since both data and model are fitted at the same time, it may be sensitive to over-fitting if not used with care. As far as we know the method is little used in this area.

Other methods that are useful in this area are the generalised linear models briefly mentioned in Chapter 15 and some of the methods discussed in Chapter 7.

8.9 Choice Based Conjoint Analysis

The other main class of techniques for collecting data for studying factor effects is the choice based methods (see e.g. Haaijer and Wedel, 2003; Louviere *et al.*, 2000; Elrod *et al.*, 1992). Instead of letting the consumers give a rating score for a number of attribute combinations, the consumer has to choose one of several alternatives. For each so-called choice set, consisting of a small or moderate number of individual factor combinations from the full set of possibilities, each consumer has to choose just one representative. Some arguments that have been put forward in favour of this methodology as compared to rating based studies are:

- The situation is more similar to the situation that the consumer is in when he/she is in a real buying situation.
- The method is simpler for the consumer since it requires less mental processing.
- Since only one choice is made, a larger number of profiles can be presented.
- The models used for analysis make it easier to do market simulations

It is, however, not obvious that all these arguments are valid in all possible situations. For instance the first point is questionable since the real choices one has to make in a buying situation are seldom presented simultaneously on a table in front of the consumer. It is often equally likely that a consumer is presented with one alternative at a time when walking around in the store. This is particularly true for context studies where there is always only one alternative available at a time. The third point is also questionable, in particular when there are true products involved. In this case, a large number of products need to be tested since several choice sets have to be used for each consumer in order to provide enough statistical power to the tests. Note that choice data are essentially a type of ranking data where only the highest rank is asked for and analysed. For further discussion of the relations between different ways of collecting data in conjoint analysis, we refer to Boyle *et al.* (2001) and Jaeger and Rose (2008).

Another problem with the choice based method is the extra complexity that it introduces both in the experimental design procedure and in the analysis. The two-step procedure of first setting up all combinations and then selecting sensible choice sets, introduces several methodological challenges that are not fully solved yet. The area is complex and the concept of fractional factorial and optimality are no longer as simple as they appear in for instance Chapter 12 for the more classical situation. Modelling by the use of multinomial regression (see Chapter 15), which is the method most frequently used here, is also more complex than the standard model fitting used for rating based studies. This is true both when concerns model estimation, segmentation and model checking. In other words, the analysis method is less flexible and makes the analysis less intuitive and more difficult to understand from a user's point of view.

In the literature about choice based experiments, there is a distinction between so-called labelled and nonlabelled choice studies. Here we will only consider the nonlabelled case which means that none of the combinations is given a name. In the same literature, there is also a distinction between so-called stated preference and revealed preference studies. The latter is more realistic, but the former is easier to conduct. As argued in Jaeger and Rose (2008), the possible problems with the stated preferences may sometimes not be so large. Methods for analysis of data and design are more or less the same for both situations.

8.9.1 Design of Choice Based Studies

In choice based conjoint analysis, each consumer is presented with a number of so-called choice sets (usually the same sets for all consumers). These choice sets consist of a subset of the full set of possible factor combinations or profiles. For each choice set the consumer is asked to select the one he/she prefers. The design problem in choice based conjoint analysis thus consists of two steps, first designing the factor combinations to select from and then selecting the choice sets to be tested. The first step is simple since regular factorial or fractional factorial design can be used. The second step is, however, much more complicated with respect to estimability of the regression coefficients and also with respect to statistical power and confounding pattern of the effects. As mentioned above, the theory is complex and appears fragmented and much less developed than regular design procedures. We recommend that in all cases where choice based conjoint analysis is used, one should simulate some data prior to the real test and check if the actual model is possible to estimate by the design generated. We refer to Street and Burgess (2007) for an overview of the most important methods for design of choice studies.

In choice based conjoint analysis, a so-called base alternative is often generated and used in each choice set in addition to the attribute combinations one is interested in. The purpose is (according to Gustafsson *et al.*, 2003) to scale the utility over the choice sets. This base alternative can be a combination of the basic attributes, but can also be a nonchoice alternative. The use of this strategy gives more complex modelling. An alternative that has been put forward is to add a column of ones to the design matrix (Gustafsson *et al.*, 2003, p. 384).

Selecting choice sets can be done in various ways and a number of alternatives are presented and discussed in (Kuhfeld *et al.*, 1994; Louviere and Woodworth, 1983). The method proposed in Louviere and Woodworth (1983) is probably the simplest and is based on generating a regular factorial (or fractional factorial) design in the actual variables to be investigated. Then one lists all these alternatives in a row. The next step is to allocate two levels to each of the profiles, absent or present. If the number of factor combinations or profiles is equal to J, the number of combinations of the absent/present combinations will then be equal to 2^J. A subset of these alternatives can be selected by using fractional factorial design theory. The selected alternatives will constitute the choice sets. This procedure creates a type of balance since the different profiles appear equally often. Variants with the same number of elements in each choice set are available (Gustafsson *et al.*, 2003).

8.9.2 Analysing Choice Based Experiments

For choice based experiments, it is difficult to use the individual modelling procedure suggested for the rating data. Only a joint and a segmentation based alternative exist. In the present chapter we only discuss the first of these alternatives. The segmentation approach is discussed in Chapter 10.

Choice based analysis methodology is based on modelling the probability of choice as a linear function of the experimental factors (see also model (8.1) and discussion around it). The most common model used for modelling choices is the multinomial logit model (McCullagh and Nelder, 1989) which assumes that the probability p_{km}, of choosing the alternative m from the choice set k can be written as

$$p_{km} = \frac{\exp(\mathbf{x}_{km}^T \beta)}{\sum_{n=1}^{M} \exp(\mathbf{x}_{kn}^T \beta)} \tag{8.8}$$

This model can be argued for by assuming a linear utility model with independent random errors that follow a Weibull distribution (Gustafsson *et al.*, 2003, p. 386). The log likelihood for this choice model can be found in for instance Gustafsson *et al.* (2003).

This model is based on the so-called IIA assumption (independence of irrelevant alternatives) which arises from the assumption of independent random errors in the utility model. The assumption implies that the odds of choosing one alternative over another is constant regardless of what other alternatives are present (see Louviere and Woodworth, 1983, p. 387). It implicitly also assumes that all consumers have the same structure regarding their choice pattern. Techniques based on so-called latent classes have been developed for avoiding these restrictions. These techniques are essentially segmentation methods which assume a subdivision of the consumers into homogeneous subgroups (see Chapter 10). For other modifications based on random regression coefficients we refer to for instance Gustafsson *et al.* (2003).

Again, the factors can be either categorical or continuous in the model with the advantages and disadvantages being more or less the same as indicated above. As in the ANOVA approach above, additional consumer attributes can be added as extra variables in the model (8.8).

Validation of this type of model, which belongs to the class of methods called the generalised linear models, is most often done by using the AIC (Akaike, 1973) or other criteria for testing adequacy of the model (see e.g. Wedel and Kamakura, 1998; McCullagh and Nelder, 1989). Residuals for outlier detection are also developed for this type of generalised linear models (McCullagh and Nelder, 1989), but these are more complex and not so directly interpretable as the residuals for regular rating based ANOVA models.

8.10 Market Share Simulation

In some cases one is not only interested in finding the factor effects and the utilities, but also to get some information about the market potential of the products involved in the test. This is often referred to as market share simulation (see Wittink *et al.*, 1994; Baier, 1999; SAS institute report, 1993; Baier and Gaul, 2003). This type of analysis is relevant both for rating and choices based experiments.

For rating based experiments there are three different techniques that are given main attention in the literature. The three methods differ in the way the estimated probability that the n'th consumer will buy the j'th product ($\hat{p}_{jn}$) is defined. All of them are based on estimated utilities $\hat{y}_{jn}$ for product j and consumer n. If individual ANOVA models are used (above, Section 8.4.2), these estimates come out directly by putting in parameter estimates in each of the models. If the joint model is used (above, Section 8.4.3), one can use the population values plus estimates of the individual random effects.

The simplest method is the maximum utility model which simply defines the $\hat{p}_{jn}$ equal to 1 for the product with the maximum utility for that consumer and 0 elsewhere. The second method is sometimes referred to as the Bradley-Terry-Luce method and this defines the probability ($\hat{p}_{jn}$) equal to

$$\hat{p}_{jn} = \frac{\hat{y}_{jn}}{\sum_{m} \hat{y}_{jm}} \tag{8.9}$$

Because of the shape, this is also called the relative utility model. Finally, the logit model is defined by simply computing the probabilities as

$$\hat{p}_{jn} = \frac{\exp(\hat{y}_{jn})}{\sum_{m} \exp(\hat{y}_{jm})} \tag{8.10}$$

For all the three methods the final market share estimate (referring to the products tested) is defined as the average of the probability values for all the consumers in the study. All these methods are important, but they also have some problems (Elrod and Kumar, 1989; Green and Krieger, 1988). Improvements are therefore proposed in Baier and Gaul (2003).

References

Akaike, H. (1973). Information theory and an extension to the maximum likelihood principle. In: Petrov, B.N., Csaki, F. (eds). *Second International Symposium on Inference Theory*, Budapest, Akademiai Kiado, 267–81.

Bagozzi, R.P. (1994). *Advanced Methods of Marketing Research* (ed. R. Bagozzi). Mass, USA: Blackwell Business.

Baier, D. (1999). Methoden der conjointanalyse in der marktforschungs- und Marketingpraxis. In (Gaul, W., Schader, M. eds.). *Mathematisce methoden der wirtschaftswissenschaften*. Heidelberg, Germany.

Baier, D., Gaul, W. (2003). Market simulation using probabilistic ideal vector model for conjoint data. In (Gustafsson, A., Herrmann, A., Huber, F. eds.). *Conjoint measurement. Methods and Applications*. Berlin: Springer.

Box, G.E.P., Hunter, W., Hunter, S. (1978). *Statistics for Experimenter*. New York: John Wiley & Sons, Inc.

Boyle, K.J., Holmes, T.P., Teisl, M.F., Roe, B. (2001). A comparison of conjoint analysis response formats. *Amer. J. Agr. Econ.* 83(2), 441–54.

Elrod, T., Kumar, K. (1989). Bias in the first choice rule for predicting shares. *Proceedings of the 1989 Sawtooth Software Conference*, 259–71.

Elrod, T., Louviere, J.J., Davey, K.S. (1992). An empirical comparison of ratings-based and choice based conjoint models. *Journal of Marketing Research* 29, 368–77.

Gacula, M.C. Jr., Singh, J., Bi, J., Altan, S (2009). *Statistical Methods in Food and Consumer Science*. Elsevier, Amsterdam, NL.

Green, P.E. (1973). On the analysis of interactions in marketing research data. *Journal of Marketing Research* 10, 410–20.

Green, P.E., V.R. Rao (1971). Conjoint measurement for quantifying judgemental data. *Journal of Marketing Research* 8, 355–63.

Green, P.E., Srinivasan, V. (1978). Conjoint analysis in consumer research: Issues and outlook. *Journal of Consumer Research* 5, 103–23.

Green, P.E., Krieger, A. (1988). Choice rules and sensitivity analysis in conjoint simulators. *Journal of the Academy of Marketing Science* 16, 114–27.

Green, P.E., Srinivasan, V. (1990). Conjoint analysis in marketing. New developments with implications for research and practice. *Journal of Marketing* 54, 3–19.

Gustafsson, A., Herrmann, A., Huber, F (2003). *Conjoint Measurement: Methods and Applications*. Berlin: Springer.

Haaijer, R., Wedel, M. (2003). Conjoint choice experiments: General characteristics and alternative model specifications. In A. Gustafsson, A. Herrmann, F. Huber (eds), *Conjoint Measurement: Methods and Applications*. Berlin: Springer.

Hand, D., Crowder, M. (1996). *Practical Longitudinal Data Analysis*. London: Chapman & Hall.

Helgesen, H., Solheim, R., Næs, T. (1998). Consumer purchase probability of dry fermented lamb sausages. *Food Quality and Preference* 9(5), 295–301.

Hersleth, M, B-H. Mevik, Næs, T. Guinard, X. (2003). The use for robust design methods for analysing context effects. *Food Quality and Preference* 14, 615–22.

Jaeger, S.R., Rose, J.M. (2008). Stated choice experimentation, contextual influences and food choice. A case study. *Food Quality and Preference* 10, 539–64.

Johansen, S., Næs, T., Øyaas, J., Hersleth, M. (2010a). Acceptance of calorie-reduced yoghurt: Effects of sensory characteristics and product information. *Food Quality and Preference* 21, 13–21.

Johansen, S., Hersleth, M., Næs, T.M. (2010b). The use of fuzzy clustering for segmentation in linear and ideal point preference models. *Food Quality and Preference* 21, 188–96.

Kendall, M.G. (1948). *Rank correlation methods*, London: Charles Griffin.

Kuhfeld, W.F., Tobias, R.D., Garratt, M. (1994). Efficient experimental design with marketing research applications. *Journal of Marketing Research*, 545–57.

Lange, C. Martin, C., Chabanet, C., Combris, P., Issanchou, S. (2002). Impact of the information provided to consumers on their willingness to pay for Champagne: comparison with hedonic scores. *Food Quality and Preference* 13, 597–608.

Letsinger, J.D., Myers, R.H., Lentner, M. (1996). Response surface methods for bi-randomisation structures. *J. Qual. Techn.* 28(4), 381–97.

Louviere, J.J., Woodworth, G. (1983). Design and analysis of simulated consumer choice or allocation experients: An approach based on aggregated data. *Journal of Marketing Research* 350, 367.

Louviere, J.J., Hensher, D.A., Swait, J.D. (2000). *Stated Choice Methods: Analysis and Applications*. Cambridge: Cambridge University Press.

McCullagh, P., J. Nelder (1989). *Generalized Linear Models* (2nd edn). London: Chapman & Hall/CRC.
MacFie, H.J.J., Bratchell, N., Greenhoff, K., Vallis, L.V. (1989). Designs to balance the effect of order of presentation and first-order carry-over effects in Hall tests. *J. of Sens. Stud.* 4, 129–48.
Martens, H., Næs, T. (1989). *Multivariate Calibration*. Chichester: John Wiley & Sons, Ltd.
Montgomery, D.C. (2005). *Design and Analysis of Experiments* (6th edn). New York: John Wiley & Sons, Inc.
Moskowitz, H.R., Silcher, M. (2006). The application of conjoint analysis and their possible uses in sensometrics. *Food Quality and Preference* 17(3–4), 145–65.
Myers, R.M., Montgomery, D.C. (1995). *Response Surface Methodology: Process and Product Optimisation Using Designed Experiments*. New York: John Wiley & Sons, Inc.
Næs, T., Aastveit, A., Sahni, N.S. (2007). Analysis of split-plot designs: an overview and comparison of methods. *Quality and Reliability Engineering International* 23, 801–20.
Næs, T. Lengard, V. Johansen, S.B., Hersleth, M. (2010). *Food Quality and Preference* 21(4), 368–378.
Olsen, N.V., Menichelli, E., Grunert, K, *et al.* (2010). Consumer acceptance for apple juice based on novel processing techniques: A conjoint study incorporating values and consequences. Submitted.
Rødbotten, M, Martinsen, B.K. Borge, G.I., *et al.* (2009). A cross-cultural study of preference for apple juice with different sugar and acid contents. *Food Quality and Preference* 20, 277–84.
SAS Institute Inc. SAS Technical report R-109. (1993). Conjoint Analysis Examples. Cary, NC: SAS Institute Inc. p. 85.
Sattler, H., Hensel-Börner, S. (2003). A comparison of conjoint measurement with self-explicated approaches. In A. Gustafsson, A. Herrmann, F. Huber (eds), *Conjoint Measurement. Methods and Applications*. Berlin: Springer.
Street, D.J. and Burgess, L. (2007). The Construction of Optimal Stated Choice Experiments: Theory and Methods. J. Wiley and Sons, Inc.
Wakeling, I.N. and MacFie, H.J.H. (1995). Designing consumer trials balanced for first and higher orders of carry-over effect when only a subset of *k* samples from *t* may be tested. *Food Qual. Pref.* 6, 299–308.
Wedel, M., Steenkamp, J.-B.E.M. (1989). A fuzzy clusterwise regression approach to benefit segmentation. *International Journal of Research in Marketing* 6(4), 241–58.
Wedel, M., Steenkamp, J.-B.E.M. (1991). A clusterwise regression method for simultaneous fuzzy market structuring and benefit segmentation. *Journal of Marketing Research* 28(4), 385–96.
Wedel, M., Kamakura, W.A. (1998). *Market Segmentation. Conceptual and Methodological Foundation*. New York: Kluwer Academic Publishers.
Weisberg, S. (1985). *Applied Linear Regression*. New York: John Wiley & Sons, Inc.
Wittink, D.R., Vriens, M., Burhenne, W. (1994). Commercial use of conjoint analysis in Europe: Results and critical reflections. *International Journal of Research in Marketing* 11, 41–52.
Young, F. (1981). Quantitative analysis of qualitative data. *Psychometrika* 46(4), 357–88.

9

Preference Mapping for Understanding Relations between Sensory Product Attributes and Consumer Acceptance

The differences between this chapter and Chapter 8 is that here we focus directly on relations between sensory profiles and consumer acceptance data. This means that the products can come from any type of investigation as long as their properties are measured by sensory analysis. The methodology for the two cases have a lot in common, but differ also at important points due to the differences in data structure. The main goal of the methodology presented here is to identify the main sensory drivers of liking and the most liked products. Focus will be on individual differences in acceptance rather than on population averages, in line with the main tradition in the area. Relations between the individual differences in acceptance and additional consumer attributes, such as gender, age and habits, will be studied for the purpose of improved insight. We will study linear preference mapping methods as well as ideal point methods which allow for nonlinear relations between sensory and consumer data. All methods are based on PCA related methodology. Strong emphasis will be given on how to use graphical interpretation tools.

The present chapter is related to Chapter 8 which is about conjoint analysis and Chapter 10 which describes various segmentation methods. The chapter is based on methodology presented in Chapters 14 and 15.

Statistics for Sensory and Consumer Science Tormod Næs, Per B. Brockhoff and Oliver Tomic

9.1 Introduction

Preference mapping is defined as methodology for investigating consumer liking for a series of products by the use of statistical mapping methods (PCA, PCR, PLS, see Chapters 14 and 15, Thybo *et al.*, 2003; McEwan, 1996; Hough and Sanchez, 1997; Faber *et al.*, 2003; Greenhoff and MacFie, 1994; Elmore *et al.*, 1999; Guinard *et al.*, 2001; Helgesen *et al.*, 1997). Consumer liking data can be analysed without taking other data sets into account, but main interest usually lies in understanding their relations to sensory attributes or other properties of the products. In this chapter main attention will be given to relations between consumer acceptance data and sensory profiles, but other types of data such as chemical profiles, can be handled more or less the same way (see Chapter 6). Important application areas for the methodology are product development and product improvement, but the methods can also be used for more general purposes (Helgesen and Næs, 1995).

Typical questions that one may be interested in and which can be answered by preference mapping methods are:

- Which samples/products in the sample set are the most and least liked?
- Which sensory attributes characterise the most and least liked samples (what are the drivers of liking and disliking)?
- Which consumers prefer which samples (segmentation) and what characterise the different consumer segments (in terms of demographics, habits and attitudes)?

Note that these questions are very similar to those discussed in the previous chapter, the main difference being that here the external data are sensory attributes which are typically many and continuous and also often highly collinear. They are also generally impossible to design directly by an experimental design, even though an experimental design may be the basis for the construction of the samples themselves. These particular characteristics require special attention and other methods than those discussed in Chapter 8. We refer to Næs and Nyvold (2004) for further discussion on this.

Since consumers may have quite different acceptance pattern, main attention will here be given to methods that analyse each consumer separately. This is also in line with the main tradition in the area (McEwan, 1996). Some attention will, however, also be given to how all consumers can be analysed simultaneously and how segmentation techniques can be applied (see Chapters 8 and 10). Advantages and disadvantages of the different approaches will be discussed.

The most important data set for preference mapping is as in Chapter 8 the matrix of consumer acceptance scores for all the products involved. This gives a data matrix with dimensions equal to the number of samples (J) times the number of consumers (N). The sensory data set has the same number of rows (samples) as the consumer data set, but the columns of the table here correspond to the sensory attributes (K). The third data set, containing information about consumer demographics and attitudes has dimension equal to the number of consumer attributes (M, attitudes, demographics etc) times the number of consumers (see Figure 9.1).

When concerns the number of samples to test in preference mapping studies, there is always a trade-off between the need for statistical power of the estimated models and the need for reliable data.. The main problem lies in the fact that there is a limit as to how many samples a consumer can test; and still produce reliable data. This aspect depends strongly

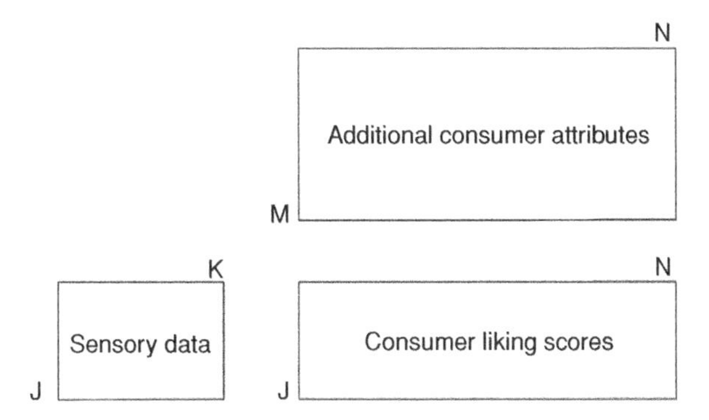

Figure 9.1 ***Data sets involved and their relation.*** *Three blocks of data are relevant in this chapter, sensory data, individual consumer acceptance data and additional consumer attributes such as demographic variables, attitude and habits. Note that the number of samples (J) coincide for the two first data sets and that the number of consumers (N) coincide for the two other data sets. Note the similarity between this setup and the one in Figure 8.1. The difference lies in the fact that here we considered the properties of the products, not the design of the products.*

on the type of samples tested, for instance samples with a very strong taste such as spicy sausages are more problematic than less tasty products such as crackers or different waters. Usually, it is quite difficult to go beyond 8–12 products (Faber *et al.*, 2003) and 5–6 is often a more realistic alternative in many preference mapping studies (Helgesen *et al.*, 1997). Using less than 5 samples is not recommended because of problems with statistical power when estimating individual regression models. When concerns the number of consumers to use, very many applications use a number of consumers between 100 and 150. It is our general experience that this is enough for providing interesting results. Even a lower number may in some cases be used if necessary, but lower than 50–60 is not recommended.

Consumers should as always be selected in such a way that they represent the population one is interested in studying, for instance a certain age group, the whole country or only a city. Random sampling from the actual group is a simple alternative, but one should always, if possible, check the representativity in terms of important demographic variables such as gender and age group (see Chapter 8.2).

Usually, the same products will be used for all consumers, but in some cases it is also possible to give different samples to different consumers (Johansen *et al.*, 2010b). For the external preference mapping below, this is generally no problem, but for the internal preference mapping this is more difficult to handle.

Main attention in this book will be given to situations where the consumers rate the products according to degree of liking, but some discussion will also be given to situations where the consumers rank the products. For a discussion of the relative merits of the different ways of collecting data we refer to Hein *et al.* (2008).

9.2 External and Internal Preference Mapping

The primary interest in preference mapping lies in finding good predictive and interpretable relations between the consumer data on one side and the sensory data on the other

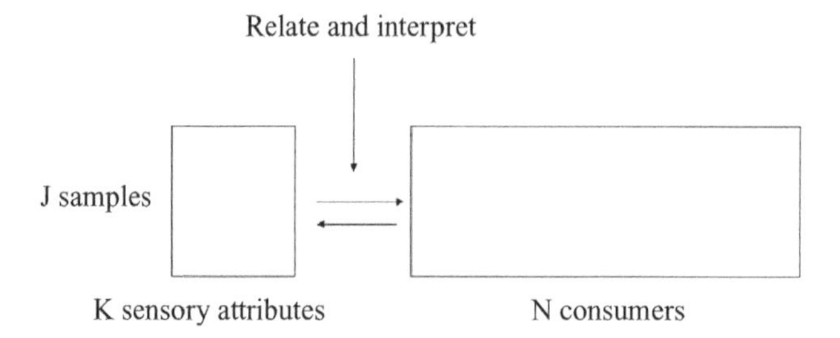

Figure 9.2 Finding and interpreting relations between sensory and consumer data. *The illustration emphasises that the main interest lies in predicting and interpreting the relation between the sensory panel data and the consumer acceptance data.*

(Figure 9.2). Experience has shown that the multivariate statistical methods described in the Chapters 14 and 15 are very useful for this purpose since they provide both predictions and interpretable plots of the data.

The most commonly used preference mapping methods are those that are based on a linear model for the relationship between the sensory and preference data for each consumer. These models assume that the preferences for each consumer either decrease or increase linearly for each sensory attribute. For instance, if a consumer likes sweetness, the increase in liking for the product increases with increasing degree of sweetness within the experimental area. This can be a valid assumption, but for some attributes and in particular if the experimental region is large, this may represent an over-simplification. The consumers may for instance like an attribute up to a certain point, but not beyond. A more advanced class of methods that handles this problem is the class of so-called ideal point models which allow for peaks and valleys in the map. Usually these models are based on quadratic polynomial models. For the sweetness example, this means for instance that lower than a certain limit, the liking increases as a function of sweetness while above this limit, the liking decreases. In many cases this is a better model for describing the reality and for finding optimal values, but in practice there are also drawbacks as will be discussed below. For all methods available, the results are presented in maps of the sensory attributes (loadings), of the samples (scores) and of the consumer liking (either loadings or max/min values in the ideal point case). Sometimes two of the plots are presented together in so-called biplots (Gower and Hand, 1995) or also in triplots if all three of them are considered together (see L-PLS methods, Martens *et al.*, 2005).

In most cases, one will concentrate on interpreting two components. Experience has shown that 2 or 3 components is usually enough for describing the most important variability in sensory and consumer data (Baardseth *et al.*, 1992; Rødbotten *et al.*, 2009; Ellekjær *et al.*, 1996).

9.2.1 Linear Preference Mapping

Linear preference mapping methods can broadly be split in two different traditions or methodologies (Figure 9.3 and 9.4), based on the same underlying techniques, PCA and linear regression.

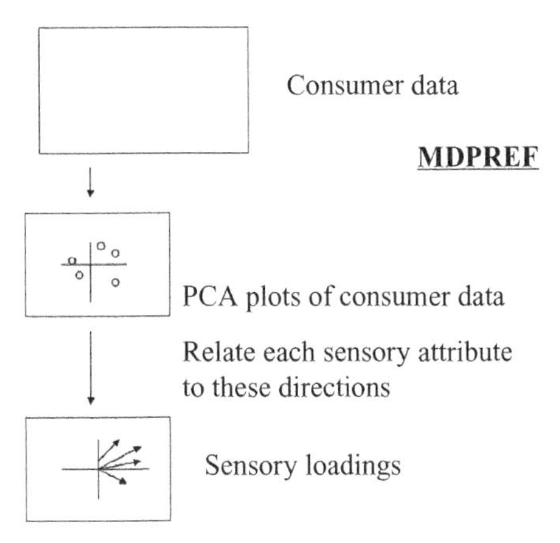

Figure 9.3 Illustration of MDPREF. *Illustration of the information flow in internal preference mapping (MDPREF). First the consumer acceptance data are analysed by PCA. Then the individual sensory attributes are regressed onto the principal components.*

9.2.1.1 Internal Preference Mapping (MDPREF)

This method is based on first applying PCA on the consumer data in order to interpret the consumer liking for the different products. This gives a PCA scores plot and a PCA loadings plot with the samples as scores and the individual consumer preferences as loadings. The next step is to regress all the sensory attributes y_k (k denotes attribute) onto the estimated PCA scores from the consumer data using the linear model

$$y_k = \beta_{ok} + \hat{t}_1\beta_{1k} + \hat{t}_2\beta_{2k} + \varepsilon_k \tag{9.1}$$

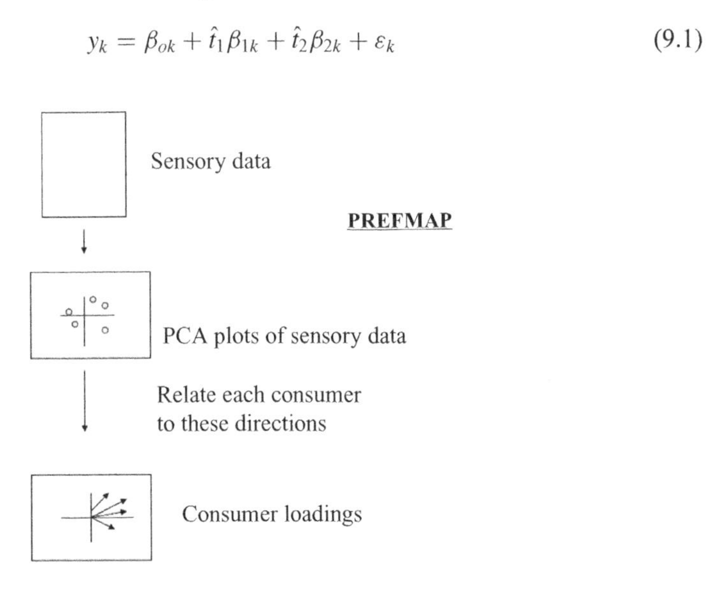

Figure 9.4 Illustration of PREFMAP. *Illustration of the information flow in external preference mapping (PREFMAP)). First the sensory data are analysed by PCA. Then the acceptance values for the individual consumers are regressed onto the principal components.*

As can be seen, this model assumes that each sensory attribute is a linear function of the consumer liking. The regression coefficients β_1 and β_2 represent the strength of the relations between the two consumer scores and the sensory attributes. The regression coefficients β_1 and β_2 for each attribute are plotted in the same way as for the consumer loadings, either in a separate plot or in the same. This means that one ends up with three plots, one score plot and two loadings plots, the consumer PCA loadings and the sensory loadings (i.e. the regression coefficients) obtained from the regression model. The three plots will always have to be considered together in order to respond to the typical questions mentioned in the introduction of Chapter 9. Interpretation is discussed further below in Sections 9.2.2 and 9.3.

Note that no intercept is needed if the sensory data table is centred, which is often the case (see below for more details). Three principal components can easily be incorporated by just adding an extra term $\hat{t}_3\beta_{3k}$in model (9.1).

The advantage of this method is that it concentrates on the consumer data first. These data represent the most important information and this approach thus ensures that the most important liking dimensions are represented in all further analyses. A drawback is that nonlinear preference mapping methods for identifying ideal points ('peaks' or 'valleys') is more difficult in this case. Some recent suggestions for how to solve this problem can, however, be found in the literature (Meullenet *et al.*, 2008). Another limitation is that testing of different products by different consumer groups can be difficult since this can give missing value configuration that PCA can not handle (see Chapters 14 and 17 and Figure 9.10).

9.2.1.2 External Preference Mapping (PREFMAP).

This method is based on first using PCA of the sensory data and then plotting the scores and loadings the usual way for interpretation. The sensory scores are then related to the acceptance data for each of the consumers, using the linear model

$$y_n = \hat{t}_1\beta_{1n} + \hat{t}_2\beta_{2n} + \varepsilon_n \tag{9.2}$$

Here the y's are the liking values and the $\hat{t}'s$ are the principal components from the sensory data. Adding an intercept is necessary if the consumer data are not centred (which they usually are, see below). In the terminology of Chapter 8, the β's are the part worths for the principal components dimensions.

In this case, each consumer (n) gives rise to a set of regression coefficients to be plotted in the same way as the attribute loadings for the sensory data. Thus, the method provides the same three plots as for the internal mapping, sensory PCA loadings, consumer loadings (obtained from regression) and a scores plot for the samples.

One advantage of external preference mapping is that it can easily be used both for linear and ideal point models. A possible problem with the method is that attributes that are unique (for instance attributes related to spice) and which thus may appear only in components with small variance, may be important for liking and this will be missed if focus is only given to the two first components. Another important aspect of the method is that since individual models are used for each consumer, different samples can be tested by each consumer, under the assumption that the sets of samples share the same underlying structure dimensionality (see Chapter 14 and Section 9.5, below).

9.2.1.3 General Considerations

As can be seen, for the linear approaches both methods are essentially PCR methods with the order of **X** and **Y** switched. Note that for both methods the collinearity problem is handled since PCA is used for data compression before the regression is performed. For internal preference mapping, the consumer loadings are the regular PCA loadings, while the sensory loadings represent regression coefficients. For the external preference mapping the situation is the opposite. Note, however, that there is no differences between true loadings and regression coefficients since the true loadings can also be interpreted as regression coefficients of the manifest variables onto the scores as discussed in the PCA chapter (Chapter 14). In our experience, despite the advantages and disadvantages listed above, it usually matters little in practice whether one uses internal or external preference mapping (see also Helgesen *et al.*, 1997).

In some publications, PLS regression is used instead of PCR for relating sensory and consumer data. The idea is to be able to extract more of the interesting information in the first few components. For external mapping this is not likely to represent any improvement because the sensory data have usually much more structure than the consumer data and they will therefore dominate the extraction of components. For the internal mapping the stability of the model may be improved, but possibly at the cost of using too much of the information in the sensory data for determining the components and thus losing the possible advantage of the internal mapping described above. In Helgesen *et al.* (1997) both methods were used, but from a practical point of view, the two approaches gave similar results.

Note that there also exist other methods for handling relations between multivariate data sets than PCR and PLS. One of them is the canonical correlation method (Mardia *et al.*, 1979). This method is, however, not suitable here due to problems with collinearity and the fact that there are more variables than samples, which is always the case in preference mapping applications.

For the linear mapping method, the direction of liking is always opposite to the direction of disliking. If a high level of sweetness is liked, low level of sweetness is disliked. This is the case also for the attributes that lie opposite to the sweetness in the attribute plot. This may be an over-simplification which in some cases can be solved by using the more advanced ideal point mapping methods to be described below.

9.2.1.4 Three-Way Components Models in Preference Mapping

All the methods above are based on the average sensory profile. It is, however, also possible to use the individual sensory profiles and relate these individual values to the consumer acceptances. This represents a clear methodological complication and in most cases it is not necessary unless there are strong individual differences in detection levels. In such cases it may be of interest to analyse individual differences in the sensory data set for improved understanding of consumer differences in acceptance. This theme will not be pursued here, but we refer to Chapter 6 for a discussion of methodology for relating three-way data to two-way data in another context (see also Chapter 17).

9.2.2 Plotting and Interpreting the Consumer Loadings

The scores and the sensory and consumer loadings are plotted and interpreted the usual way as shown in Chapter 14. This means for instance that one consumer that lies in the direction of a certain sensory attribute likes this particular attribute and dislikes those that lie in the opposite direction.

The regular loading values can be used (see Rødbotten *et al.*, 2009), but in some applications the loadings are linked to the centre by straight lines and in some cases also scaled to have the same (unit) length (Helgesen *et al.*, 1997). The reason for this is ease of interpretation since the directions of the acceptance values are the most important. Note, however, that scaling the loadings always represents loss of information.

Another possibility is to use correlation loadings. Using this plot it is also easy to identify and possibly also eliminate those assessors that have the least structure common with the rest. This is done by for instance eliminating those consumers that fall within the 50 % explained circle, and redo the analysis with the rest of the consumers. This type of consumers can either be confused, nonconsistent or they may simply have a totally different acceptance pattern than the rest of the consumers. Our experience is that eliminating consumers with low explained variance does not necessarily change much. Another way of identifying consumers who have a weak relation between the consumer acceptances and the sensory data is to use regression F-tests as described in Chapter 11.

Note that consumers falling close to the middle/origin (for both internal ad external mapping) are consumers with no specific direction for their preference. The regression coefficients are close to 0 and this means that they have no clear direction for their preferences. Note in particular that it does not mean that they prefer samples in the middle.

In standard applications of internal preference mapping the consumers data are always centred. This is normal for all applications of PCA since modelling variability with respect to the centre is in most cases more natural than modelling with respect to the origin. It should, however, be mentioned that this means that only the 'relative' difference between samples is used in the analysis (see chapter on PCA) and not the absolute acceptance values. For instance, if one person uses the higher end of the scale while another uses the lower end, this difference will be neglected in the analysis. For external mapping this type of differences can be analysed by adding an intercept for each consumer model and then interpreting the intercepts as differences in use of the scale.

Scaling of the consumers by dividing the data for each of them by their standard deviation, is also possible. This is seldom done in practice, but may be natural if one believes that the different span of the scores has no relation to differences in the liking pattern, i.e. it is only related to different use of the scale. In some cases, it may be difficult to make a decision. The common way of solving the problem is to use no scaling at all.

Since one generally uses only few samples for preference mapping, validation of the models (both the sensory model and the consumer model) and selecting the number of significant components by the use of cross-validation (Chapter 15) is often difficult. Typically, each sample is unique and cross-validation will give unreliable results. One can also question the role of cross-validation in situations where it is not natural to think of the samples as representing a true population. In the example related to apple juice discussed below, it was used with success, but generally this is not the case.

As usual, however, one can consider PCA simply as a descriptive method that presents the data in a compressed and graphical way. If the first few components describe a substantial amount of variation before the explained variance curve flattens out, these few components are good candidates for reliable interpretation. Regression analyses (both internal and external mapping) accompanied by model F-tests (Chapter 15) can also be useful for assessing the importance of the principal components and thus to determine how many to rely on for interpretation.

An additional and quite interesting aspect of preference mapping is that it can also sometimes be used for providing ideas for new product development. If for instance one identifies an attribute direction with many consumers and this is an area with few or no samples in it, this may be an indication of a possible region for further study and new development. It is likely that a new product in this region with the attributes indicated by the sensory loadings plots may lead to high liking among many consumers.

9.2.3 Other Possible Ways of Modelling the Data

In addition to modelling each individual separately as done above, it is also sometimes possible to model all or groups of consumers simultaneously.

9.2.3.1 Average Preference Data

The simplest example of this is to model the average preferences directly, but this can have important drawbacks if the consumers are very different. Imagine for instance that there are two different groups of consumers with opposite opinion about two different products. The two products will then end up with the same average score and no effects will be found (see Tang *et al.* (2000) for an interesting application). This type of modelling is therefore most suitable when one expects the group of consumers to be relatively homogeneous. Its importance for practical studies should, however, not be neglected due to its simplicity.

9.2.3.2 Simultaneous Modelling

A better method is to model all consumer acceptance values as functions of the sensory principal component dimensions in one single regression model. The simplest possibility is to assume

$$y_n = \beta_1 \hat{t}_1 + \beta_2 \hat{t}_2 + Cons_n + \varepsilon_n \tag{9.3}$$

for each sample. Here the y's now represent acceptance values, the $\hat{t}'s$ the principal components of the sensory data fo the actual sample, the β's are the regression coefficients and the extra term *Cons* represents a systematic random consumer effect. An additive consumer effect will, however, only act as an intercept and account for level differences (see also discussion above). It is therefore more realistic to add interactions between consumer and the principal components using for instance the model:

$$y_n = \beta_1 \hat{t}_1 + \beta_2 \hat{t}_2 + Cons_n + \beta_{1n} \hat{t}_1 + \beta_{2n} \hat{t}_2 + \varepsilon_n \tag{9.4}$$

where both *Cons* and the individual regression coefficient β_{1n} and β_{2n}, are considered random effects. Note the similarity between this model and the models (8.3) and (8.7).

Note that the estimates of the regression coefficients of the fixed effects (β_1, β_2), will as for the average data approach, be related to the average liking of the consumers in the population. The advantage of this approach is, however, that the individual differences can be investigated afterwards both individually by plotting of residuals and by comparing the variance components. This aspect will not be pursued further here, but an example of how this is done for a similar situation in conjoint analysis is given in Chapter 8. In that example, both the individual differences and the variance components are computed and interpreted. The same aspects regarding cautious interpretation of R^2 as discussed in Chapter 8 are valid also here.

As can be noted this approach represents a kind of compromise between the individual model approach and the modelling of the average data. It essentially models the average, but opens also for a study of the individual differences.

9.2.3.3 Segmentation

Another interesting approach which represents a compromise between the two methods above is to segment the data accompanied with modelling of averages within each segment. This means that one collects the consumers with similar response pattern in separate segments and fits the segment models separately by using a standard regression models without individual consumer effects. A number of approaches may be used for segmentation, based on for instance cluster analysis. A broad discussion of these methods will be given in Chapter 10 where both conjoint analysis and preference mapping will be considered at the same time within the same clustering framework.

9.2.3.4 Analysing Ranking Data

If the consumer data are collected as rank data, the sum of the scores for each consumer is equal to the sum of the ranks and some of the above analyses are not longer natural. This holds in particular when concerns ANOVA and regression analysis. Pragmatic use of PCA may, however, still be quite useful since this method does not make any assumption about residual distributions.

A possible method which can sometimes be used for this type of data is the so-called optimal scaling (OS, Young, 1981) approach which transforms the data before modelling (see Chapter 17 for a short description). The transform is defined in such a way that the data after the transform still satisfy the same ranking as the raw data, but now the data are considered as regular rating data that can be fitted to any type of model. The two parts of the fitting procedure, the optimal scaling/transform and the model estimation, are done at the same time using the principle of alternating least squares ALS (Chapter 17). This is a very flexible strategy and has as far as we know never been tested within the area of preference mapping. Since it is flexible it is also be sensitive to over-fitting and it must be used with care.

9.3 Examples of Linear Preference Mapping

9.3.1 Apple Juice Example

The example to be considered first focuses on the effect of varying sugar and acid content of apple juice (see Rødbotten *et al.*, 2009). The sample set consists of 6 samples, prepared according to an experimental design (a 2*3 factorial design) with two levels of acid concentration and three levels of sugar concentration. The three sugar levels correspond to regular level of sugar in commercial juice, 20 % reduction and 40 % reduction as compared to the standard juice. The two acid levels are both within the range found in regular commercial apple juice. The samples are all based on the same concentrate and manipulated by adding water, sugar and acid. The study was conducted as a part of an EU funded project with focus on reducing the sugar content in apple juice (Rødbotten *et al.*, 2009). The samples were analysed by sensory analysis and then tested by 126 consumers using a standard rating system anchored in 'like very much' and 'dislike very much'.

The results in Figure 9.5 are based on external preference mapping. As can be seen, a very large portion of the variance in the sensory attributes is collected in the first two principal components (97 %).The components are also valid according to cross-validation as shown by the lower explained variance curve. The scores plot is very similar to the design of the samples, with the samples with the highest degree of sugar to the right, those with the lowest to the left and the medium ones in the middle. The samples with the high acid level lie above the horizontal axis and the rest below. The samples with the higher sugar content have the highest values of sweetness etc. It seems, however, that when concerns acid content, the bitterness plays a more prominent role than does acidity for describing the effect of adding acid. Most of the other attributes have a weak relation to the two sensory dimensions. This was also to be expected since the samples were based on the same concentrate.

The consumer loadings plot from linear preference mapping tell us first of all that almost all the consumers fall to the right of the vertical Y-axis. This means that the acceptance values for these persons go in the direction of the sweetest samples with only a few lying in the opposite direction. If we look at the correlation loadings plot it can be seen that a very large portion of the consumers fall outside of the 50 % explained circle. This means that almost all consumers have a rather strong tendency of more or less the same type, namely of preference for the sweetest samples.

If we look at the second component, it is clear that there are representatives at both sides of the horizontal X-axis. This means that there are persons who prefer samples with high acid content and those who prefer samples with low acid content.

This example shows that sweetness is the most important driver of liking for the apple juices considered (not very surprising) and that this was the case for almost all the consumers. The study has also shown that when concerns amount of acid in the samples, about half of the consumers prefer the low level and half of them prefer the high level. There were also consumers close to the horizontal line indicating an indifferent attitude regarding this dimension.

The data were also analysed by ANOVA for each of the sensory attributes to confirm the findings.

9.3.2 Sausage Example

The second example concerns consumer liking of dry fermented lamb sausages. The study was conducted in order to understand the entire Norwegian market of this type of products. First, 14 different products covering more or less the whole market were selected and analysed by sensory profiling. Then six samples were selected from the PCA scores plot of the sensory data. The selection process is illustrated in Figure 9.6 and commented on in more detail below. As can be seen, the samples are selected in such a way that they cover the whole region well. The six samples, renamed 1–6, were then assessed by all the consumers. The sensory and consumer data sets were analysed by external preference mapping. The consumer loadings are scaled and linked to the centre in such a way that the direction, not the strength, is highlighted.

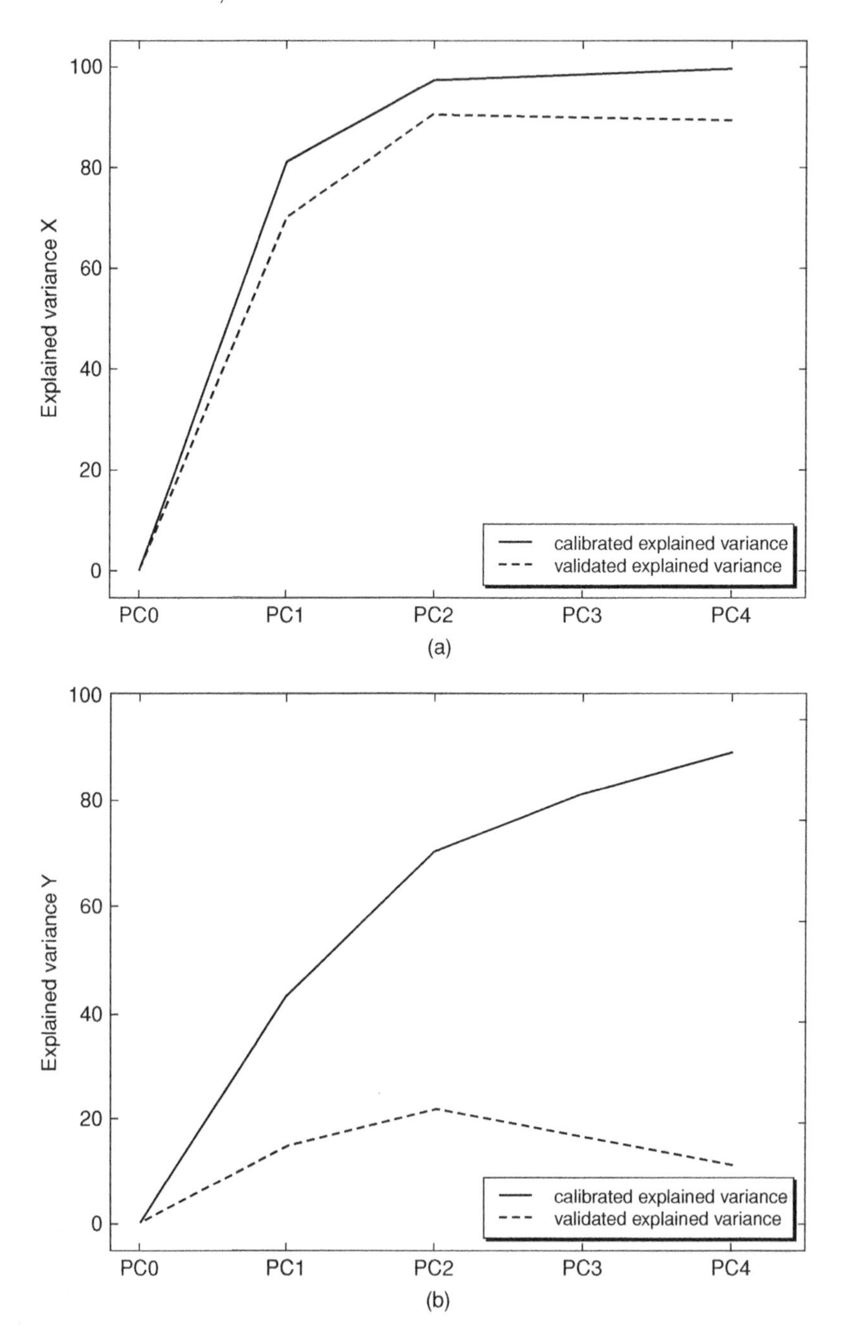

Figure 9.5 ***Preference mapping results for the apple juice study****. Explained variances for* ***X*** *and* ***Y*** *(both calibration/estimation and validation) are presented in figures a) and b). The scores, sensory loadings, consumer loadings and correlation loadings are presented in c), d), e) and f). The first digit in the symbols in the scores plot refer to sugar level (L-low, M-medium, H-high) and the last digit to the acid level (L-low, H-high) (See also Rødbotten* et al. *(2009).*

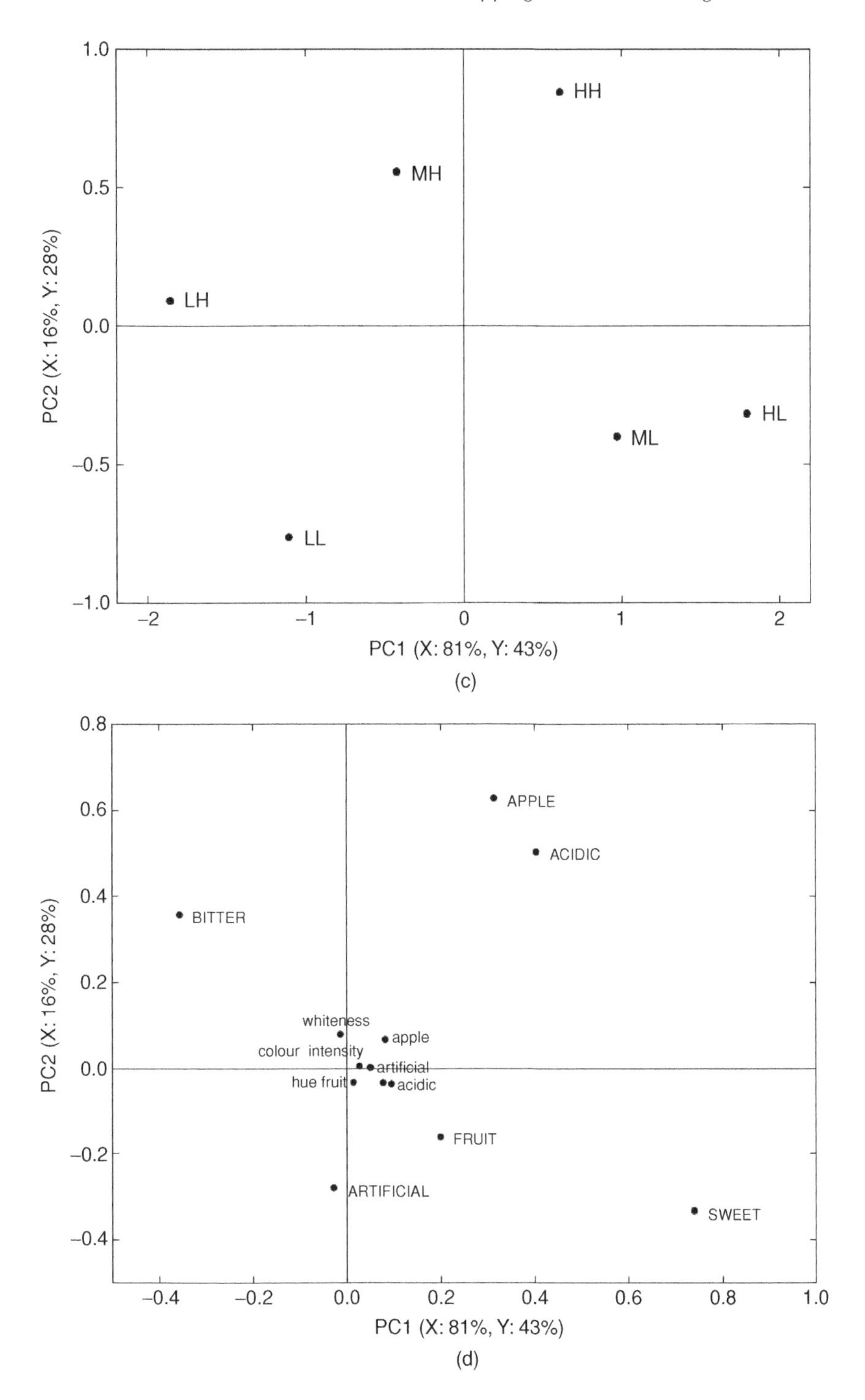

Figure 9.5 *(Continued).*

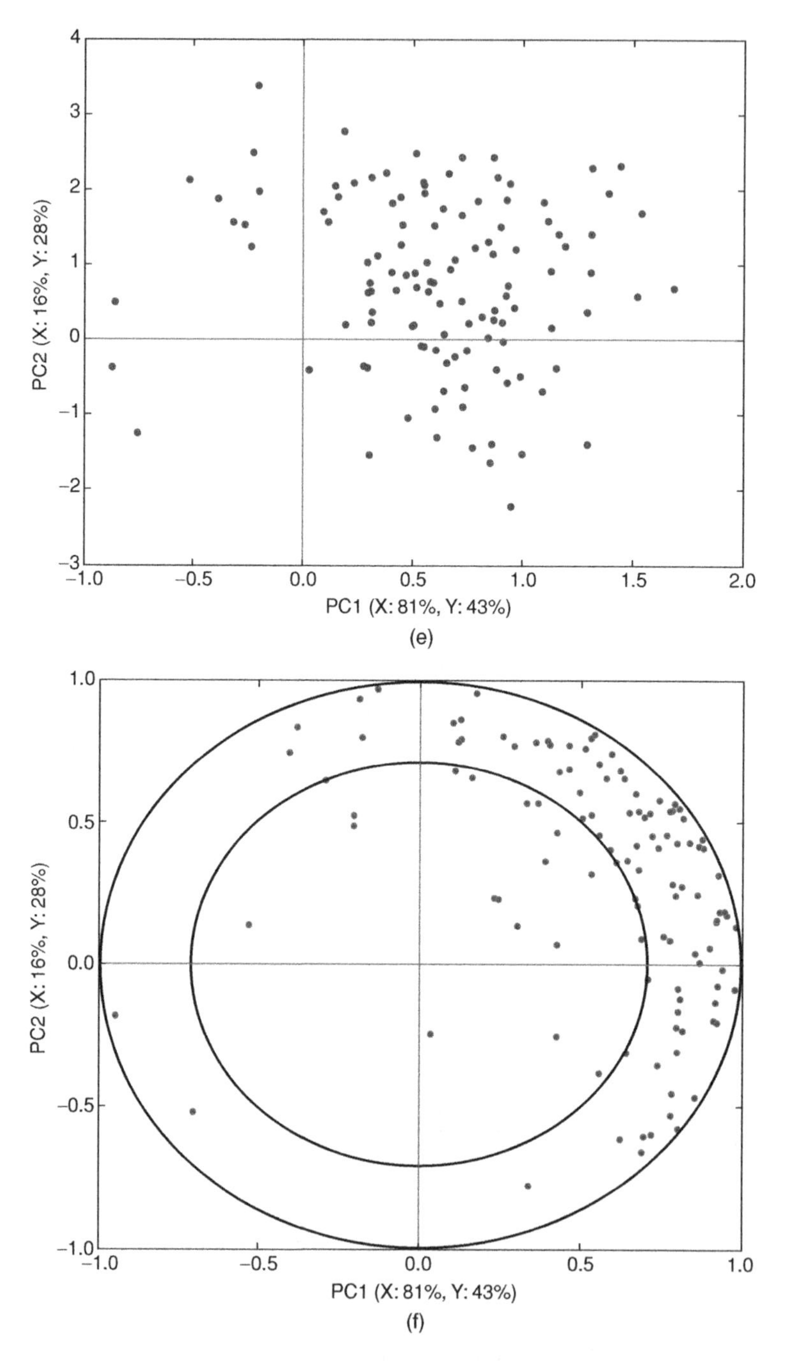

Figure 9.5 (Continued).

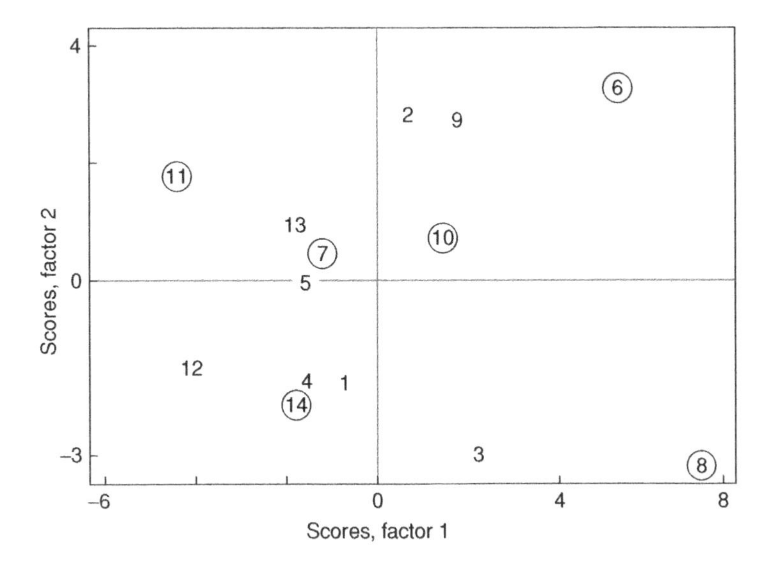

***Figure 9.6** **Sample set selection in dry fermented lamb sausage example (Helgesen and Næs**(1995)). All 14 products are plotted and the circled products are selected for the consumer study. The criteria for selection are described in the text. Reprinted from Food Quality and Preference, 20, Helgesen and Næs, Selection of dry fermented lamb sausages fopr consumer testing, 109–120, 1995, with permission from Elsevier.*

As can be see from Figure 9.7a, most of the consumers go in the direction of products 2, 5 and 6 while very few consumers seem to prefer sample 3. This is in good correspondence with the average liking results reported in Helgesen *et al.* (1997). The sensory properties which seem to be the most liked (Figure 9.7b) are juiciness, whiteness, lamb flavour, greasiness, acidic odour and acidic flavour. The same tendencies were found after elimination of the consumers with the least contribution to the explained variance.

As can also be seen, it seems that some of the directions which are liked according to the consumer loading plot have no clear product representatives. It might therefore be tempting to propose product development in this area. For instance along the vertical axis, there are several consumers that focus on this dimension, but there are large open areas in the score plot to be explored by new product development. The sensory loadings indicate that this axis is dominated by spiciness vs. smoke.

9.4 Ideal Point Preference Mapping

One of the advantages of the external preference mapping is that it can easily be used for handling other types of relations than the classical linear one. This is particularly important when a certain attribute is liked, but too much of it is disliked. Typical examples of attributes of this type are sweetness and saltiness.

In principle there are many models that can be used for this, but the simplest and most well established approach is to use second order polynomials based on the principal components (Chapter 15). One of the advantages of this type of models is that they can be

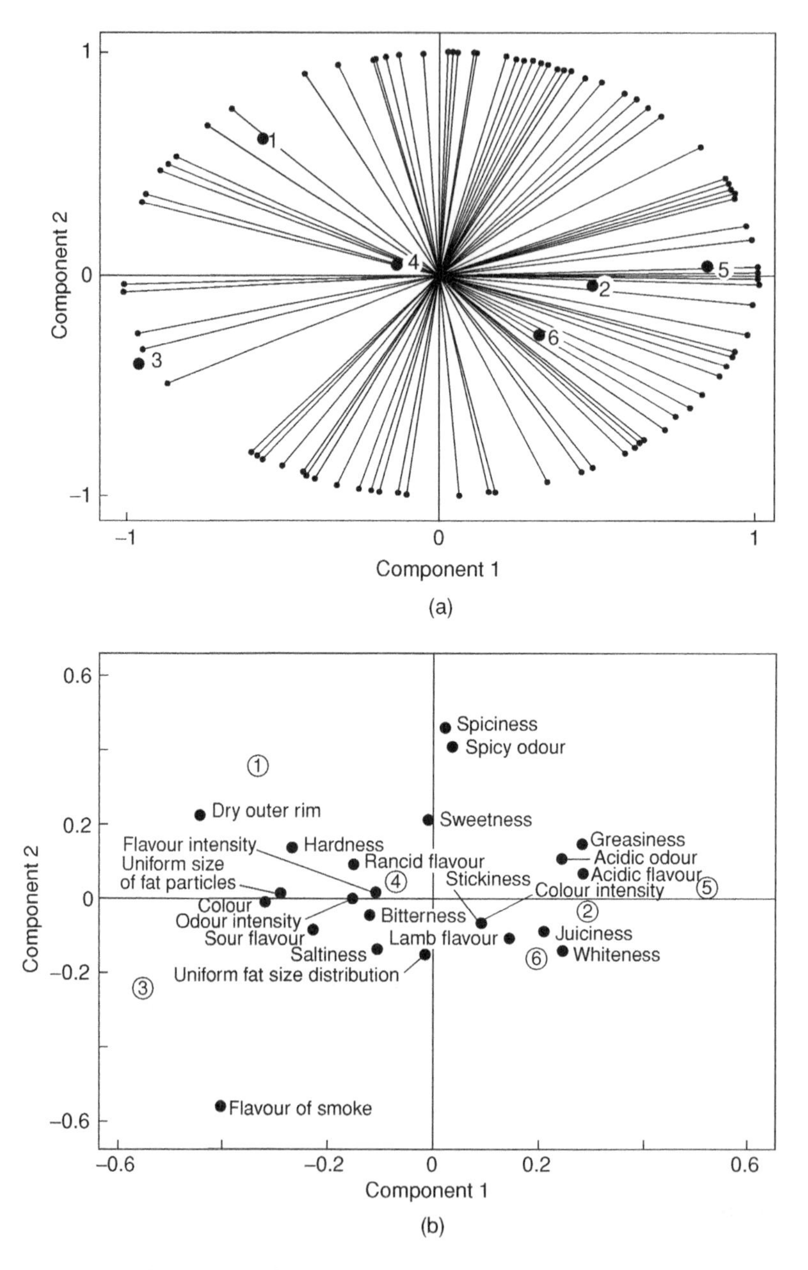

Figure 9.7 Results from external prefmap of dry fermented lamb sausage data (Helgesen et al. (1997). *The product scores and the consumer loadings are presented in figure a) and the product scores and the sensory loadings are given in figure b. The product scores are the same in both cases. The consumer loadings are scaled and joined to the centre by a straight line in order to focus on direction of the liking. Reprinted from Food Quality and Preference, 8, Helgesen* et al., *Consumer preference mapping of dry fermented lamb sausages, 297–109, 1997, with permission from Elsevier.*

handled by standard linear regression techniques as discussed in Chapters 11 and 15. The full polynomial model based on two principal components can be written as

$$y = \beta_0 + \beta_1\hat{t}_1 + \beta_2\hat{t}_2 + \beta_{11}\hat{t}_1^2 + \beta_{22}\hat{t}_2^2 + \beta_{12}\hat{t}_1\hat{t}_2 + \varepsilon \tag{9.5}$$

As can be seen, this involves 6 terms meaning that it can not be used for less than 6 samples if an individual models approach is used. Even for 6 samples, the model will be very unstable and no residuals will be available for testing model adequacy. Used for individual fitting this model may therefore have limited applicability.

Note that model (9.5) allows for both positive and negative peaks as well as saddle points (Myers and Montgomery, 1995). It is also important to note that the linear model is a special case of model (9.5). An intercept is always needed for polynomial regression model even though the consumer acceptance values and sensory scores are centred. The reason is that even for centred data, the quadratic terms are noncentred.

A couple of simplifications have been proposed for reducing the number of parameters in the model. One of these is to eliminate the interaction term (so-called elliptic model) and the other one is to assume the same size of the coefficients for the two quadratic terms (so-called circular model). The first suggestion gives a model with 5 parameters and the latter one a model with 4, in both cases at the expense of loss of flexibility. The elliptic model gives contours with principal axes parallel to the two principal component axes, while the circular model gives circular contours if the two axes are standardised prior to modelling.

In practice, the ideal point model is usually tested for curvature for each individual separately. The nonlinear models are then optimised and the max or min values are indicated in the scores plot. The 'linear consumers' can be presented in a separate loading plot using the same techniques as described above.

A comparison between a linear model and an ideal point model with elliptic contours is given in Figure 9.8 (see also McEwan (1996) for more detail).

For situations with 3 principal components in the preference model, the full second degree polynomial with all three quadratic terms and all three interactions has as many as 11 parameters to be estimated. This means that more than 11 samples are required for each consumer if individual models are used.

This discussion implies that even though the ideal point model is more realistic than the linear one in some cases, it may be difficult to use for individual modelling. The best approach in such cases is probably to use some type of segmentation combined with joint fitting within each segment. Such an approach will be discussed in Chapter 10 where several possible approaches to segmentation will be discussed within a joint framework.

An important aspect to mention is that ideal point models may, as opposed to linear models, be used also for situations with different drivers of liking and disliking. Assume for instance that an 'acceptance peak' is positioned in the first quadrant. The less liked samples may then lie in all different directions and not necessarily in the third quadrant (opposite to the first). Such a situation can not be handled by linear models, but a quadratic polynomial may still be adequate (see also Kano *et al.*, 1984; Riviere *et al.*, 2006).

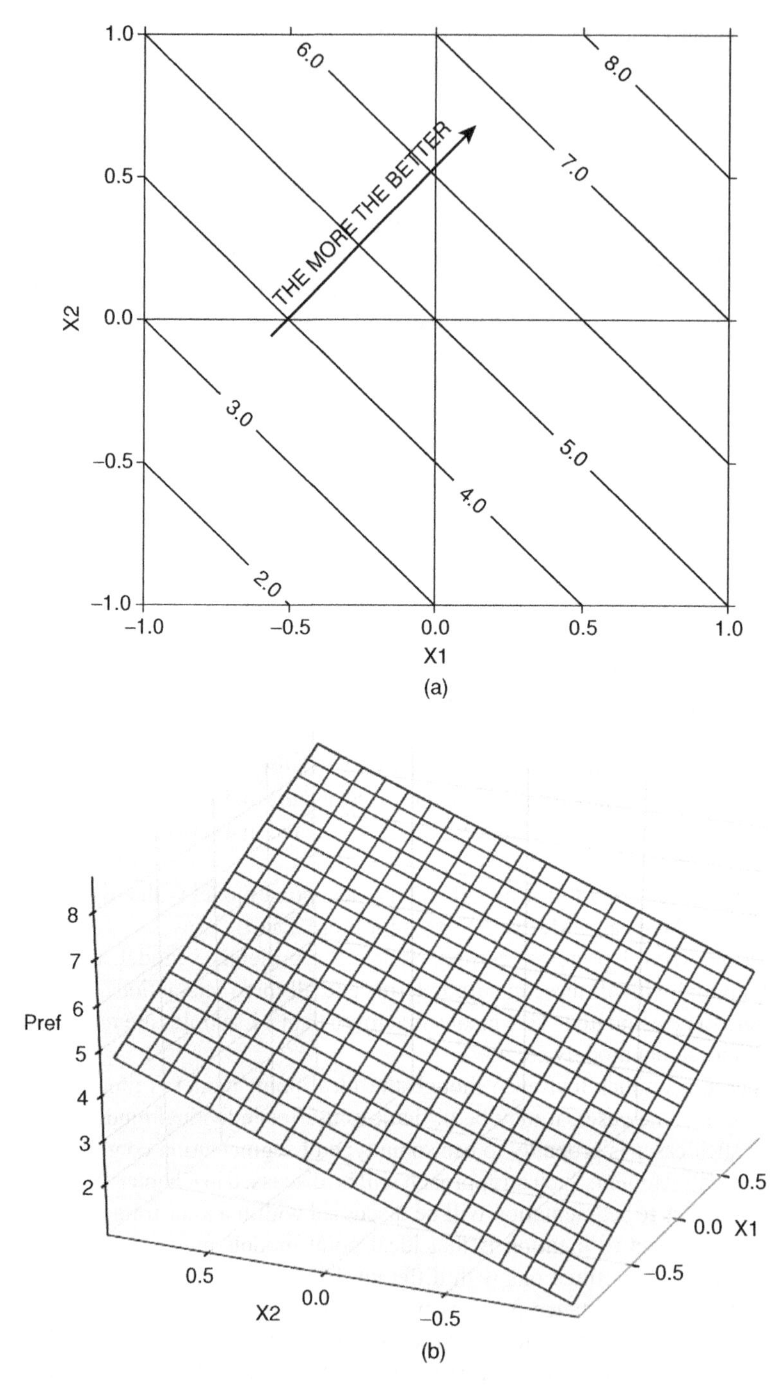

Figure 9.8 Linear and ideal point models illustrated. *The plots in figures a) and b) show the contour plot and three-dimensional plot for the linear model. The figure c) and d) show the contour plot and three-dimensional plot for a polynomial model.*

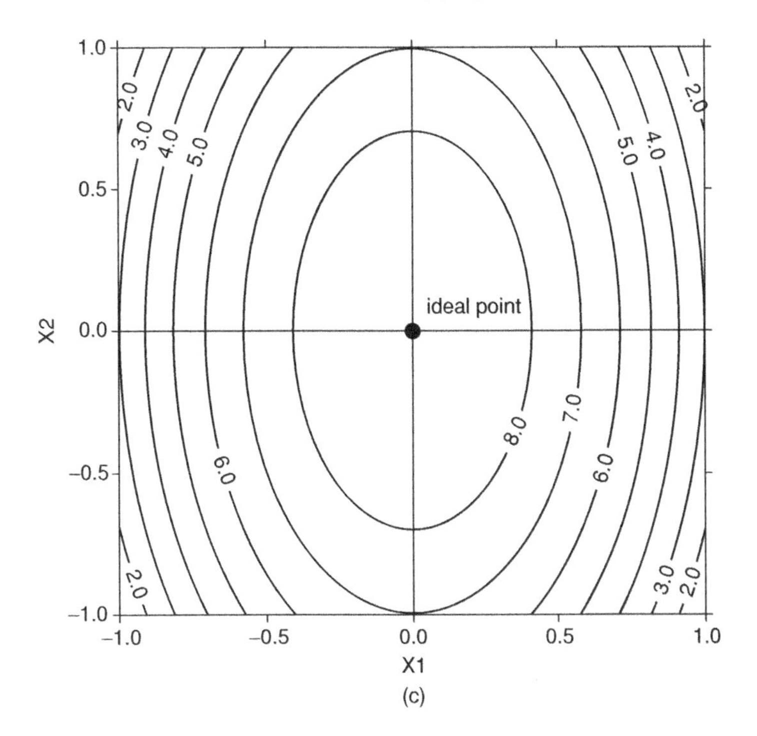

(c)

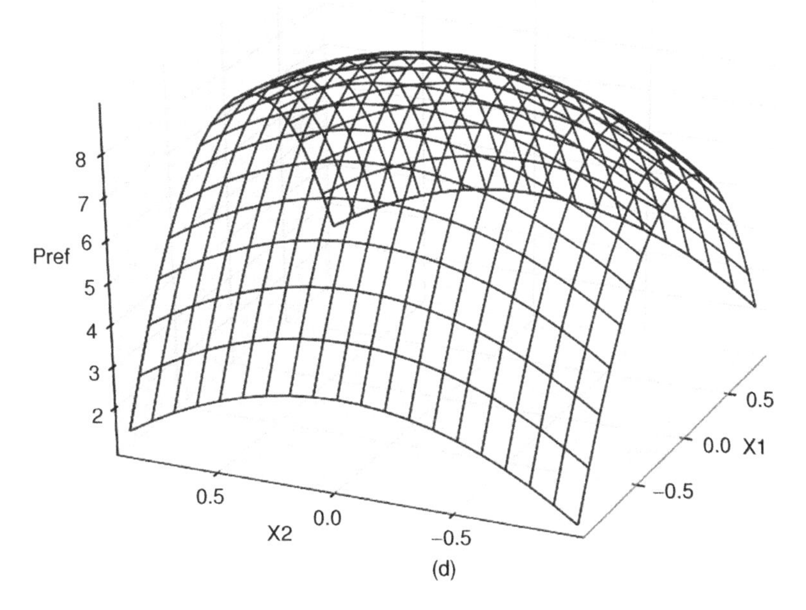

(d)

Figure 9.8 (Continued).

9.5 Selecting Samples for Preference Mapping

As stated above, in consumer acceptance studies there is always a trade-off between the statistical need of having as many samples as possible for estimation of coefficients and the fact that consumers can only test a few samples each. Usually, one ends up with a compromise between 5 and 10 samples for each consumer, depending on the type of samples used. If the samples are particularly strong in taste or flavour it is in most cases better to be closer to 5 than to 10.

In many cases, it is clear from the context exactly which samples that are of interest to test, but in other cases one may have a choice among a larger set of possible samples. An example of the latter was given in Helgesen and Næs (1995) where the purpose was to characterise the Norwegian market of a special type of sausage product and also to investigate the consumer acceptance values within the market for this product category. In this example, 14 different samples were selected (covering more or less the whole market of the product) and analysed by sensory analysis. In this case 6 samples were selected and tested by the consumers using the following idea:

The basis for the selection process is the PCA scores plot based on the sensory data of all samples available (Helgesen and Næs, 1995). In order to ensure good model precision and also the ability to detect model adequacy it was proposed to select samples that cover the whole region of samples (the whole PCA scores plot) as evenly as possible (see Figure 9.9, and Næs and Isaksson, 1989). It was shown in Zemroch (1986) and Næs (1989) that an even spread of points also has a robustness aspect in the sense that it will give reasonable fit and predicted values even if the model used is not the best.

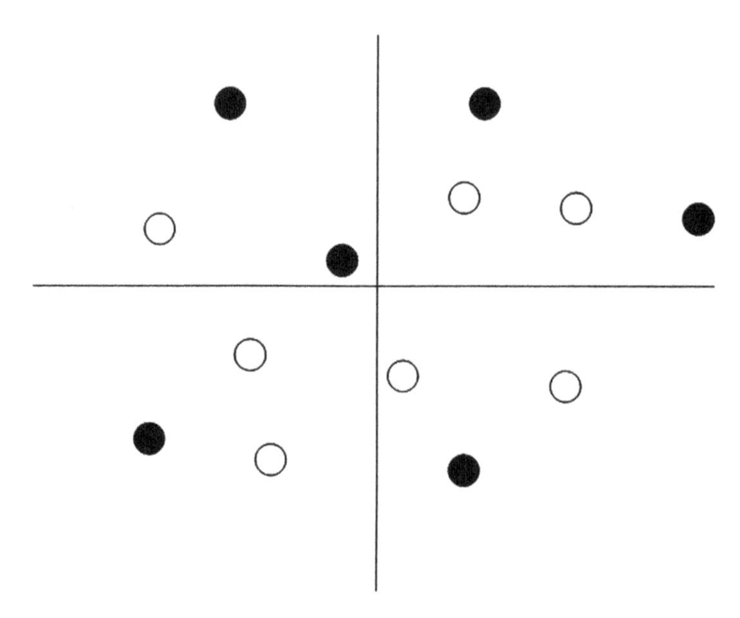

Figure 9.9 Illustration of principle for selection of samples. *The black circles are those selected according to the criterion given in the text.*

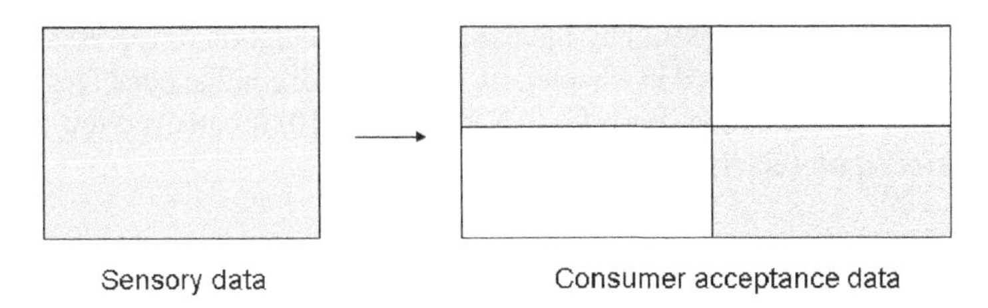

Figure 9.10 ***Consumer data sets set with many missing values.*** *The figure depicts a situation with two groups of consumers having tested two different groups of products. For both sets there are many missing values in the data set (illustrated by the white regions).*

Note that this is a general principle and one should not hesitate to overrun it if there are other criteria that need to be satisfied. Such a case was reported by Johansen *et al.* (2010a) where two of the samples were of special interest. A slight modification of the ideal criterion was therefore implemented.

If a more formal and fully automated procedure is wanted, which may be of some interest if there are more than two principal components to select from, a regular cluster analysis can be used (Zemroch (1986) and Næs (1989). The procedure is based on using cluster analysis, for instance a hierarchical method, of all sensory samples and stop the clustering when the number of clusters is equal to the number of samples to be used for consumer testing. The samples within each of the clusters are similar and samples in different clusters are different. In order to satisfy the general criterion, one can then select one sample from each cluster, preferable the one lying furthest away from the centre in order to obtain the best possible spread. An alternative is to use the one closest to the centre in each cluster.

When using data selected according to the criterion above for preference mapping, one has a choice between repeating the PCA on the reduce data or using the original scores obtained by PCA of all the samples. In most cases it would not matter very much which of them is chosen.

If different samples are used for different consumers, one ends up with a consumer preference data set which contains missing values. An extreme case with two consumer groups having tested two different sets of samples is presented in Figure 9.10. Using PCA for data sets with missing values is possible (see Chapter 15), but the situation in Figure 9.10 is too extreme for this. This means that internal preference mapping can be problematic in such cases. As usual, external mapping can handle such situations easily since all columns in the preference data set are used separately in separate regression equations.

9.6 Incorporating Additional Consumer Attributes

A preference map can be very interesting in itself, but in many cases one is also interested in further interpretation of the individual differences detected. A number of different techniques can be envisioned for this. A possible strategy is to collect the consumers into relatively homogeneous groups, by for instance visual inspection of consumer loadings plots or by more formal segmentation methods and then relate the segments to additional consumer attributes (see Figure 9.1) by the use of either tabulation or regression analysis. This approach will be considered in Chapter 10. Here we will confine ourselves to methods that do not require segmentation. We refer to Næs *et al.* (2010) for an overview of methods. See also Endrizzi *et al.* (2009).

9.6.1 Finding Relations Using Regression Methods

The simplest and most straightforward way of relating a preference mapping solution directly to external consumer attributes is to use regression analysis with the consumer loadings as Y-variables and the external data as X-variables. It is generally most useful also here to use a PCR or a PLS approach, but if the external data set only contains a few variables like for instance age group and gender, a regular LS regression is a good alternative. If two principal components are used in the preference map, the analysis ends up with two regression equations, one for each dimension.

An example of this approach can be found in Table 9.1 and Figure 9.11 based on the same data as used in the apple juice example above. In Table 9.1 are given both the description of the consumer attribute used (frequency of use) and the ANOVA table for the analysis. In this case the loading for the second preference component, i.e. the one related to acid level, is regressed onto frequency of use. As can be noted from the plot in Figure 9.11, the more often the consumer drinks apple juice, the more he/she seems to prefer the juice with high acid content. The effect is, according to Table 9.11, not significant, so one should be careful about making too strong statements in this particular case. The fact that the effect is systematic, however, supports the validity of the tendency.

Note that for certain structures in the loadings plot, this type of analysis may fail to detect important results. Imagine for instance that quadrant 1 and 3 in the loadings plot have the same relation to the external data, and that the same is true for quadrants 2 and 4. This pattern will then be impossible to detect if the two principal components are considered separately, simply because the two effects will cancel each other. We believe, however, that such situations are rare.

Table 9.1 Relation between consumption and liking for the apple juice example. *First, the four different response possibilities for the consumers are given. Next, the ANOVA table for the analysis where the second loading in Figure 9.5e is used as the response and frequency of use (FU) is used as the explanatory variable.*

How often do you drink apple juice (four levels)?

1. **2 times per month or more seldom**
2. **3 times per month**
3. **1–3 times per week**
4. **4 times or more per week.**

Source	DF	SS	MS	F	p-value
FU	3	2.33	0.78	0.58	0.626
Error	122	161.83	1.33		
Total	125	164.16			

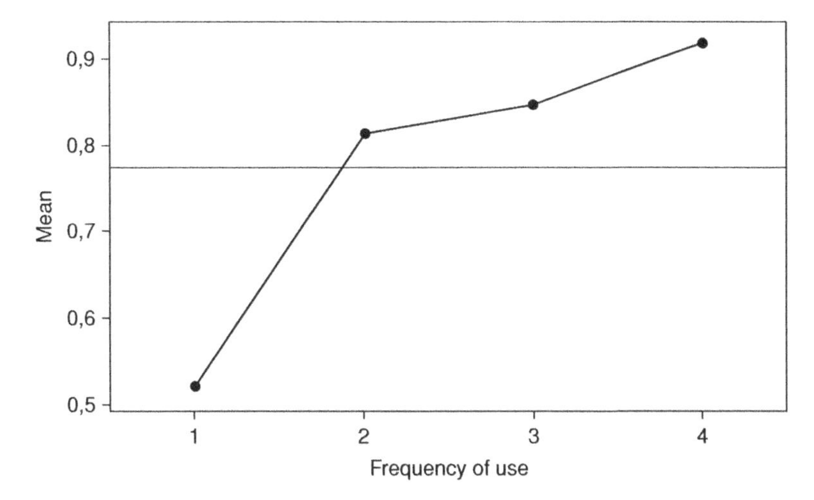

Figure 9.11 ***Relation between consumption and liking in the apple juice example.*** *The plot illustrates the relation between loading 2 and frequency of use of the juice. The more the consumers drink, the more they seem to like the juices with high level of acidity. The numbers on the horizontal axes represent the four levels in Table 9.1.*

9.6.2 Finding Relations Using L-PLS Regression

An alternative approach is the newly developed L-PLS regression (L-PLSR) method (Martens *et al.*, 2005), which analyses all three data sets simultaneously. The L-PLS is a very elegant extension of regular PLS regression methodology for data structure of this type (with an L-shape, see Figure 9.1), but how this approach relates to interpretation of the two-step approach above is still not fully explored. A possible advantage of the method is that everything is done in one single analysis.

The method ends up with two-dimensional plots where both the sensory variables, the consumer attributes, the scores and the additional consumer attributes are represented, either separately or in one single plot. Some more details about the method are presented in Chapter 17.

9.7 Combining Preference Mapping with Additional Information about the Samples

In a buying situation the modern consumer emphasises a number of aspects of the product in addition to the sensory properties (Enneking *et al.*, 2007). Important examples of such additional or extrinsic attributes are information about health benefits, user-friendliness of the product and impact on the environment. For a product developer it is important to have techniques available, both for design and analysis, for investigating the relative importance of these additional factors and the sensory attributes and also for evaluating how they interact with each other.

The simplest way of setting up a design in such cases is to consider the different products as individual levels of one of the experimental factors and then use the same types of methods as those explained in Chapter 8 for full factorial designs. Such an approach was taken in for instance Helgesen *et al.* (1997) where the focus was on evaluating a number of sausages in combination with information about fat level and price. The advantage of the method is that it is conceptually simple and fits well into established ANOVA methodology.

If the number of samples is large, however, one will need to use some type of fractional factorial design for multilevel problems, which is considerably more complex than situations with two levels only. Another possible problem with this approach is that it does not provide much information about what are the most important sensory drivers of liking and how they interact with the additional attributes. In order to obtain this type of information one needs to incorporate sensory analysis of the samples and use this information as covariates in the analysis. A third problem is that the samples used are not necessarily spread out in such a way that they span the relevant sensory space adequately.

In the following we will discuss an alternative strategy first proposed by Johansen *et al.* (2010a) that can be used to solve these problems.

9.7.1 An Alternative Design Strategy for Combining Products and Additional Information

The situation considered here is again one in which there exist a large set of possible product candidates and the problem is to select the best subset. This type of situation is typical in for instance product development where a relatively large number of possible prototypes are available, but only a few of them can be tested in the market, either for economic reasons or for reasons related to consumer fatigue.

As above one can select samples based on the PCA scores of the sensory data of a larger set of samples, but in this case one should use a slightly different criterion than the one used in Chapter 9.6. The main focus is to select samples that first of all represent all the relevant combinations of the sensory attributes, but also in such a way that it is easy to combine the products selected with other extrinsic variables in the experimental setup. A possible way of doing this is to select samples in the PCA scores space such that the corresponding geometric shape resembles a rectangle (see Figure 9.12) with the two rectangle axes interpreted as new so-called 'meta-attributes'. These new attributes can then easily be combined with the other extrinsic attributes in for instance a factorial design.

Let us as an example assume that there are two significant principal components for the sensory data and two additional attributes, for instance related to price and a health aspect, with two levels each. Let the two sensory meta-attributes be the two illustrated in Figure 9.12. We then have four variables with two levels each, two meta-attributes and two additional extrinsic attributes. In Figure 9.12, the point marked by '1' represents low level of both meta-attributes, the point marked by '2' represents low level of meta-attribute 1 and high level of the other etc.

Using standard factorial design theory, one can then set up a full factorial design in the four variables with 16 experimental runs in total. If a half fraction is selected, one ends up with 8 runs. An example of such a design is given in Table 9.2. As can be seen, the first run

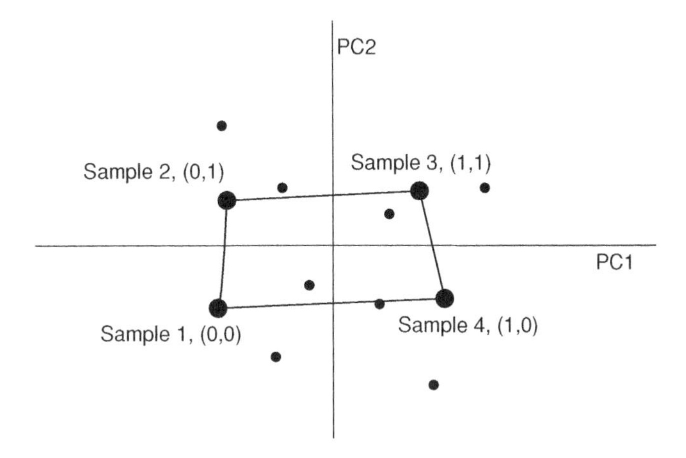

Figure 9.12 The idea behind selection of samples in a situation where they have to be combined with extrinsic attributes in the design. *The products are selected in such a way that they represent the corners in a structure similar to a rectangle. The directions of the two axes in the rectangle are used as meta-attributes to be combined with extrinsic attributes.*

in this design consists of combining sample marked 1 with low level of extrinsic attribute 1 (price) and high level of extrinsic attribute 2 (health related). A similar description can be given for the next run etc.

The PCA loadings can be used to interpret the meta-attributes, which is always to be recommended for better understanding. Since a rectangle is used, all major combinations of the sensory attributes will be represented and spread out and as can be seen, it is easy to combine the samples with other attributes interpreting the rectangle axes as new attributes.

Table 9.2 Design of experiment for combined study. *The − and + signs represent "low" and "high" level respectively. The third attribute is generated as a product of the other three, i.e. the generator is $D = -ABC$. The resolution is IV, which means that some two factor interactions art confounded, but the main effects are not confounded with any two-factor interactions.*

Run	A Meta-attribute 1	B Meta-attribute 2	C Extrinsic attribute 1 (price)	D Extrinsic attribute 2 (health related)
1	−	−	−	+
2	+	−	−	−
3	−	+	−	−
4	+	+	−	+
5	−	−	+	−
6	+	−	+	+
7	−	+	+	+
8	+	+	+	−

Thus both the two goals are reached, good spread of the samples as well as a simple design procedure.

If no additional requirement concerning selection of samples exists, the rectangle can be constructed in such a way that the axes are parallel to the two principal components axes. If, however, there are other criteria that need to be satisfied, for instance certain attributes that are of special interest, the rectangle can be tilted as was done in for instance Johansen *et al.* (2010a). If a rectangle shape is difficult to find, one will try to approximate it as much as possible.

9.7.2 Analysis of the Combined Data

Since the sensory meta-attributes are combined with the other variables in a regular factorial design, a possible way of analysing the data is to simply use ANOVA with all variables treated as categorical attributes using for instance the model (8.3). With a good interpretation of the meta-attributes this may be a useful approach. Another possible approach is to use the sensory variables directly as covariates in the model (ANCOVA, Weisberg, 1985). If the rectangular approximation above is poor, this will usually be the best strategy.

Since sensory variables are usually many and collinear, it is necessary to reduce the dimensionality by the use of PCA (or use LS-PLS as proposed in Jørgensen *et al.*, 2004, 2007) of the sensory variables prior to combining them with the other design variables. Since interactions may be important here, using individual regression coefficients for the different levels of the external factors may be the most natural approach.

References

Baardseth, P., Naes, T., Mielnik, J, Skrede, G., Hølland, S., Eide, O. (1992). Dairy ingredients effects on sausage sensory properties studied by principal component analysis. *J. Food Science.* 57(4), 822–8.

Ellekjær M.R., Ilseng M.R. and Næs, T. (1996). A case study of the use of experimental design and multivariate analysis in product improvement. *Food Quality and Preference* 7(1), 29–36.

Elmore, J.R., Heymann, H., Johnson, J., Hewett, J.E. (1999). Preference mapping: Relating acceptance for 'creaminess' to a descriptive sensory map of a semi-solid. *Food Quality and Preference* 10, 465–75.

Endrizzi, I., Gasperi, F., Calo, D.G., Vihneau, E. (2009). Two-step procedure for classifying consumer in a L-structured data context. *Food Quality and Preference* (in press).

Enneking, U., Neumann, C., Henneberg, S. (2007). How important intrinsic and extrinsic product attributes affect purchase decision. *Food Quality and Preference* 18, 133–8.

Faber, N.M., Mojet, J., Poelman, A.A.M. (2003). Simple improvement of consumer fit in external preference mapping. *Food Quality and Preference* 14, 455–61.

Gower, J.C, Hand, D.J. (1995). *Biplots*. London: Chapman & Hall.

Greenhoff, K., MacFie, H.J.H. (1994). Preference mapping in practice. In H.J.H MacFie, D.M.H. Thompson (eds), *Measurements of Food Products* (1st ed, pp 137–66). Glasgow: Blackie Academic and Professional.

Guinard, J., Uotani, B., Schlich, P. (2001). Internal and external mapping of preference for commercial lager beers: comparison of hedonic ratings by consumers blind versus with knowledge of brand and price. *Food Quality and Preference*, 12, 243–55.

Hein, K.A., Jaeger, S.R., Carr, T.B., Delahunty, C.M. (2008). Comparison of five common acceptance and preference methods. *Food Quality and Preference* 19, 651–61.

Helgesen, H., Næs, T. (1995). Selection of dry fermented lamb sausages for consumer testing. *Food Quality and Preference* 6, 109–20.
Helgesen, H., Solheim, R., Næs, T. (1997). Consumer preference mapping of dry fermented lamb sausages. *Food Quality and Preference* 8(2), 97–109.
Hough, G., Sanchez, R. (1997). Descriptive analysis and external preference mapping of powdered chocolate milk. *Food Quality and Preference* 9(3), 197–204.
Johansen, S., Næs, T., Øyaas, J., Hersleth, M. (2010a). Acceptance of calorie-reduced yoghurt: Effects of sensory characteristics and product information. *Food Quality and Preference* 21, 13–21.
Johansen, S., Hersleth, M., Næs, T.M. (2010b). The use of fuzzy clustering for segmentation in linear and ideal point preference models. *Food Quality and Preference* 21, 188–96.
Jørgensen, K., Segtnan, V., Thyholt, K., Næs, T. (2004). A comparison of methods for analysing regression models with both spectral and designed variables. *J. Chemometrics* 18, 451–64.
Jørgensen, K. Mevik, B-H., Næs, T. (2007). Combining designed experiments with several blocks of spectroscopic data. *Chemometrics and Intelligent Laboratory Systems* 88(2), 143–212.
Kano, N, Seraku, N, Takahashi, F., Tsuji, S. (1984). Attractive quality and must-be quality. *Journal of the Japanese Society for Quality Control* 14(2), 39–48.
McEwan, J.A. (1996). Preference mapping for product optimization. In T. Næs, E. Risvik (eds), *Multivariate Analysis of Data in Sensory Science* (Vol. 16, Data Handling in Science and Technology, pp. 71–102): Amsterdam: Elsevier Science B.V.
Martens, H., Anderssen, E., Flatberg, A., *et al.* (2005). Regression of a data matrix on descriptors of both rows and of its columns via latent variables: L-PLSR. *Comp. Stat. and Data Analysis* 48, 103–23.
Meullenet, J.F., Lovely, C, Threfall, R., Morris, J.R., Striegler, R.K. (2008). An ideal point density plot method for determining and optimal sensory profile for Muscadine grape juice. *Food Quality and Preference* 19, 210–19.
Myers, R.M. and Montgomery, D.C. (1995). Response surface methodology. Process and product optimisation using designed experiments. New York: John Wiley & Sons, Inc.
Næs, T. (1989). The design of calibration in near infra-red reflectance analysis by clustering. *Journal of Chemometrics* 1, 121–34.
Næs, T., Isaksson, T. (1991). Splitting of calibration data by cluster-analysis. *Journal of Chemometrics* 5(1), 49–65.
Næs, T., Nyvold, T. (2004) Creative design. *Food Quality and Preference* 2, 97–104.
Næs, T., Lengard, V., Johansen, S.B. and Hersleth, M. (2010) Alternative methods for combining design variables and consumer preference with information about attitudes and demographics in conjoint analysis. *Food Quality and Preference* 21(4), 368–78.
Riviere, P., Monrozier, R., Rogeaux, M., Pages, J., Saporta, G. (2006). Adaptive preference target: Contributions to Kano's models of satisfaction for an optimized preference analysis using a sequential consumer test. *Food Quality and Preference* 17, 572–81.
Rødbotten, M, Martinsen, B.K. Borge, *et al.* (2009). A cross-cultural study of preference for apple juice with different sugar and acid contents. *Food Quality and Preference* 20, 277–84.
Tang, C., Heymann, H., Hsieh, F-h. (2000). Alternatives to data averaging of consumer preference data. *Food Quality and Preference* 11, 99–104.
Thybo, A.K., Kuhn, B.F., Martens, H. (2003). Explaining Danish children's preference for apples using instrumental, sensory and demographic/behavioural data. *Food Quality and Preference*, 15, 53–63.
Weisberg, S. (1985). *Applied Linear Regression*. New York: John Wiley and Sons, Inc.
Young, F. (1981). Quantitative analysis of qualitative data. *Psychometrika* 46(4), 357–88.
Zemroch, P.J. (1986). Cluster analysis as an experimental design generator, with application to gasoline blending experiments. *Technometrics* 28, 39–49.

10

Segmentation of Consumer Data

For both the problem areas considered in Chapters 8 and 9, the analysis of individual differences among the consumers is an important issue. Various methods for studying individual differences and also how to relate them to additional consumer attributes were considered. In some cases, it may, however, be more natural and useful to collect together the consumers which show a similar acceptance pattern and analyse these different groups or segments separately. In this chapter we will consider both segmentation methods based on hierarchical clustering and methods based on partitioning. Special attention will be given to fuzzy clustering. It will also be demonstrated how segments can be related to additional consumer attributes using regression analysis or tabulation.

This chapter is closely related to Chapters 8 and 9. The methodology is based on techniques presented and discussed in Chapters 15 and 16.

10.1 Introduction

Individual differences among consumers is mentioned and analysed in several chapters in this book, in particular in Chapters 8 and 9 in connection with conjoint analysis and preference mapping. The present chapter is devoted specifically to methodology for studying patterns among individual differences using segmentation or clustering methods (see also Næs *et al.*, 2001; Westad *et al.*, 2004). The importance of segmentation lies in improved insight which can be useful both for product development and for developing better marketing strategies. For a more detailed description of statistical methodology for segmentation, we refer to the cluster analysis chapter (Chapter 16, see also Wedel and Kamakura, 1998).

At an overall level, there are two different types of segmentation; a priori segmentation and a posteriori segmentation (see Næs *et al.* (2001) for a comparison). The former has to do with determining segments from the additional consumer attributes or characteristics,

Statistics for Sensory and Consumer Science Tormod Næs, Per B. Brockhoff and Oliver Tomic

for instance gender, age or differences in attitude. These segments can then be analysed separately or by the use of ANOVA techniques that combine design variables and consumer attributes in one single model (see Chapter 8). The other type of segmentation is on the other hand based only on consumer acceptance data. The advantage of this approach is that it is unsupervised in the sense that the segments are determined without external influence and is then more open to new and unexpected results. Subsequent analysis of relations between segments and external data is usually done by regression analysis (Næs *et al.*, 2001, or by tabulation (Helgesen *et al.*, 1997).

It is important to mention that for a posteriori segmentation of consumers, one will seldom be able to identify clearly separated clusters. In most cases, there will be a continuum of different types of acceptance patterns, thus leading to segments with no sharp border or split between them. Consumers in the border zone between two segments will not have any clear membership to any of the segments. This aspect is one of the reasons why fuzzy clustering methods play such an important role in the area.

In this chapter focus will be on rating data, but choice data will also be covered briefly. Ranking data will not be covered as a particular topic here but for the purpose of clustering, ranking data can in some cases be treated as regular rating data. We will focus on both hierarchical and criterion based partitioning methods, although it is claimed in Wajrock *et al.* (2008) that partitioning methods are generally superior to hierarchical methods in this context.

The data structure considered in this chapter is basically the same as the ones discussed in Chapters 8 and 9; consumer data that describe the acceptance for all the samples, an additional data set of sensory measurements or experimental design variables and then additional information about the consumers for interpretation purposes.

In Hottenstein *et al.* (2008) is stated that in some cases serving order may be an important reason for cluster structures in consumer data. This emphasises the need for using different randomisations in the data set and testing out afterwards that the segments found are not overlapping with differences in serving order.

10.2 Segmentation of Rating Data

We will in the following distinguish between three different types of input data to use for segmentation:

- the direct use of the acceptance profiles for each consumer;
- regression coefficients (factor effects, part worths, either all or some of them) from analysis of individual acceptance patterns (either by conjoint analysis or prefmap, see Chapters 8 and 9);
- residuals obtained from relating acceptance values to sensory or other attributes.

The first of these cases is only relevant for situations with the same samples presented to all consumers, i.e. the classical and simplest situation discussed in Chapter 8 and 9. The second approach can sometimes also be used for other situations, for instance when incomplete block designs are used, as long as the regression coefficients are estimated without confounding with other factors that vary from consumer to consumer (see Chapter 8). The most important advantage of the last approach is that it can easily be used also for the situation where different samples are tested by each consumer and also for situations with a very small number of samples for each consumer (see e.g. Wedel and Steenkamp, 1989, 1991).

10.2.1 Segmentation Based on Original Acceptance Data

For the situation where the original acceptance data are used for clustering, there are basically two different approaches that can be taken, one is pragmatic, intuitive and visually oriented and the other one is more automatic and requires more advanced statistical methodology. The former approach is based simply on using PCA of the consumer acceptance data matrix with subsequent inspection of the consumer loadings plot (see also Chapter 9). Consumers that lie close to each other are determined to belong to the same segment. If there is no tendency of grouping, only a relatively even spread of consumer loadings, one can still obtain useful segments by for instance setting the limits between the segments such that the segments are about equal in size, either in number of consumers within each group or with respect to the sixe of the regions they cover. The PCA approach is most suitable for applications where the number of useful components is limited to 2 or 3. Low-dimensional solutions are very common for sensory data and standard preference mapping situations, but for conjoint studies where the number of dimensions is often deliberately set higher, the number of PCA dimension may sometimes become too high.

Note that for prefmap studies the primary focus is on direction of the acceptance (see Chapter 9). It may therefore be more natural also here to cluster according to direction than according to exact closeness in the PCA space.

An application of this type was discussed in Rødbotten *et al.* (2009), where the consumer loadings were commented on according to whether the consumers like the juice samples with the lowest or the highest acid content. In that case this corresponds to segmentation according to whether the consumer loadings lie above or below the X-axis (Figure 10.1).

A more automatic approach is to use a statistical clustering method (Chapter 16). A simple possibility is to use a hierarchical clustering method based on a matrix of distances between the consumers. The results are usually presented in so-called dendrograms, which are useful for identifying the number of clusters in the data set. Another possibility is to use a partitioning method based on optimising one single criterion. For these methods one usually has to decide on the number of clusters before the algorithm starts. One will typically consider solutions with 2, 3 and 4 clusters and then compare the solutions afterwards. A method of special interest here is fuzzy clustering since it also provides membership values for each consumer to each of the clusters. These values can be used for determining the number of meaningful clusters.

Even though an automatic approach can be useful, we recommend that one always looks at the PCA loadings for improved interpretation. As always, one should not hesitate to overrun the algorithm if the results are not satisfactory. A fact that emphasises the importance of this is that for consumer segmentation, different clustering methods will often give somewhat different results and it is important to use prior knowledge and personal judgment to decide on the most natural solution.

In Helgesen *et al.* (1997) both methods were applied and compared (see Figure 10.2). In that case a regular hierarchical clustering was run first and stopped at 4 clusters. The

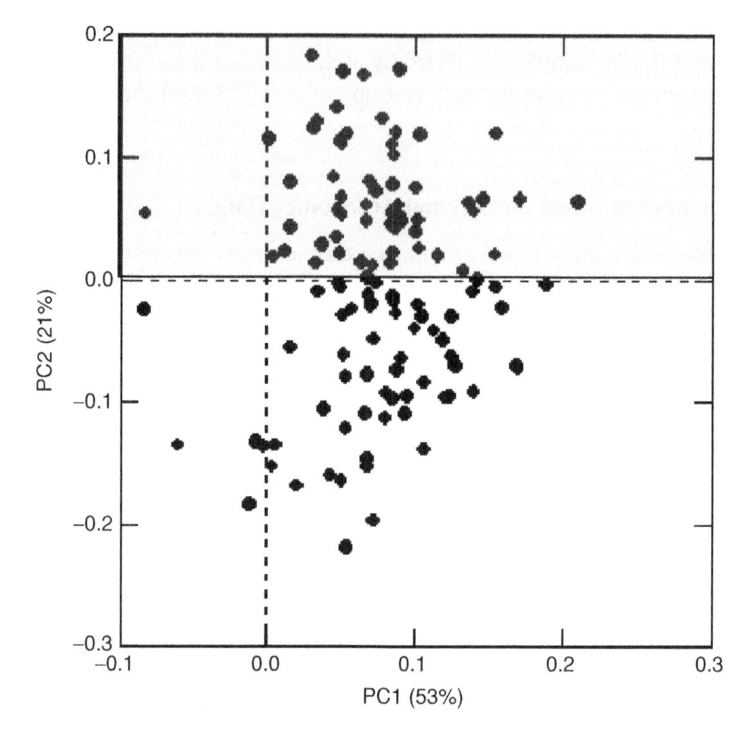

Figure 10.1 Illustration of how to split according to loading value for the apple juice data. *The plot shows the consumer loadings from the PCA of the consumer acceptance data. The consumer group can for instance be split according to whether the consumer lies above or below the horizontal line. In this case this corresponds to whether the consumer prefers the juices with the higher or the lower acid content. Reprinted from Food Quality and Preference, 20, Rødbotten et al. A cross-cultural study of preference for apple juice with different sugar and acid contents. 277–284, 2009, with permission from Elsevier.*

4 clusters were then indicated in the consumer loadings plot and they were found to correspond quite well with the four quadrants in the PCA plots.

As usual, one has to decide whether to centre the data or not. If data are not centred, the general level of acceptance values for a consumer will play a part in the clustering. If the consumer data are centred for each consumer, only the relative differences will be analysed in the segmentation process. As was discussed in Chapters 8 and 9, this is not an obvious decision. If in doubt it may also be possible to test out both alternatives and compare the results.

10.2.2 Segmentation Based on Individual Regression Coefficients (Part Worths)

The cluster analysis treated here is based on the regression coefficients determined when fitting acceptance data to either the experimental design (Chapter 7) or sensory profiles (Chapter 8) of the samples. In both cases one ends up with a set of regression coefficients for each individual.

It is important to note that if a linear model is used and the number of samples is the same as the number of parameters, the model is so-called saturated, and the coefficients

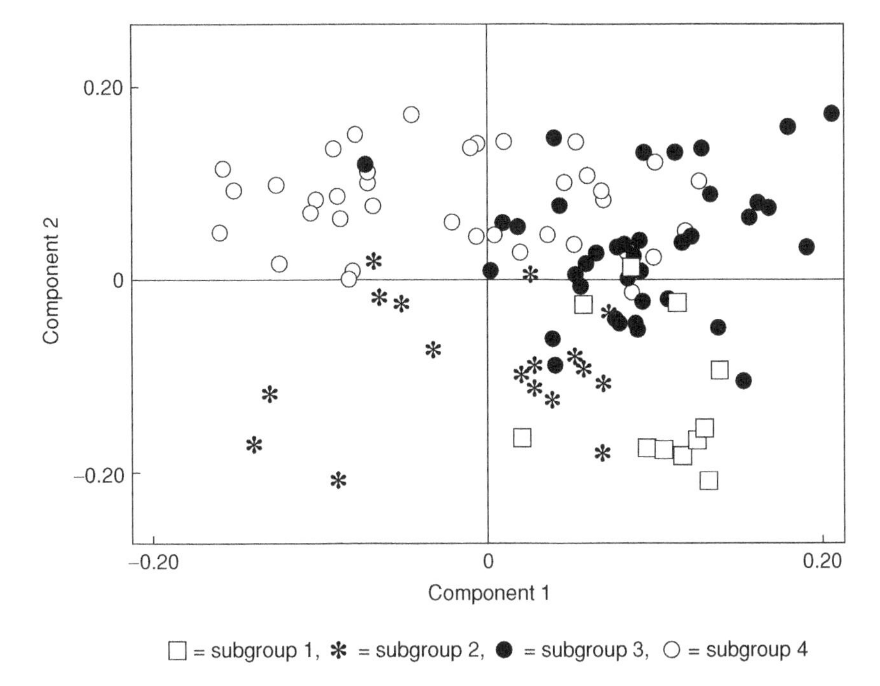

Figure 10.2 Hierarchical clustering of raw data compared to visual inspection of PCA loadings. *The plot compares the results from hierarchical clustering as indicated by the different symbols with what is obtained by visual inspection of the loadings.(Helgesen* et al. *(1997)). Reprinted from Food Quality and Preference, 8, Helgesen* et al.*, Consumer preference mapping of dry fermented lamb sausages, 297–109, 1997, with permission from Elsevier.*

represent simply a linear invertible transform of the raw data. This means that clustering will then give essentially the same information as if the raw data were used.

For external linear preference mapping, the regression coefficients to be used for clustering are simply the consumer loadings discussed in Chapter 9. Also for more complex models, like for instance ideal point models, the regression coefficients can be used for clustering, but here it is also possible to use the position of max or min points. For categorical design data (see Chapter 8), a possible approach is to look at line plots, i.e. plots of coefficient values vs. the coefficient number. This provides a line for each consumer with each point representing a coefficient value. With many coefficients and consumers, however, this may be difficult, so a more formal procedure may be required. Note that for this type of modelling it is also possible to base the clustering on a subset of the coefficients. As an example, consider a situation with both price and a number of additional product attributes in the model

$$y = price + attributes + \varepsilon \tag{10.1}$$

but one is only interested in the price. One may then simply decide to cluster the consumers according to the level of the coefficients for price. This is a very simple thing to do since the estimate is a scalar value.

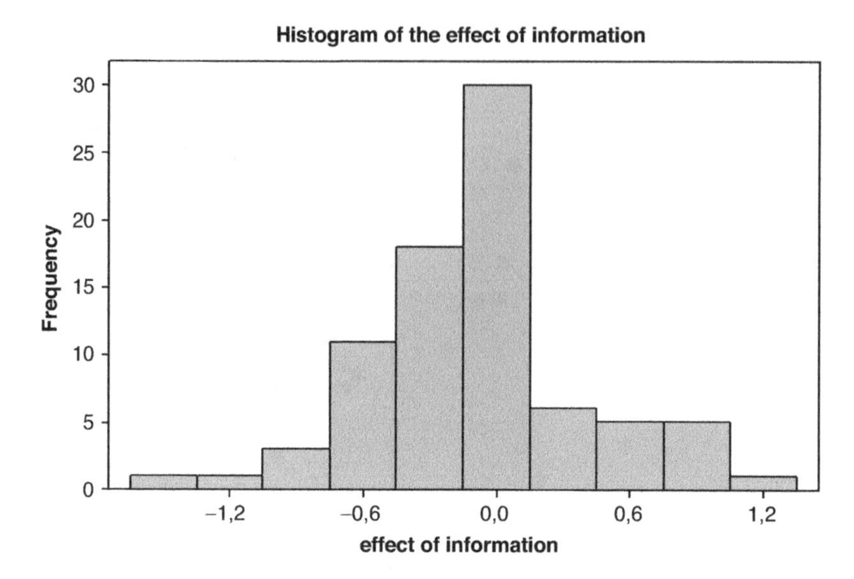

Figure 10.3 ***Histogram of effect estimates from individual model fitting.*** *Histogram for the effect of information in the ham study reported in. The ANOVA model is fitted for each consumer separately and the effect value (the differences between the two levels) for the information factor is plotted along the horizontal axis.*

In the ham study discussed in Chapter 8, there were two conjoint factors, information at 2 levels and product at 4 levels. In Figure 10.3 is presented a histogram of the information effect computed for each of the individuals in the study. As we can see, the distribution resembles a normal one and there is no sign of well separated clusters. One could, however, in this cases decide to segment according to whether the effect of information is positive or negative and proceed with an interpretation of the two segments using one of the methods described in Section 10.3 below.

10.2.3 Simultaneous Segmentation and Model Fitting

The idea behind this approach is essentially the same as in Section 10.2.2 above, namely to base the clustering on similarity of regression coefficients for the individual consumers. The main difference lies in the fact that here all consumers are investigated simultaneously, which leads to a number of important advantages. First of all, more complex models like ideal point models can safely be used. In addition it provides better opportunities for letting consumers test few and possibly different products (Wedel and Steenkamp, 1989; Steenkamp and Wedel, 1993).

The reason for these advantages is that the clustering is based on the residuals obtained from regressing the acceptance values onto either design variables (for conjoint analysis) or principal components of sensory data (for preference mapping). This means that the method does not depend on the explanatory variables directly, only implicitly through the residuals. For applications of the approach we refer to Wedel and Steenkamp (1989), Næs *et al.* (2001) and Johansen *et al.* (2010b).

10.2.3.1 Fuzzy Clustering

The optimisation criterion used for the fuzzy clustering (FCM) approach with the use of the residual distance can according to Chapter 16 be written as

$$J = \sum_{c=1}^{C} \sum_{n=1}^{N} u_{nc}^{m} (y_n - \mathbf{x}_n^{T} \boldsymbol{\beta}_c)^2 \tag{10.2}$$

Here the u_{nc}'s are the membership values, and the distance d_{nc} in the general FCM criterion is given as a residual $d_{nc} = |y_n - \mathbf{x}_n^T \boldsymbol{\beta}_c|$. Furthermore, m is the fuzzifier, N is the number of objects in total, C is the number of clusters, the y's are the acceptance values, the $\boldsymbol{\beta}$'s are the regression coefficients and the $\mathbf{x}$'s are either design variables or the principal components of sensory measurements (or functions of them, for instance polynomial functions).

Minimising the criterion J with respect to $\mathbf{U}=\{u_{nc}\}$ and $\mathbf{D}=\{d_{nc}\}$ will favour combinations of large values of u and small values of d and *vice versa*, corresponding to as clearly separated clusters as possible. The J is optimised using an iterative algorithm that has nice convergence properties. More details about the method and how to find the solution can be found in Chapter 16.

The main results from the FCM method are the membership values u_{nc} and the regression coefficients $\boldsymbol{\beta}_c$ for each cluster. The membership values are all between 0 and 1 and can be used to interpret which consumers are related to which group. A consumer is usually allocated to the cluster for which it has the highest membership value. The regression coefficients are use to interpret how the consumers in the different clusters respond to the various stimuli in $\mathbf{x}$.

For assessing the validity of clusters and choosing an appropriate number of them, several alternatives exist. A simple approach is to sum the residual variances after fitting of the regression equations and to compare them for different values of C (Johansen *et al.*, 2010b). Another possibility is to use the membership values and look at how natural the splitting is; for instance membership values close to the average membership value (1/C) is an indication that the splitting is not very clear. A more sophisticated approach is to compute an index that for instance measures the ratio between the within vs. between cluster variability. A third possibility is to use discriminant analysis accompanied with cross-validation (Steenkamp and Wedel, 1993). One should, however, always keep in mind that determining the number of clusters in segmentation is often a rather subjective matter since one can seldom expect to find nicely separated clusters of consumers. Therefore, aspects such as for instance having a reasonable number of consumers within each segment also become important.

The FCM method has the additional advantage that it can easily be extended to handle outliers by the use of the noise cluster method (Dave, 1991). A strategy has also been developed for identifying the most obvious cluster first and then continuing with the next most obvious cluster etc, until no cluster structure is left in the data set. This is called sequential clustering and can be very important for many practical situations (Berget *et al.*, 2005).

Table 10.1 *Regression coefficients for three different groups based on results in Næs* et al. *(2001). The results are obtained with FCM the use of FCM. Standard errors are presented parenthesis. The product 1 with no health information and information about low fat content is the reference in this case (i.e it represents the intercept). The R-squares for the three groups are 0.29, 0.18 and 0.06. Næs* et al. *2001.*

Effect	Segment 1	Segment 2	Segment 3
Intercept (ref)	2.72 (0.11)	1.89 (0.11)	3.01 (0.011)
Health claim	−0.13 (0.09)	−0.13 (0.09)	0.21 (0.09)
Fat content	0.20 (0.09)	0.51 (0.09)	0.16 (0.09)
Product 2	0.27 (0.13)	1.23 (0.13)	0.46 (0.13)
Product 3	−1.29 (0.13)	0.71 (0.13)	−0.24 (0.13)
Product 4	−0.99 (0.13)	0.55 (0.13)	0.11 (0.13)

In Næs *et al.* (2001), the method was used for the analysis of data from a conjoint study of margarine with additional information about price and fat level. A designed experiment was set up based on 2 levels of price, 2 different health claims related to fat content and 4 different margarines (2*2*4=16 combinations in total). In this case all consumers tested all combinations and gave an acceptance score for each. Three segments were identified. The regression coefficients for the design variables within each segments were used for interpretation. The regression coefficients and their standard errors for this example can be found in Table 10.1. As can be seen, the regression coefficients are quite different within the groups.

Another application of the method can be found in Johansen *et al.* (2010b) where it was used for ideal point mapping of consumers in their acceptance for low-fat cheese. The consumer group was split in two and each group was given 6 samples each. There were 12 samples in total and the two groups tested two completely different sets. The technique was used to reveal three different segments with different acceptance patterns. One of the segments was linear, while the other two were highly nonlinear. No serious convergence problems related to the algorithm were found, but it was recommended to always test out different starting values and to select the solution with the smallest criterion value.

The finite mixture model clustering described in Chapter 16 can if wanted be used to replace the FCM. A possible advantage of this approach is that it can be extended to a random coefficients model allowing for individual differences within each segment instead of essentially assuming that all individuals have the same acceptance pattern within each group. This is a generalisation which can make the model more realistic. A possible drawback of this approach as compared to the FCM is the explicit need for a distribution assumption. The FCM opens up the possibility of more intuitive modifications of the distance measure (see Næs and Mevik, 1999). We refer to Vigneau and Qannari (2002) and Carbonell *et al.* (2008) for descriptions of similar methodologies.

10.2.3.2 Segmentation of Choice Based Data

The standard segmentation methodology used for this type of data is usually based on the multinomial logit model. One starts by assuming that the observed choices are independent

Table 10.2 *Table from Helgesen* et al. *(1998). The p-value for this consumer attributes is here equal to 0.3. Other consumer attributes in the same study were significant.*

Gender	Total %	Cluster 1, %	Cluster 2, %	Cluster 3, %	Cluster 4, %
Male	56	36	67	51	62
Female	44	64	33	49	38

and follow a multinomial distribution and also come from a population which is a mixture of *C* different clusters or segments. Again one ends up with the so-called mixture model which is described in Chapter 16.

10.3 Relating Segments to Consumer Attributes

As soon as consumer segments have been identified, one is usually interested in understanding more about the clusters using external consumer data (see also Chapter 8). There are essentially two different approaches to this; tabulation and regression.

It may in some cases be advantages for interpretation to remove some of the consumers which have a low membership to all the segments. This means for instance consumers with all membership values close to the average *1/C*. The analysis thus proceeds with the most typical consumers for each of the segments.

Tabulation of additional consumer attributes is the simplest approach. For each cluster one simply creates a table that contains the percentages of the consumers in the different categories of the consumer attributes. If for instance age is an important consumer attribute, one can split age into age categories and compute the percentages within each category for each of the segments. The different tables can then be compared visually or by the use of a homogeneity testing as described in Chapter 11. The table of percentages for different age groups in the study in Helgesen *et al.* (1997) is given in Table 10.2. As can be seen, there are quite large differences between age groups within the 4 clusters (Figure 10.2) identified, although they are not significantly different here.

The other possible approach is to use regression or discriminant analysis (Chapter 15). One can for instance define a dummy variable for each cluster and use discriminant PLS

Table 10.3 *p-values for the consumer attributes for the three groups in Table 10.1 (see Næs* et al. *(2001)).*

Effect	Group 1	Group 2	Group 3
Gender	0.003	0.060	0.044
Age	0.802	0.552	0.903
Education	0.040	0.358	0.010
Gender*Age	0.005	0.804	0.007
Gender*Education	0.944	0.491	0.848
Age*Education	0.372	0.906	0.247
Gender*age*education	0.382	0.879	0.227

directly on the dummy variables versus the external data. Alternatively one can use one of the other discriminant analysis methods described for instance Ripley (1996). As advocated in Wedel and Steenkamp (1989) and also tested out in Næs *et al.* (2001) it is possible to use the membership values from the fuzzy clustering directly as the response in regression equations. In this may one also incorporates the degree of membership into the analysis, which may be a natural thing to do. It was advocated in Wedel and Steenkamp (1989) to use a logistic transform of the external consumer attributes before regression.

Some of the results obtained in Næs *et al.* (2001) are presented in Table 10.3. The *p*-values are presented for each group and each external variable. As can be seen, gender is significant in all, while education and interaction between age and gender are significant only for two of the segments.

References

Berget, I., Mevik, B.-H., Vebø, H., Næs, T. (2005). A strategy for finding biological relevant clusters in microarray data, *Journal of Chemometrics* 19, 482–91.

Carbonell, L. Izquierdo, L., Carbonell, I., Costell, E. (2008). Segmentation of food consumers according to their correlations with sensory attributes projected on preference spaces. *Food Quality and Preference* 19, 71–8.

Dave, R.N. (1991). Characterization and detection of noise in clustering, *Pattern Recognition Letters* 12, 657–64.

Helgesen, H., Solheim, R., Næs, T. (1997). Consumer preference mapping of dry fermented lamb sausages. *Food Quality and Preference* 8(2), 97–109.

Hottenstein, A.W., Taylor, R., Carr, B.T. (2008). Preference segments. A deeper understanding of consumer acceptance or a serving order effect? *Food Quality and Preference* 19, 711–18.

Johansen, S., Hersleth, M., Næs, T.M. (2010b). The use of fuzzy clustering for segmentation in linear and ideal point preference models. *Food Quality and Preference* 21, 188–96.

Næs, T., Mevik, B.H. (1999). The flexibility of fuzzy clustering illustrated by examples, *Journal of Chemometrics* 13, 435–44.

Næs, T., Kubberød, E., Sivertsen, H. (2001). Identifying and interpreting market segments using conjoint analysis. *Food Quality and Preference* 12(2), 133–43.

Ripley, B.D. (1996). *Pattern Recognition and Neural Networks*. Cambridge: Cambridge University Press.

Rødbotten, M, Martinsen, B.K. Borge, G.I. *et al.* (2009). A cross-cultural study of preference for apple juice with different sugar and acid contents. *Food Quality and Preference* 20, 277–84.

Steenkamp, J-B, E.M., Wedel, M (1993). Fuzzy cluster-wise regression in benefit segmentation: Applications and investigation into its validity. *Journal of Business Research* 26, 237–49.

Vigneau, E., Qannari, E.M. (2002). Segmentation of consumers taking account of external data. A cluster of variables approach. *Food Quality and Preference* 13, 515–21.

Wajrock, S., Antille, N., Rytz, A, Pineau, N. and Hager, C. (2008). Partitioning methods outperform hierarchical methods for clustering in preference mapping. *Food Quality and Preference* 19, 662–669.

Wedel, M., Steenkamp, J.-B.E.M. (1989). A fuzzy clusterwise regression approach to benefit segmentation. *International Journal of Research in Marketing* 6(4), 241–58.

Wedel, M., Steenkamp, J.-B.E.M. (1991). A clusterwise regression method for simultaneous fuzzy market structuring and benefit segmentation. *Journal of Marketing Research* 28(4), 385–96.

Wedel, M., Kamakura, W.A. (1998). *Market Segmentation: Conceptual and Methodological Foundation*. New York: Kluwer Academic Publishers.

Westad, F., Hersleth, M., Lea, P. (2004). Strategies for consumer segmentation with application on preference data. *Food Quality and Preference* 15, 681–7.

11

Basic Statistics

In this chapter we will discuss some of the most basic concepts and methodologies of statistics and also indicate how they are related to methodologies treated elsewhere in the book. In the first part, the basic concepts of randomness, population, sample and distribution will be defined. The next step is to discuss a number of important properties of a distribution and show how these best can be estimated from data. The most important of these are the mean, the standard deviation and the variance. Hypothesis testing and confidence intervals and how these concepts relate to important and simple situations in basic statistics will be illustrated. The last part of the chapter is devoted to relations between two or several variables. Correlation and simple linear regression will be defined and discussed. The final sub-chapters present some simple tools for analysing categorical data tables.

11.1 Basic Concepts and Principles

The most basic principles and concepts of statistics are the following:

Randomness: This is a philosophically quite complex concept, but in most cases it is satisfactory to think of it as lack of predictability in an 'experiment' with many possible outcomes. An experiment with randomness in it is sometimes referred to as a stochastic experiment.

Population vs. data: In order to apply statistics properly, it is necessary to distinguish between the population which is thought of as the underlying 'reality' that we want to understand and the data (in statistics often called a sample) that are collected in order to tell us something about this reality. We can seldom or never obtain exact information about the population, but the data can, if collected in a reasonable way, be used to estimate or approximate important characteristics of it. The larger the sample is, the more precise

Statistics for Sensory and Consumer Science Tormod Næs, Per B. Brockhoff and Oliver Tomic

information we get about the population. The sample can be taken at random or in a more systematic way depending on situation. This will be discussed more thoroughly in Chapter 12.

Stochastic variable: The term stochastic variable (or just variable) refers to a quantity that is measured in a stochastic experiment. The value of a stochastic variable can not be predicted exactly; it will always have a component of randomness in it.

Sample space (or domain): This is the set of possible values/outcomes of a stochastic variable.

The **distribution** of a stochastic variable describes how likely the different outcomes of the stochastic variable are.

It should be mentioned that the statistical term sample is a bit problematic in this type of text since it is used very differently in statistics and in applied science. In statistics, it is the term used for a set of observations taken from a certain population while in other disciplines, like for instance in chemistry, it is a term used for a physical object for which the stochastic variable is measured. In the rest of this book we will mainly use the term in the latter meaning of the word. When confusion is possible, a clarification will be given.

11.2 Histogram, Frequency and Probability

Let us now assume that we consider N independent random measurements of the stochastic variable y and that we call them $y_1, \ldots . y_N$. The most common way of estimating the distribution of the variable y is to use the so-called histogram. The histogram is a sample analogue to the population or true underlying distribution.

For continuous variables, i.e. variables which can take on any numerical value, the histogram is found by dividing the actual interval into sub-interval of equal length and by counting the number of objects within each interval. The number of objects in an interval divided by the total number of objects in the data set is called the relative frequency for that particular interval. A plot of the relative frequencies for the different sub-intervals is called a histogram.

The relative frequencies can be considered as estimates of the so-called probabilities (usually denoted by P) for the corresponding events (here intervals). When the number of objects goes towards infinity (N goes towards infinity) and if we let the intervals become smaller and smaller, the histogram will resemble more and more the true underlying population distribution. For variables that are not continuous (so-called discrete variables), splitting of the interval as described above has no meaning and the relative frequencies can just be presented as bars for each of the values in the sample space.

Let us now look at an example that illustrates these concepts: Let us assume that one is interested in the variability of the hardness of a certain sausage product over a whole day of production. A natural population to consider in this case is the whole production of this sausage product over the entire day. For practical reasons it is not possible to investigate the whole production, and one therefore decides to select for instance 100 sausages at random from the population/production in order to investigate its properties. The 100 sausages thus represent a sample from the population. If the 100 objects are collected at random, it is called a random sample. Sometimes this is referred to as representative sampling. The stochastic (or random) variable to be measured is the hardness of the product (here called y). Each sausage is taken at random, so the outcome of y is unknown; it is random. The sample space (or domain) of the stochastic variable is then the set of all possible values for hardness.

Some distributions play a special and central role in statistics. The most famous is the normal distribution which is a continuous distribution symmetric around the mean. If the mean is 0 and the variance is equal to 1, it is called a standard normal distribution. The normal distribution plays an important role in statistics because many real situations can be approximated by a normal distribution and because many of the theoretical aspects become much simpler for this distribution. Many of the methods presented in this book are, however, also reasonably robust to small deviations from normality. Expecting exact normality is neither realistic nor necessary. As will be discussed in Chapter 15, large deviations from normality can be detected by various types of diagnostics.

Two histograms based on 100 randomly chosen observations in each group are presented in Figure 11.1 together with the true normal distribution functions, as indicated by the solid bell-shaped curves.

Other important distributions which are used in many statistical tests (see Chapters 8 and 13) are the Student t-distribution and the F-distribution for analysis of variance purposes. The Student t-distribution is similar to the normal distribution, but has heavier tails. The F-distribution starts at 0 and goes towards infinity. A typical example of an F-distribution is given in Figure 11.2. Another distribution which will be discussed at the end of this chapter is the chi square (χ^2) distribution which also takes on only positive values. Both the χ^2-, t- and F-distributions are classes of distributions. They depend on so-called degrees of freedom, which are numbers that are adjusted according to which situation they are used in. The degrees of freedom depend for instance on the number of observations in the data

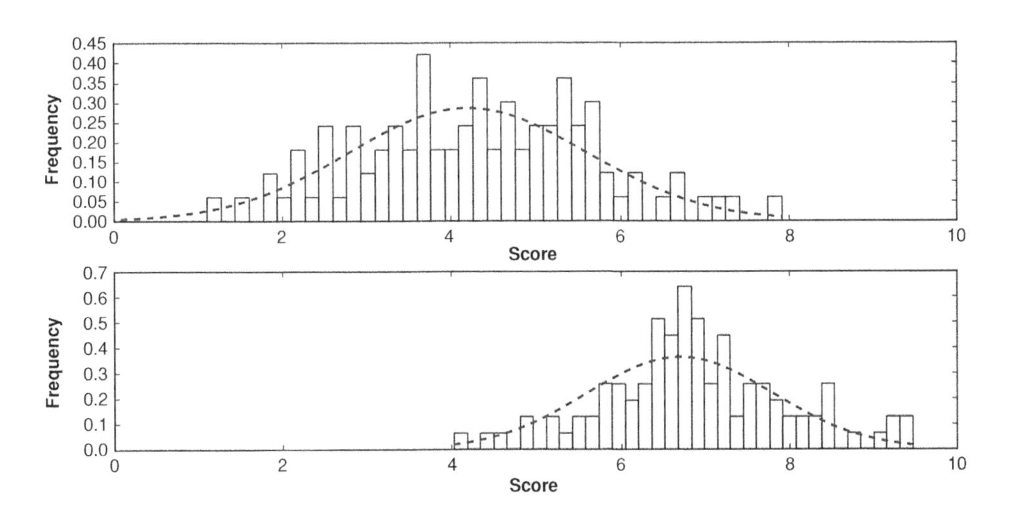

Figure 11.1 Histograms for two normal populations. *Histogram based on simulated data from two normal distributions with 100 observations in each. The theoretical normal density curves are superimposed.*

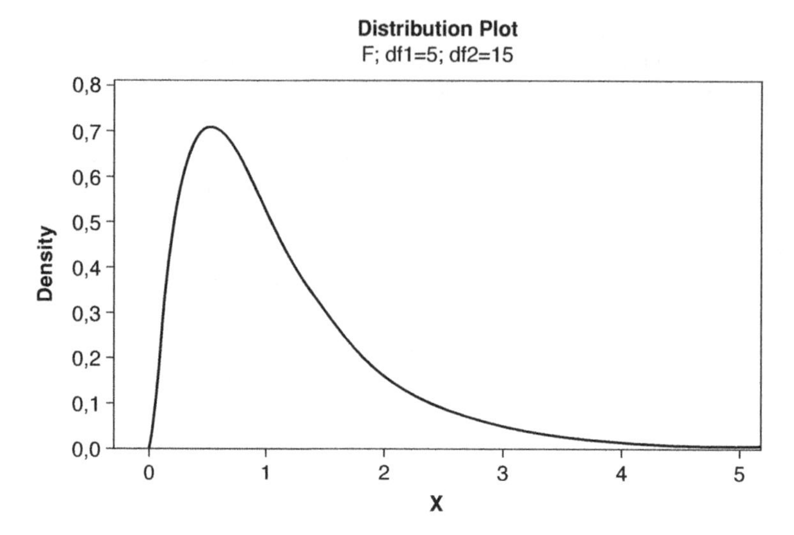

Figure 11.2 ***Shape of F-distribution***. *An example of a typical F-distribution (5 and 15 degrees of freedom).*

set and the complexity of the model. In all statistical software packages these degrees of freedom are computed automatically according to the situation considered.

11.3 Some Basic Properties of a Distribution (Mean, Variance and Standard Deviation)

Usually one is particularly interested in certain properties of a distribution, not only in the shape of the histogram. Assume again that one has a number (N) of independent measurements of a variable y (i.e. $y_1, \ldots, y_N$). The most important characteristics of such a data set are

$$\bar{y} = \frac{1}{N}\sum_{i=1}^{N} y_i \tag{11.1}$$

$$s^2 = \frac{1}{N-1}\sum_{i=1}^{N} (y_i - \bar{y})^2 \tag{11.2}$$

The first is simply the mean or the average of the observations and represents the most important summary of any type of data set. The other quantity is the variance of the observations and represents an estimate of the spread of the data. The larger the value of the variance, the more spread out are the observations. The standard deviation of the observations is defined and estimated by s, which is the square root of the estimated variance s^2. The advantage of the standard deviation over the variance is that it is given in the same units as the observations themselves.

As the number of observations (N) increases, the two quantities above will stabilise around the mean and the variance of the population, usually denoted by μ and σ^2 respectively. Population values like these are called parameters of the distribution. Sometimes one denotes the estimated versions ($\bar{y}$ and s^2) of the two parameters by $\hat{\mu}$ and $\hat{\sigma}^2$, i.e. with a hat on top of the parameters themselves. In the normal distribution, the $\mu \pm 2\sigma$ covers about 95 % of all possible values and $\mu \pm 3\sigma$ covers about 99 % of the whole distribution.

Other important parameters in a distribution are the median and the quartiles. The median is the value which divides the distribution in two in the sense that 50 % of the values fall above and 50 % below the median. The quartiles are defined as the two values that split the distribution in 25 %/75 % and 75 %/25 % respectively. As for the mean and variance, these values can be defined both for the sample and for the population. More generally, one can define a percentile of a distribution. For instance, a 30 % percentile corresponds to that place in the distribution where 30 % of the observation are to the left (lower than) and 70 % are to the right (higher than).

An advantage of the median over the mean is that it is robust to outliers and in cases with a few very high or very low observations, it may be more relevant for indicating the central tendency of the data set. The box plot is invented for illustrating the median and the quartiles in addition to maximum and minimum values.

Often one is also interested in the variability of the mean itself, i.e. variance of $\bar{y}$ (here denoted by $Var(\bar{y})$). It can be shown that this is equal to

$$Var(\bar{y}) = Var(y)/N \tag{11.3}$$

where N is the number of observations in the data set. This quantity is easily estimated by computing the s^2 above and dividing this by the number of observations N, i.e. by s^2/N. The square root of an estimate of the variance of an estimator is often called the standard error of the estimate. The square root of the estimated version of Equation (11.3) is thus the standard error of the mean.

Generally, when a parameter is estimated, for instance the mean of a distribution, one presents the results by giving the estimate as a number or as a bar and with an additional indication of the uncertainty given by the square root of the estimated version of Equation (11.3). Without indication of the precision of an estimate, it is usually very difficult to interpret the value given.

In Figures 11.3 and 11.4 illustrations are given of the mean and standard deviations and box plots for the simulated data in Figure 11.1.

11.4 Hypothesis Testing and Confidence Intervals for the Mean μ

The above section handled estimation of the mean μ and variance σ^2. Sometimes one is also interested in testing hypotheses about the mean. A typical situation is that a certain product standard has a mean equal to a fixed value and one is interested in testing whether the actual distribution has a mean which is comparable to this standard value. The estimated value is in practice always different from the standard because it is a random variable, but the question is whether it is different enough in order to be determined as so-called significantly different. This means different enough so that we can safely say that there is a very small chance that a difference of this size can happen if the hypothesis is true.

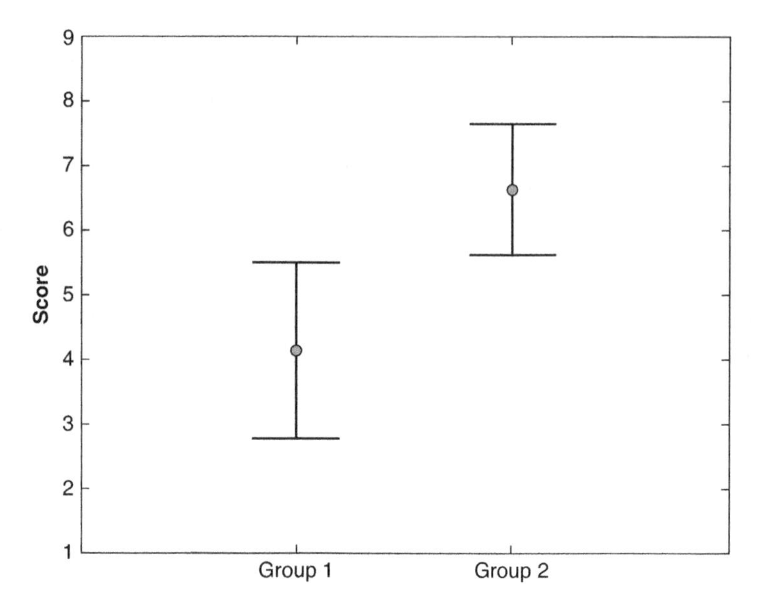

Figure 11.3 Means and standard deviations. *The values correspond to the same distributions as presented in Figure 11.2.*

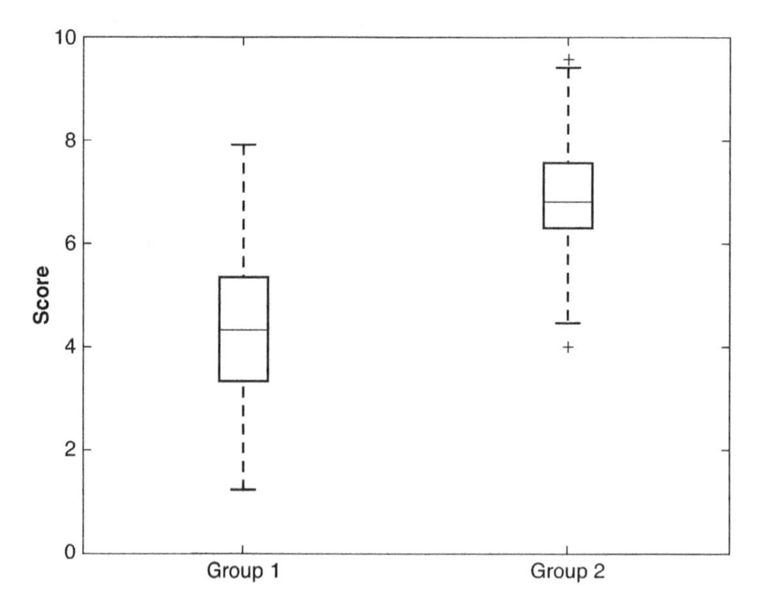

Figure 11.4 Box plot illustration. *Box plots for the same data as used in Figure 11.1 and 11.3.*

A typical way of investigating this is to first set up an hypothesis, called the null hypothesis (H_0), in this equal to H_0: $\mu = \mu_0$, where the μ_0 is the fixed reference value. Then the data are used to check if this hypothesis seems reasonable. A natural way to do this is to compute the empirical mean and compare this with the fixed value μ_0. If the difference is large, there is reason to believe that the hypothesis is not true. The problem is to decide how far away an estimated mean has to be from the fixed hypothesis value before one can conclude that the hypothesis must be rejected. The way to solve this problem in practice is to compare the difference between the measured and the hypothesis values with an estimate of the natural variability of the estimator, in this case the variability of $\bar{y}$. It can be shown that if the underlying distribution for y is the normal one and the hypothesis H_0 is true, the distribution of

$$t = \frac{\bar{y} - \mu_0}{s}\sqrt{N} \tag{11.4}$$

i.e. of the difference dividing by the standard error of the mean, is equal to a t-distribution. The hypothesis is then rejected if the absolute value of t is so large that this can only happen in for instance 5 % of all cases if the hypothesis is true. If one is very afraid of rejecting the hypothesis when if it is actually true, one can set the value of this 'significance level' (often called α) lower. Usually, however, in practice one will compute the so-called p-value of the test, which is defined as the probability of having a t-value at least as extreme as the one observed. This p-value can be used directly as a measure of the degree of correspondence of the data with the hypothesis. Small values, for instance smaller than 0.05, are indications that the hypothesis should be rejected. In such cases we say that the mean is significantly different from the μ_0.

Another way of assessing the variability of the estimated value is to compute the confidence interval. This is an interval which with a certain pre-specified probability covers the unknown parameter value with a high (usually 95 %) probability. In this case the 95 % confidence interval can be written as

$$\bar{y} \pm 1.96^* \frac{s}{\sqrt{N}} \tag{11.5}$$

The value 1.96 represents the 97.5 % percentile of the standard normal distribution. For the data used in Figure 11.3, the estimated confidence intervals are equal to [3.9, 4.4] and [6.4, 6.8].

A confidence interval can also be used to test hypotheses. One simply checks whether the actual hypothesis value lies within the limits of the interval or not. If the hypothesis value is outside of the interval, the hypothesis is rejected. This means that for this data set, for instance all hypothesis values smaller than 3.9 for the first group and smaller than 6.4 for the other, would have been rejected.

Statistical testing will play a major role in most sections of this book. In all these situations, a statistical model is defined, a natural test statistic is developed and the p-values for he actual hypotheses are computed. The same philosophy as used here is underlying all these tests.

11.4.1 Some General Comments on Hypothesis Tests, Power and Type I and Type II Risks

In all statistical tests it is possible to make a wrong conclusion. One of them is the one that all test procedures protect against, namely to reject a hypothesis when it is actually true. This is called Type I error. The other error that it is possible to make is to fail to reject a hypothesis when the hypothesis is wrong. This is called a Type II error. One minus the probability of this type of error considered as a function of how far away from the hypothesis the reality is, is called the power curve of the test. The larger the data set used for analysis is, the closer the power curve comes to 1, which means that for large data set, the power of a test is large. Most standard textbooks of statistics have a discussion of these issues.

Often one sets the significance level of a test equal to 0.05, but one must be aware that there is no obvious place to put the cut-off. If one is very afraid of making a Type I error, one can set the significance level lower. In most practical cases, one will, however, just compute the p-value and use this as a measure of the distance to the hypothesis and base the judgement and conclusions upon this value. Controlling the risk of a Type II error for a specific test is not possible without changing the number of observations or the design or the study. The concept is therefore important for comparing the properties of different design procedures.

A given test with a given significance level will, however, have a given power curve. In some simple cases it is possible to compute the power curve of a test, but in other cases it is more difficult and one must relay on statistical resampling techniques (Martens *et al.*, 2000).

In this book a pragmatic attitude towards interpretations of p-values and hypothesis tests is recommended. For instance p-values should merely be used as indications of significance, without no clear cut-off at for instance 0.05 between significant and non-significant. One should also be aware that when the number of observations increases, even small differences can show up as significant. It is therefore recommended to plot the mean values for the different factor levels (see Chapter 5 and Chapter 13) together with their standard errors in order to get an idea about the practical significance of the results. We refer to Chapter 13.9 for a discussion of p-values when the number of tests increases.

11.5 Statistical Process Control

Statistical process control (SPC, Montgomery, 1997) is a methodology applied for checking whether an actual measurement is within the normal range of variability. The SPC methodology is based on so-called control charts. These are plots of the data over time with so-called control limits superimposed. The plotting aspect is the most important, but the limits can also play a certain role for detecting drift and outliers.

The assumption behind the most basic SPC control chart, the so-called Shewart control chart, is that when the process in under control (within the normal variation range) all observations are independent identically distributed. Usually, they are assumed to have a normal distribution.

The limits used to control the process are drawn in the following way: First one computes the mean and the standard deviation of the data using all the N observations. Using the fact that over 99 % of the observation in a normal distribution lie within +/−3*s from the true mean, the control limits (LCL-lower control limit, UCL-upper control limit) are simply

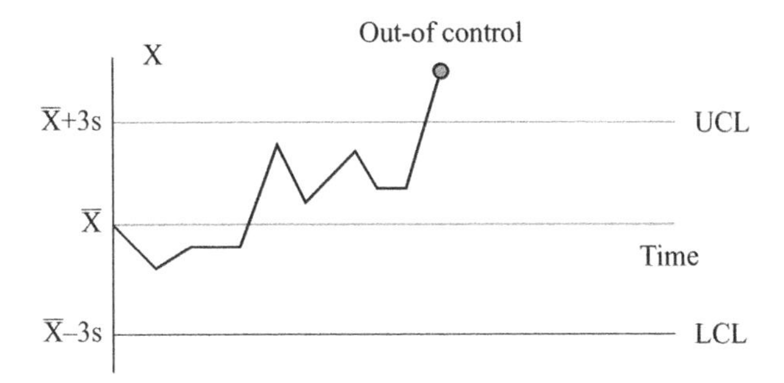

Figure 11.5 ***The principle of SPC.*** *The process is under control for several values before it suddenly goes outside of the upper control limit (UCL). The lower control limit (LCL) is indicated as equally distant from the process average.*

computed as

$$LCL = \bar{y} - 3s \quad \text{and} \quad UCL = \bar{y} + 3s \tag{11.6}$$

This interval is an estimate of an interval that contains about 99 % of all cases if the process is in control (common cause variability). Observations that are detected outside the control limits are taken as signs of abnormal variability or drift, so-called out of control situations. It is generally assumed that this is due to 'special causes'. It is simple to extend the control chart limits to hold for averages of size K, by just dividing the standard deviation in the equation by the square root of K. Note the similarity of the interval in (11.6) and the confidence interval above.

Alternative strategies have also been put forward. For instance, two consecutive observations outside the 2^*s limit is sometimes used as an indication of out of control. Extensions of the methodology exist for correlated data. One can also use empirical distributions (histograms) and base the control limits on those.

In the area of sensory and consumer science, the main relevance is within quality control of sensory panels. If for instance a standard sample is available, one can check the panel average (or individual averages) vs. the control limits developed during a calibration phase. In this way one can detect drift and other unwanted effects. An example is given in Figure 11.5.

11.6 Relationships between Two or More Variables

In the same way as for one variable, one can define a joint distribution of two or more variables. This distribution can as for the univariate case, be estimated using a generalisation of the univariate histogram.

The simplest way of studying the empirical relationship between two variables is to use a scatter plots between two and two variables. Figure 11.6 gives some examples of this.

An important measure of the degree of linear relationship between x and y is the correlation. As for the mean and standard deviation, there will be one correlation for the underlying population and one for the sample. As usual the estimate will approach the true value when the N increases. The estimated correlation coefficient is, however, in this case biased in the sense that its expectation is larger than the true value of the correlation and can therefore lead to overoptimistic assessments of the true correlation coefficient. This is particularly true when the number of objects is small.

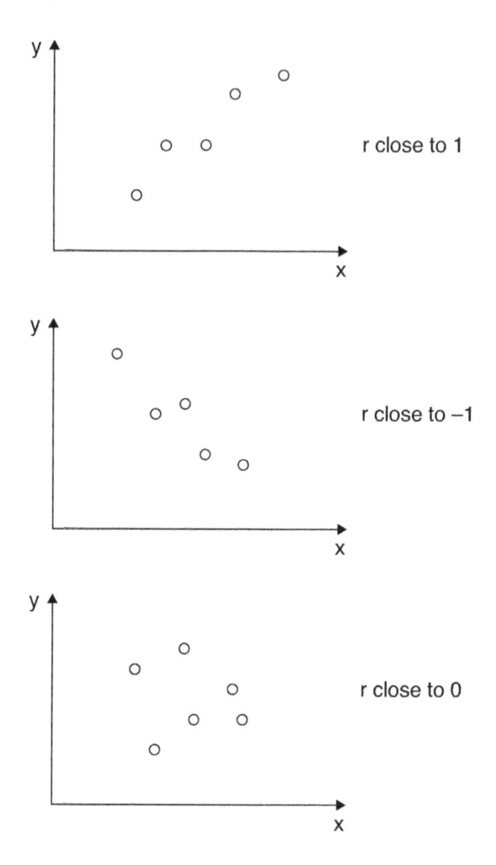

Figure 11.6 Correlation. *The three plots indicate three very different situations. In the first and the second, the correlation is strong while in the last it is very weak. In the first Figure the relation is positive and results in a positive correlation coefficient. In the second the relation is negative, which gives a correlation close to −1. The correlation is always between −1 and 1.*

The sample correlation coefficient is defined as

$$r = Cov(x, y)/s_x s_y \tag{11.7}$$

where the covariance between x and y, $Cov(x,y)$, is defined as

$$Cov(x, y) = \sum_{i=1}^{N} (x_i - \bar{y})(y_i - \bar{y})/(N - 1) \tag{11.8}$$

and the s_x and s_y are the standard deviations of x and y respectively. The population analogue of the correlation coefficient is usually denoted by ρ. The correlation coefficient r has the advantage that it is always between –1 and 1 and independent of the unit for x and y. Note that both the covariance and correlation definitions are independent on the order of the x and y.

A correlation approximately equal to 0 indicates that there is a very weak linear relationship between x and y. A correlation coefficient close to 1 or -1 indicates that there is almost a perfect linear relationship between x and y. If the correlation is positive, the relation is positive, i.e. both variables go 'up and down' at the same time. If the correlation is negative, the relation is negative. Note, however, that correlation measures degree of linear relationship, which means that nonlinear relations may have a small correlation coefficient. Illustrations of situations with low and high correlation are given in Figure 11.6.

In this case the standard hypothesis is equal to $H_0 : \rho = 0$. Significance testing of this hypothesis can be done in most statistical computer programs.

If several variables are measured, one can compute the covariances between all combinations and present them together with the variances in a so-called covariance matrix. With K different variables, this matrix is defined as

$$S = \begin{pmatrix} Var(y_1) & Cov(y_1, y_2) & .. & Cov(y_1, y_K) \\ Cov(y_2, y_1) & Var(y_2) & & \\ .. & & . & \\ . & . & & Var(y_K) \end{pmatrix} \tag{11.9}$$

As can be noted, this matrix is symmetric with variances on the diagonal. This matrix can be changes into a correlation matrix if all covariances are divided by the product of the corresponding standard deviations and the diagonal is replaced by 1's.

If no information about the distribution of a variable is available from another, the two are said to independent of each other.

11.7 Simple Linear Regression

Assume again that one measures two variables x and y and that there is a relationship between them. Assume further that x is simple to measure while y is more difficult or time consuming to measure, but also more interesting. One will then be interested in building a model between the two such that y can be predicted from measurement of x (Martens and Næs, 1989).

The simplest model that can be used for the relationship between x and y is the linear model. This is a model assumption which may be reasonable or not, but often it is a useful approximation in real applications.

The simple linear model can be written as

$$y = \beta_0 + \beta x + \varepsilon \tag{11.10}$$

This model assumes that y can be written as a constant β_0 plus β times x plus some random error ε. An illustration of a linear function and how the parameters relate to the model in a scatter plot is presented in Figure 11.7. The β_0 corresponds to the intercept of the model, which is the intersection line with the y-axis, while β represents the slope of the curve. The random error represents the part of y than can not be accounted for by x, the so-called random noise or uncertainty in the relationship. Usually, one assumes that the ε's are independent and that the variance is the same for all observations, i.e. $Var(\varepsilon_i) = \sigma^2$ is independent on i.

The parameters β_0 and β represent the true underlying relationship, which we are interested in, but which we never can find exactly. Therefore one needs measurements of both x and y for a number of objects that can be used to estimate them (see Figure 11.7).

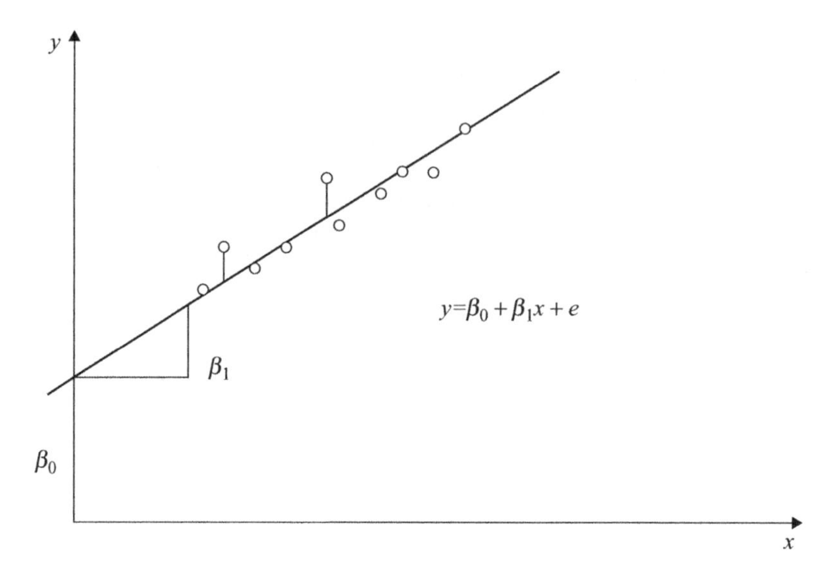

Figure 11.7 ***Linear regression illustrated.*** *The y is a linear function of x plus noise. The points indicate the observations. The problem is to fit the best possible line. This is obtained by computing the distances from each point to a potential line (indicated by the two vertical short lines), and by then "moving" the line until the sum of the squared distances is as small as possible.*

The most common criterion for estimation is least squares (LS, see e.g. Weisberg, 1985), which in the case of normal distribution is identical to the more general maximum likelihood (ML) principle. The least squares principle is also illustrated in the Figure 11.7. The parameters are found by adjusting the line in such a way that the sum of squared differences between predicted values from the model and the true measured values becomes as small as possible. In mathematical terms, this means that the criterion

$$\sum_{i=1}^{N} (y_i - \beta_0 - \beta x_i)^2 \tag{11.11}$$

is minimised over the parameter values β_0 and β. Note that while the correlation coefficient measures the strength of the relation between y and x, linear regression tries to find the best fitting model between the two.

The estimate of the error variance is based on the Equation (11.11) with the parameter values substituted by the estimated values, i.e.

$$\hat{\sigma}^2 = \sum_{i=1}^{N} (y_i - \hat{\beta}_0 - \hat{\beta} x_i)/(N-2) \tag{11.12}$$

Note that it is here necessary to divide by $(N - 2)$ in order to get the best possible estimate.

One is also in this case interested in testing hypotheses about the parameters. The most interesting hypothesis here is that there is no relationship between the two variables at all

$(H_0 : \beta = 0)$. This hypothesis corresponds to a regression line parallel with the x-axis. For testing this hypothesis it is natural to compare the differences between the estimated and given hypothesis value of β with the standard error of the estimate. This combined quantity can be written as

$$t = \frac{\hat{\beta} - 0}{se(\hat{\beta})} = \frac{\hat{\beta}}{\hat{\sigma}}(\sqrt{\sum_{i=1}^{N} (x_i - \bar{x})^2} \qquad (11.13)$$

Under the null hypothesis and when the error distribution of ε is normal, this t-variable is distributed with a t-distribution with $(N - 2)$ degrees of freedom.

In regression analysis x is usually called the independent, the explanatory or the predictor variable, and y the dependent or the predicted variable. In some cases when analysing the relation between two sets of variables, it is not always obvious what to use as x and what to use as y. Examples also exist where both ways are of interest, as for instance in preference mapping (Chapter 9).

These methodologies will be extended in Chapters 15 to more realistic situations with several X- and Y-variables, which are directly relevant both in Chapter 6, 9 and at many other places in the book.

11.8 Binomial Distribution and Tests

The binomial distribution plays an important role in certain areas in statistics. In this book it will be of particular importance in Chapter 7.

The binomial distribution is a so-called discrete distribution with the sample space consisting of only a number of integer values. The situation leading to the distribution is a variable x which measures the number of 'successes' in a number (N) of independent and identical trials with two outcomes in each case, 'success' or 'no-success' (or alternatively the event A or not A). The sample space in this case consists of the integer values between 0 and N and the distribution function can be written as

$$P(x = x_0) = \binom{N}{x_0} p^{x_0}(1 - p)^{N - x_0} \qquad (11.14)$$

where p is the probability of success in each of the trials and the quantity in front of the first p denotes the number of ways that one can select x_0 out of N different units. Sometimes the binomial distribution is denoted by bin(N,p) indicating both the probability (p) and the number of trials (N).

In this distribution, a main focus is on the estimation and testing the parameter p. The most natural way of estimating p is simply to compute the number of successes and divide by the number of trials in total, i.e. $\hat{p} = x/N$. The standard deviation of this unbiased estimator is equal to

$$\sqrt{\frac{1}{N} p(1 - p)} \qquad (11.15)$$

This can be estimated by substituting the parameter p by its estimate. It can be shown that the binomial distribution can be approximated by the normal distribution if N is large.

This can be used to construct confidence intervals and also tests for hypotheses of the type $H_0 : p = p_0$ where the p_0 is a specified value of the probability p. One can test against one-sided or two- sided alternatives. For large values of N, it can be shown that the confidence interval (with confidence 1-α) for p can be approximated by

$$\left(\hat{p} - \lambda_{1-\alpha/2}\sqrt{(1/N)\hat{p}(1-\hat{p})},\quad \hat{p} + \lambda_{1-\alpha/2}\sqrt{(1/N)\hat{p}(1-\hat{p})}\right) \tag{11.16}$$

where $\lambda_{1-\alpha/2}$ is equal to the upper $\alpha/2$-percentile (α often set equal to 0.05) of the standard normal distribution (with mean 0 and variance = 1). The confidence interval can as above be used to test an hypothesis by checking whether the hypothesis value p_0 lies within or outside the interval. If it is outside, the hypothesis must be rejected.

11.9 Contingency Tables and Homogeneity Testing

A variable may be categorical because categories like for instance different colour are investigated, but may also have been made categorical by splitting a continuous axis up into subgroups. The latter may be natural in some analyses as is discussed in for instance Chapter 8. In the former case, the variable is called nominal and in the latter it is ordinal since there is an order to the classes (see also Chapter 1). Questionnaire data are typical examples of categorical data. One asks the consumer a number of questions and the consumer responds by indicting which of a small number of categories he/she belongs to. Examples are age group, gender, type of work etc.

When the stochastic variable is categorical, other analysis techniques may be more suitable than when data are continuous. In these cases it is for instance of no interest to compute means and standard deviations. Instead, one is interested in other properties of the distribution such as percentages in different categories and their relation to external variables (see e.g. McCullagh and Nelder, 1989). A simple way of presenting results from such studies both when interest is in one variable and when it is in the relationship between several, is to use tables, so-called contingency tables (Agresti, 1990). If two-way tables are used, one is usually interested in relationships between the two variables. The tables can be presented using true counts or percentages that are obtained either horizontally or vertically.

A very simple example of such a table is given in Table 11.1. Here 220 different adult consumers between 20 and 60 years from a certain population are asked to respond to which age group they belong to; 20-29 years, 30–39 years, 40–49 years or 50–59 years. As can be seen, there are quite large differences between the groups.

A simple chi-square test can then be set up for testing hypotheses about fixed values of the percentages. The definition of the test statistic normally used is

$$\chi^2 = \sum_{j=1}^{J} \left(y_j - Np_j^0\right)^2 / Np_j^0 \tag{11.17}$$

where the y_i's are the counts in J different groups/categories and the p^0's are the percentage values (probabilities) under the specified null hypothesis. It can be shown that under the null hypothesis, this test statistic has a chi square (i.e. χ^2) distribution with J-1 degrees

Table 11.1 Chi-Square Goodness-of-Fit test for the categorical variable "Age Group". True values, expected values under the null hypothesis of equally sized groups and the group percentages are given. The p-value for the chi square test of equal percentages is equal to 0.04.

	20-29	30-39	40-49	50-59
Observed	70	40	57	53
Expected	55	55	55	55
Percent	32%	18%	26%	24%

of freedom. If we test the hypothesis that all percentages are equal to each other, the chi-square test for the data in Table 11.1 is equal to 8.33 with a p-value equal to 0.04 (DF = 3). Note also that compared to the expected value which is 55 for each group under the null hypothesis, the main contribution to rejecting the hypothesis comes from the first two categories (see also Figure 11.8). The low p-value indicates that the age distribution in the actual population is not uniform in this case.

If one is for instance interested in comparing the age distribution between the genders, one can split accordingly and compute percentages for both genders. This is done in Table 11.2. As can be seen the two genders have a somewhat different distribution. It is then of

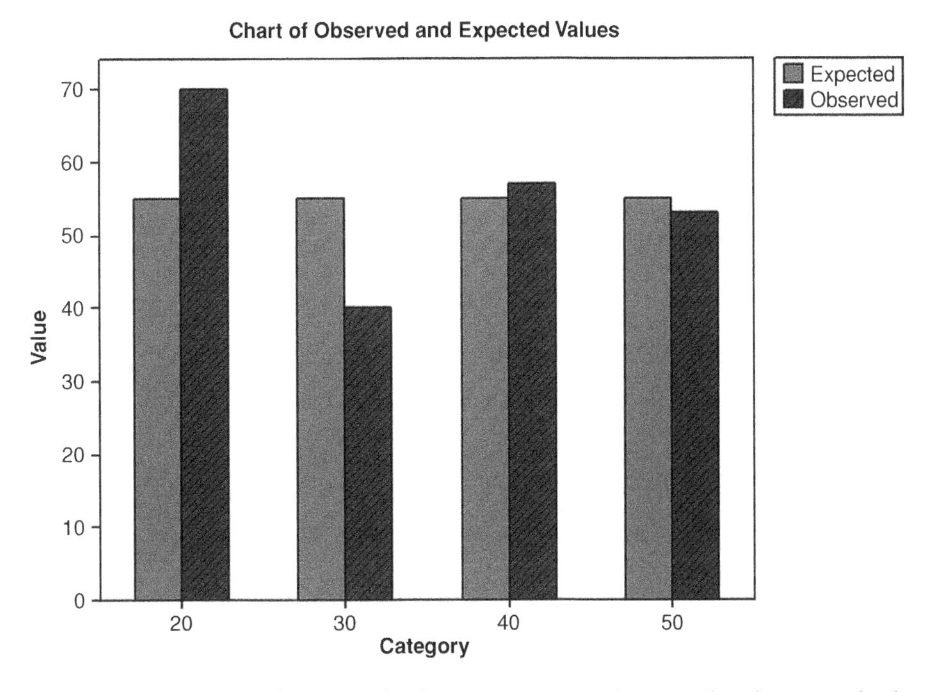

Figure 11.8 **Expected and computed values.** Comparison of expected and measured values for the test performed in Table 11.1. As can be seen, the largest differences are for the first two categories, which shows that these two are the most different from the hypothesis.

Table 11.2 The data for a homogeneity test for age group percentages for men and women. The real observed values and the percentages (taken horizontally) are presented together with the row sums. p-value = 0.048 (chi square test).

	20-29	30-39	40-49	50-59	Total
Men	29	18	28	35	110
Percentages	26%	16%	25%	32%	100%
Women	41	22	29	18	110
Percentages	37%	20%	26%	16%	100%

interest to investigate whether the differences are large enough to be determined significant and not only due to random noise. This type of test can be conduced using a so-called homogeneity test which again is based on the chi-square distribution. The statistic used for this purpose is equal to

$$\chi^2 = \sum_{j=1}^{J}\sum_{k=1}^{K} \frac{(y_{jk} - y_{j.}y_{.k}/N)^2}{y_{j.}y_{.k}/N} \tag{11.18}$$

Here the y_{jk} is the entry in cell j,k in the table (with J and K levels for the two ways), the $y_{j.}$ and $y_{.k}$ are row and column sums respectively. The N is the total number of observations. Under the null hypothesis of no differences between the groups, the statistic has a chi square distribution with $(J\text{-}1)(K\text{-}1)$ degrees of freedom. For this data set, the test gave a value equal to 7.9 with 3 degrees of freedom and with a p-value equal to 0.048. This means that for this data set, the two groups have a significantly different age distribution at 5 % level.

It should be mentioned that the methods discussed here are closely related to methods for generalised linear models discussed in Chapter 15.

References

Agresti, A. (1990). *Categorical Data Analysis*. New York: John Wiley & Sons, Inc.

McCullagh, P., Nelder, J. (1989). *Generalized Linear Models* (2nd edn). London: Chapman & Hall/CRC.

Martens, H., Næs, T. (1989). *Multivariate Calibration*. Chichester: John Wiley & Sons, Ltd.

Martens, H., Byrne, D.V., Dijksterhus, G. (2000) Power of experimental designs, estimated by Monte Carlo simulation. *J. Chemometrics* 14, 441–62.

Montgomery, D.C. (1997). *Introduction to Statistical Quality Control* (3rd edn). New York: John Wiley & Sons, Inc.

Weisberg, S. (1985). *Applied Linear Regression*. New York: John Wiley & Sons, Inc.

12

Design of Experiments for Sensory and Consumer Data

This chapter presents the most basic ideas related to experimental design for sensory and consumer science. The chapter starts with some important and general ideas, distinctions and principles behind design of experiments. Then some of the most important building blocks are presented, comprising factorial designs, randomisation and replicates. More complex designs such as fractional factorial designs, nested designs and split-plot designs will also be covered.

The theory described here is related to many chapters in this book, in particular Chapters 3, 5 and 8.

12.1 Introduction

The planning or design of experiments (DoE-design of experiments) is a large and important area within statistics, see e.g. Cochran and Cox (1957) or Coleman and Montgomery (1993), MacFie (1986) and John (1971). The main purpose of all these methods and strategies is to set up experimental plans that provide as much information as possible in the most efficient way. The present chapter is devoted to a discussion of some of the most important and useful principles and methods with relevance for sensory and consumer experiments. It will be discussed how the different techniques relate to concrete problem areas and references will be given to books and articles where the methods are applied and also described in more detail.When planning sensory and consumer experiments there are always two important and quite different aspects that have to be taken into account. The first of thesc is the initial choice of products and conditions for the study (see e.g. Coleman and Montgomery, 1993). The other one is how to select the sensory assessors and consumers and also how to present the products to them.

Statistics for Sensory and Consumer Science Tormod Næs, Per B. Brockhoff and Oliver Tomic

Some typical situations where good design strategies are needed are:

1. Profiling the current status of a product category and potentially viewing this in light of a competitor's product (see Chapter 5).
2. Learning about the key factors influencing the sensory properties of a product (see Chapter 5).
3. Identifying the most important factors for either preference and/or buying intent (see Chapter 8).
4. Optimising the recipe with respect to either sensory profile or preference (see Chapter 5, 8, 9).

The toolbox of DoE provides what is necessary to cope with both the selection of products and with individual differences in order to provide valid, relevant and affordable product information. It is important to stress that in all these cases, subject matter knowledge is equally important as statistical competence and methodology for providing useful solutions.

In the following, the concept of an experiment will both be used to describe the full set of experimental runs in the full design and also for one single run. When confusion is possible, the exact meaning will be emphasised.

12.2 Important Concepts and Distinctions

12.2.1 Varying One Factor at a Time vs. Varying Several Factors

One of the most important motivating factors for DoE is that even in the simplest of all cases the simultaneous consideration of all experimental factors at the same time is superior to a sequential varying one-factor-at-time approach. Consider as an example the two small experiments illustrated in Figure 12.1. The situation is that one is interested in testing the effect of adding an ingredient A and an ingredient B at two different levels (Low-L and High-H) to a certain food product.

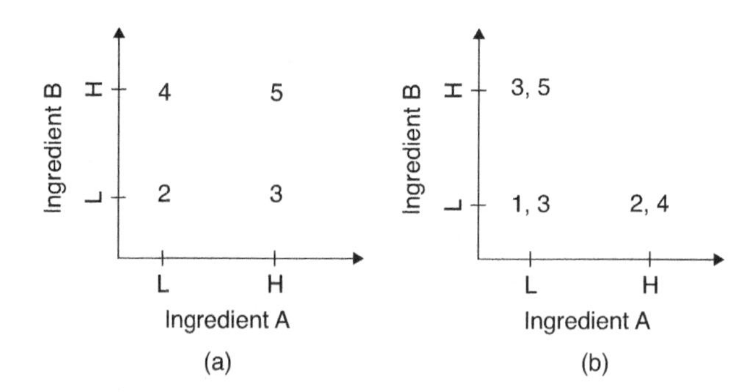

Figure 12.1 Two different design strategies. *Comparing a a) 2-factor factorial design with a b) varying 1 factor at a time experiment. The ingredients A and B are tested at two levels each (Low-L and High-H). The outcome of the experiment is given in the Figure. In Figure a) there are 4 experiments in total and in Figure b) there are 6 experiments in total (2 and 2 values separated by a comma).*

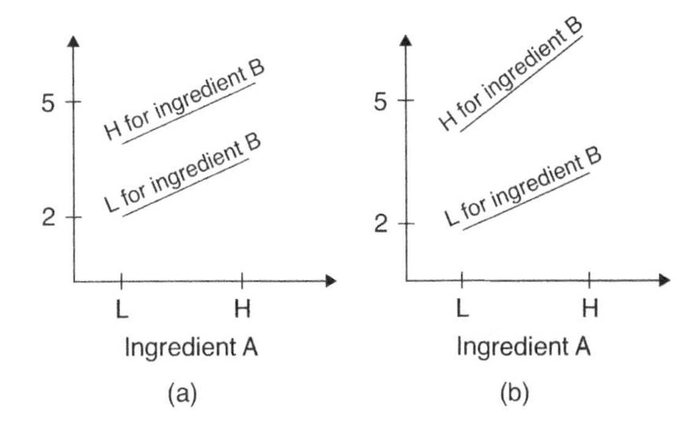

Figure 12.2 The concept of interaction in factorial experiments. *Comparing the no-interaction with the interaction case. In Figure a), the lines correspond to the observations obtained in the factorial design in Figure 12.1.a). As can be see, the two lines are parallel, which means that there is no interaction here. In Figure b), the observation for the constellation H/H (high level for both ingredients), is changed from 5 to 7. In this case, there is an interaction between the two factors since the effect of varying ingredient A is dependent on the level of ingredient B.*

Experiment 1: Two factors with two levels each are investigated by testing each of the four combinations. ($N = 4$ observations); low and high levels for both of the ingredients A and B.
Experiment 2: The same two factors are investigated in two separate sub-experiments: ($N = 6$ observations)
Experiment 2a: Two replications of adding ingredient A at two levels (L and H).
Experiment 2b: Two replications of adding ingredient B at two levels (L and H).

Experiment 1 will provide information about the potential presence of the so-called interaction as illustrated in Figure 12.2, i.e. whether the potential effect of adding ingredient A depends on whether B is added or not. In the second experiment, this type of information is simply not available since the combination of adding both A and B is not present in the experiment. Secondly, even in cases where no interaction is expected, it can be shown that these two experiments will provide the same precision/knowledge of the important effects (see Chapter 13) of A and B even if Experiment 2 is 50 % larger than experiment 1. The two-factor factorial design presented in Experiment 1 is then said to have a relative efficiency of 6/4 = 1.5. For more factors, the relative efficiency of a full factorial experiment (see Section 12.3) as compared to the one-factor-at-a-time approach will increase. A third advantage of Experiment 1 is that since a factor is tested at more than one level of the other(s) there will be support in the full experiment (in the case of no interaction) for a broader validity of the main effects of each factor.

For further discussion of efficiency and the importance of interaction we refer to Box *et al.* (1978) and Montgomery (2005).

12.2.2 Important Building Blocks for DoE

In situation (1) in Section 12.1 above, a natural way of selecting products would be to take a completely random sample from the population of products of each of the categories tested. If no further information is connected to the products, this would lead to a so-called one-way (single factor) product structure. For situation (2), however, there could be a few, one, two or maybe three factors (Chapter 5) to investigate. It would then be natural to utilise the so-called full factorial experiments where all combinations of all factors are tested (see Chapter 13). For situation (3) the full factorial design strategy of testing all combinations of factors rapidly explodes in size. Instead a so-called fractional factorial design can be used, where an assumption about no (higher order) interactions are traded in for the ability to study more main effects (below, Section 12.4). The most well known are the fractional 2^K-designs, where K factors on two levels are investigated in a fraction of either one-half, one-fourth, one-eighth or even less, depending on the size of K. In these designs the main effects computed will be confounded with some of the interaction effect, meaning that it is really impossible to know whether a potential effect observed in the data is due to the factor in question or due to some of the higher order interaction effects. The success of these designs rely on the fact that often main effects are larger than higher order interaction effects, such that the observed effects indeed are stemming (primarily) from the main effects. However, such fractional designs are primarily to be used for screening purposes.

Experimental factors may have purely categorical levels or they may have quantitative levels. If one starts out with a problem with many quantitative factors and the purpose is optimisation (Situation 4 in Section 12.1, above), a typical experimental strategy would be to start with a screening design based on a fractional factorial design method, then continue with a small factorial design in the most important factors in the region identified as the most promising, before finally adding some 'star' points to form a central composite design for optimisation purposes. For more details about this strategy (response surface optimisation) we refer to Box *et al.* (1978). When the factors are all ingredient factors where they must add up to 100 %, special designs, the so-called mixture designs (Cornell, 1990), have been developed for this optimisation process.

12.2.3 Replicates, Randomisation and Blocking

One of the most important concepts in any type of statistics is the concept of *replication.* All useful statistical analyses are based on some kind of aggregation of more than a single

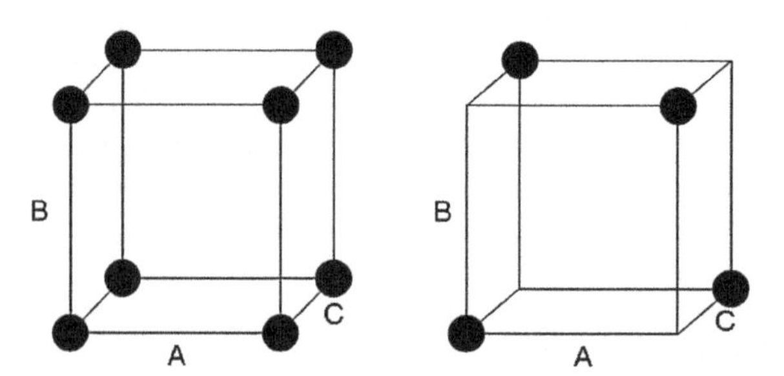

Figure 12.3 A full factorial and a fractional factorial design illustrated. *Examples of a full factorial and a fractional factorial design in the situation with only three design variables, A, B and C. The fractional design is obtained by confounding factor C with the interaction between factor A and B.*

piece of information: Using a single observation as an estimate of an unknown parameter μ gives low precision and provides no information about the uncertainty of this estimate. With N independent replicated observations, the standard error of the mean $\bar{y}$ used as an estimate of μ is equal to

$$\sigma_{\bar{y}} = \frac{\sigma}{\sqrt{N}} \tag{12.1}$$

where σ is the standard deviation of the individual observations. We see how the precision of the estimate increases with $\sqrt{N}$, the square root of the number of independent pieces of information. An estimate of the standard error can be obtained by replacing the σ by an estimate s (Chapter 11).

It is, however, not always obvious how to define an independent replicate. An important distinction is the distinction between two different observations of the same sample and two observations taken on two different samples selected at random. The former is usually called *repeated measurements* and the latter *true replicates*. The repeated measurements would then be so-called *nested* within the true replicates, since for each true replicate one can take several repeated measurements. However, in certain cases this terminology has its limitations as demonstrated in for instance Chapter 5 where a number of different replicate structures are defined and analysed.

Randomisation is another key principle in experimental design. In our context this is usually defined as presenting samples to the sensory assessors or to the consumer in a random order. The idea is that all effects of no interest should be made an integral part of the random error. If not done properly, one can possibly detect effects which are only due to an unexpected relation between the actual effect and an underlying nuisance factor, for instance an order effect. A proper randomisation will ensure that the effects of extraneous factors (factors not controlled nor observed in the experiment) are averaged out in the long run. In some cases one will split a consumer group into subgroups and use a different randomisation in each.

Blocking is another important principle used for controlling unexpected and unwanted variation. One simply tests out the different products, or a subset of them, under a number of selected combinations of possible influential factors, for instance time, serving order, place etc. In sensory and consumer science the individual human being is often considered a block when testing out different products or attribute combinations.

12.3 Full Factorial Designs

In the full factorial design all combinations of all factors are studied in the same experimental setup. In Chapter 5 (Figure 5.1) a 4x2x2 full factorial is exemplified. The 16 products in the design are listed in Table 12.1. The benefit of this design is that both the main effects and all interactions between the three factors can be identified. In Chapter 5 is shown how the product differences can be decomposed into components due to each of the three main effects, the three two-factor interactions and the three-factor interaction. Two-factor interactions capture synergistic and/or antagonistic effects with the straightforward interpretation that the effect of one of the factors is dependent on the level of the other. The three-factor interactions are a bit harder to interpret, but measure essentially how the

Table 12.1 The 16 products used for the pea experiment. The design is a full 4*2'2 design.

	Sucrose level	Colour	Size		Sucrose level	Colour	Size
1	Low	Light	1	9	High	Light	1
2	Low	Light	2	10	High	Light	2
3	Low	Light	3	11	High	Light	3
4	Low	Light	4	12	High	Light	4
5	Low	Dark	1	13	High	Dark	1
6	Low	Dark	2	14	High	Dark	2
7	Low	Dark	3	15	High	Dark	3
8	Low	Dark	4	16	High	Dark	4

interaction between two of the factors vary with the level of the third. For experiments with more than 3 factors, higher order interactions become more challenging to comprehend. Often higher order interactions will be checked merely as a kind of lack-of-fit for the more simple structure given by main effects and first order interactions. Often it can be justified that higher order interaction terms are negligible and can be regarded as a random part of the ANOVA model.

In industrial experiments and in particular for sensory and consumer experiments it is relatively rare that much larger experiments than the one above, in terms of the number products tested, are carried out. In Table 12.2 is given the number of products needed for various combinations of the number of factors and the number of levels used for each factor. It is clear that full factorial designs rapidly become unrealistically large. This is why a large part of DoE theory is dedicated to reducing the size of factorial designs in an intelligent way that maintains the ability to detect the main effects and two-factors interactions while sacrificing higher order interactions.

The two-level full factorial 2^K-designs are often given special attention as they represent the smallest possible designs to jointly investigate K factors and their potential interactions. Often the factors are quantitative and in such cases one also has to decide on the levels of the factors to use.

Note that two-level designs leave us with the problem that we can only estimate linear effects. Adding a center point to the design, however, will enable the estimation of quadratic effects without affecting the estimates of the linear effects. A centre point is a design point where each factor is set at exactly the midpoint (average) of the two levels employed in

Table 12.2 Number of products needed for various full factorial designs.

	Levels pr factor			
No factors	2	3	4	5
2	4	9	16	25
3	8	27	64	125
4	16	81	256	625
5	32	243	1012	3125

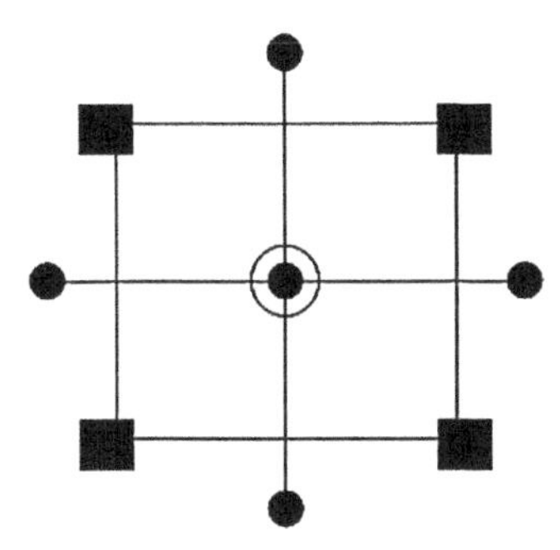

Figure 12.4 **Central composite design.** *An example of a Central Composite Design (CCD) in the situations with only two factors. The factorial design (indicated by squares) is augmented by so-called axial or star points (indicated by filled circles) and a number of center points.*

the 2^K-design. Adding more than one centre point will moreover allow for an independent estimate of the random error. If one or more of the factors are categorical, center points may be added for each combination of the categorical factors (Ellekjær *et al.*, 1996).

Adding center points will not enable the estimation of a full quadratic response surface model (see Chapter 15). To obtain this, usually a Central Composite Design (CCD) is employed. A CCD can be seen as a 2^K-design augmented with so-called axial or star runs/points for each factor, see Figure 12.4. CCDs are frequently used within the more general approach of Response Surface Methodology (RSM, Box *et al.*, 1978) mentioned above.

12.4 Fractional Factorial Designs: Screening Designs

In most cases one will use 2^K-designs as points of departure for reducing the number of products to be tested, but other possibilities also exist (John, 1971). Typical examples are half fractions and quarter fractions of the full designs leading to $2^{K-1}$and 2^{K-2} points respectively, but the fraction needed depends heavily on the number of factors K. In Montgomery (2005) a number of fractional factorial designs are tabulated. The most sparse one given is a 1/2048 fraction of a 2^{15}-design, that is, a 2^{15-11}-design, where the effect of 15 two-level factors are investigated in only 16 runs. The full factorial design would have required $2^{15} = 32.768$ runs. Clearly such an extreme reduction comes with a price: Each main effect is confounded with 7 different two-factor interactions (and many higher order interactions).

If main effects are confounded, i.e. impossible to separate in the estimation, with two-factor interactions, such a design is called a resolution III design (see Box *et al.*, 1978) for explanation and discussion). A better design is a resolution IV design where main effects are only confounded with 3-factor interactions. For resolution IV designs, some of the two-factor interactions can be confounded with each other. An even better design is therefore the resolution V design where also the two-factor interactions are only confounded with higher order interaction. Higher order interactions can, as also mentioned above, often be neglected. This implies that for resolution V designs all the relevant effects can in many cases be estimated without confounding.

To understand the basic idea of the construction of fractional factorial designs and the concept of confounding of factors, consider the basic 2^2 full factorial design with 4 runs (factors A and B). A way to construct the full 2^3 full factorial design with 8 runs would be to run the full 2^2 full factorial design with 4 runs for each of the two levels of the third factor, see Figure 12.3. The half fraction is then obtained by leaving out the opposite diagonals of the two versions of the 2^2 full factorial design. In other words, this means that the additional factor C is added to the 2^2 full factorial design on the two diagonals: one diagonal for each level of C. By changing the levels of C with the diagonals and assuming that there is no AB-interaction, any effect (non-equality of diagonal averages) of changing the level of C would be an effect of C. But note that the observed effect of changing C is generally a sum of the effects of C and the AB-interaction. With only the half-fraction of the 2^3 full factorial design, we will not be able to distinguish between the two – they are inherently confounded. The 2^{3-1}-design with 4 runs is clearly a resolution III design (see Figure 12.3).

In situations with several factors at two levels, a fractional design is typically constructed by first fixing the number of experimental runs (a power of 2) and then forming a full factorial in the corresponding number of factors. If for instance, $16 = 2^4$ is set as the maximum, the full factorial is then constructed using four of the original factors. Then the levels of the other factors are defined by using so-called generators, which are based on defining which interaction effects the new or additional factors are to be confounded with. In the example above, the generator is C = AB, which means that factor C is confounded with the interaction between A and B. One will as indicated above seek designs with as large resolution as possible. The final confounding pattern can easily be calculated by hand, but in most cases this will be done by using some type of software package. The results of using such a package are: a reduced design of pre-determined size, the resolution of the design and also its confounding pattern. In practice one will typically try out different constellations in order to find one which is suitable.

The design strategy just indicated always ends up with a number of runs equal to a power of 2. However, there also exist similar designs for N runs whenever N is a multiple of 4 (and not only when N is a power of 2). These are the Placket-Burman designs, which are characterised by a more complex confounding (also called aliasing) structure than the standard fractional factorials. It is also possible to construct fractional/reduced designs when factors have more than two levels and/or factors have different number of levels – the mixed-level designs, but for this purpose specialised computer programs are always required. In all such cases, revealing the confounding pattern is more complex than in the standard setup.

The Taguchi name is frequently used for such extremely reduced designs as he made them famous. Another really useful technique promoted by Taguchi are the robust parameter designs and process robustness studies, see Montgomery (2005) and Hersleth *et al.* (2003).

12.5 Randomised Blocks and Incomplete Block Designs

The concept of *blocking* plays a major role for the design of sensory and consumer experiments. The individual evaluator, be it a trained panellist or a consumer, will take the role as a block whenever more than a single evaluation is carried out by the individual, which is the case for the most commonly met situations. Blocks can in addition more generally also be related to taste sessions and/or the replication structure of the products under investigation (Chapter 5). The benefit of a block design compared to a completely randomised design is that when analysing the data by ANOVA, the between-blocks variability is removed from the noise in a systematic way. If a consumer preference study of 5 products is to be carried out, a completely randomised design would be based on letting for instance 100 consumers test each product, leading to 500 consumers in total. Alternatively, the randomised block design would let 100 consumers each test all 5 products. In this way the products are compared 'within the consumer', the effect of individuals having varying overall levels in their judgement of the products is corrected for in the analysis (Chapter 13). This type of strategy generally leads to much stronger significance tests and better conclusions. Usually also, the additional costs of having individuals do more than a single evaluation is much lower than having additional individuals in the study.

We distinguish between the *complete* and the *incomplete* block designs. Both are important for sensory and consumer experiments. As the name indicates, in the complete block design all products are tested within each block. As such, complete block versions of the full and fractional factorials are straightforward to employ. In sensory profiling, the panellists will typically be complete blocks. In many cases also consumer preference studies are carried out in a complete setting (Chapter 8). However, sometimes the number of products is too large to have all products tested in a single block. One possible strategy for solving this is to use an *incomplete block designs* (Gacula *et al.*, 2009), where only a subset of the products is tested in each block. For certain combinations of number of products, number of blocks, number of products in a block (block size) and number of product replications, a *Balanced Incomplete Block Design (BIBD)* exists.

In a BIBD each product is tested the same number of times and each pair of products occurs together in a block exactly the same number of times. The practical design challenge of the incomplete situation is the allocation of products to blocks. Generally a dedicated software is needed to find a suitable BIBD design. For a one-way product structure this would simply amount to finding a design, where each pair of products occur within a block as equally often as possible. For a full factorial product structure the situation is more complex, but such cases also opens up the possibility of giving higher priority to balance for certain effects. The main problem with the incomplete block designs is that in cases with interaction between assessor and product, which can often occur, there will be confounding structure in the data that may make it difficult to analyse and to interpret the results. This problem is discussed more thoroughly in Chapter 8 in connection with conjoint studies. Therefore, unless one is an experienced user, care should be taken and a better strategy may be to reduce the number of factors or factor levels in the design and proceed with a regular complete block design.

A recurrent issue in the use of the block designs is how to handle the order of product presentation within the block. A much used solution is to simply randomise the order of presentation completely within each block. This would in the long run balance out possible order and/or carry-over effects and justify the use of the (simple) randomised block model underlying the subsequent statistical analysis. The main drawbacks with this approach are: (1) the risk of the specific experiment becoming more than just a little unbalanced in how the products are allocated to the different presentation positions leading to biased results and (2) that the potential nuisance effects related to the order of presentation will increase the size of the random error leading to a less powerful result. To remedy these problems designs that handle the order and so-called carry over effects have been developed. These types of designs are described in for instance Cochran and Cox (1957) and Schlich (1993b). In MacFie *et al.* (1989) tables of balanced designs are given for up to 16 products. In Ball (1997) the combined challenge of finding good designs to balance out the order and carry-over effects when the blocks are incomplete is discussed. Computer-generated optimal designs can be constructed for certain models according to an optimality criteria (D-optimality, A-optimality, G-optimality, V-optimality).

12.6 Split-Plot and Nested Designs

In some cases full randomisation of products for all factor levels is not possible due to practical restrictions. A typical example of this can be found in Sahni *et al.* (2001). To investigate the critical control points in a production process for the texture properties of low-fat mayonnaise a designed experiment was planned and carried out. Process factors, such as the temperatures of a heat-exchanger and cooler units could easily and rapidly be changed allowing for straightforward handling of randomised combinations of settings of these factors. But raw material (recipe) factors could only be changed for each batch which could not be changed between experimental runs. It was then necessary to vary these less frequently (so-called whole-plot factors) and then within each setting vary the process variables (so-called sub-plot factors). As can be noted, the problem is a restriction in the randomisation which may lead to systematic effects in the error structure. This type of experimental setup is usually called a split-plot design. The alternative would be to prepare new recipes for each new setting of the process factors, increasing the costs and waist of the experiment dramatically.

Another example very typical for consumer studies was published in Hersleth *et al.* (2003). In that case a number of wines produced according to a factorial designs were tested in different contexts, In this case, the wines were the whole plot effects since they were made and tested throughout all the different contexts. In consumer studies of this type, however, the random noise is often so large that the systematic structure of the random effects plays a minor role and a regular ANOVA can usually be applied.

The price to pay for this type of design if the whole-plot error is substantial is that the estimates for the whole plot effects will be less precise than the estimates of the process factors. A correct statistical analysis will have to take into account that the random variability from recipe to recipe (whole plot error) is larger than the variability for repeated experiments within the same recipe. Therefore, the different effects estimated will have to be compared to different types and level of random variability. Usually, the data from split-plot experiments are analysed by mixed ANOVA models (Næs *et al.*, 2007). If the whole-plot error is small as compared to the sub-plot error, regular ANOVA can be used without taking the split-plot structure into account (see Letsinger *et al.*, 1996). This was actually the procedure used in the paper Hersleth *et al.* (2003) mentioned above. The different replicate structure models discussed in Chapter 5 have strong similarities with split-plot models.

Nested designs (see Figure 12.5) are used if the effects of a certain factor are tested for each level of another and the experimental units for the former are physically different. A typical example is consumer studies where one is interested in testing the effects of gender

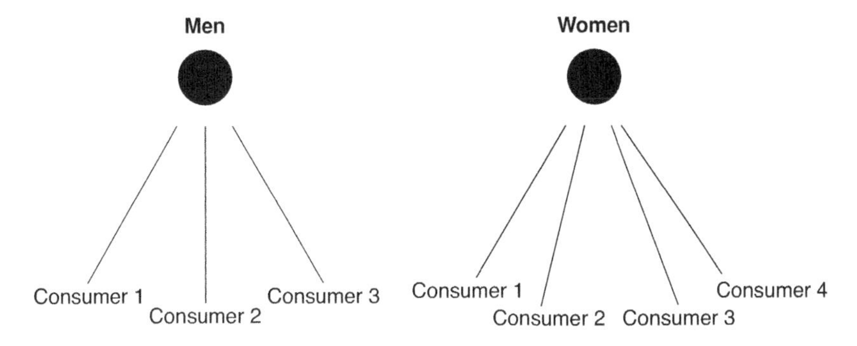

***Figure 12.5* Nested design.** *A nested design illustrated for an (simplified) consumer test with only 3 men and 4 women. As can be noted, the number of men is allowed to be different from the number of women. Each individual can only belong to one of the genders. This must be taken into account when modelling gender and consumer in an ANOVA model (Chapter 13).*

and its interaction with other factors in the designs (see Chapter 8). In this case one will also have to incorporate the consumer effect in the model and it is obvious that each consumer can only belong to one of the genders. We say that consumer is nested within or under gender. The model for analysing this type of design will be different than when analysing a factorial design where all combinations of factors are possible (so-called crossed effects).

12.7 Power of Experiments

The *power* of an experiment is defined as the chance of finding the effects which are actually there and is inherently related to the (relevant) number of observations, usually called the *sample size*. Power and sample size considerations are generally recommended prior to any kind of experimentation. Post hoc power considerations (after the experiment is carried out) is of less use, instead confidence band/interval estimation procedures should be used to quantify (lack of) information given in the data (see Chapter 11 and Hoenig and Heisey, 2001).

Generally the effects discussed above are represented by factors with several levels and the statistical tests that will be used to decide on statistical significance are often F-tests, see Chapter 13. It is possible to calculate the power of F-tests for such situations, but it is easier to understand how in the simple situation where the problem is to test the difference between two products as discussed in detail in Hunter (1996).

Software tools for power calculations are available in many statistical software packages and also available as online calculators numerous places on the internet. We refer to Zar (1999, ch. 8) for exact formulae to use. For techniques based on the Jackknife we refer to Martens *et al.* (2000).

References

Ball, R.D. (1997). Incomplete block designs for the minimisation of order and carry-over effects in sensory analysis. *Food Qual. Pref.* 8(2), 111–18.

Box, G.E.P., Hunter, W., Hunter, S. (1978). *Statistics for Experimenter*. New York: John Wiley & Sons, Inc.
Cochran, W.G., Cox, G.M. (1957). *Experimental Designs*. New York: John Wiley & Sons, Inc.
Coleman, D.E., Montgomery, D.C. (1993). A systematic approach to planning for a designed industrial experiment. *Technometrics* 35(1), 1–12.
Cornell, J. (1990). *Experiments with Mixture* (2nd edn). New York: John Wiley & Sons, Inc.
Ellekjær M.R., Ilseng M.R., Næs, T. (1996). A case study of the use of experimental design and multivariate analysis in product improvement. *Food Quality and Preference* 7(1), 29–36.
Gacula, M.C. Jr., Singh, J., Bi, J., Altan, S. (2009). *Statistical Methods in Food and Consumer Science*. Amsterdam: Elsevier.
Hersleth, M. B-H. Mevik, Næs, T. Guinard, X. (2003). The use for robust design methods for analysing context effects. *Food Quality and Preference* 14, 615–22.
Hoenig, J.M., Heisey, D.M. (2001), The abuse of power: the pervasive fallacy of power calculations for data analysis, *American Statistician* 55(1), 19–24.
Hunter, E.A. (1996). Experimental design. Chapter 2 in T. Næs, E. Risvik (eds), *Multivariate Analysis of Data in Sensory Science*. Amsterdam: Elsevier.
John, P.W.M (1971). *Statistical Design and Analysis of Experiments*. New York: Macmillan.
Letsinger, J.D., Myers, R.H., Lentner, M. (1996). Response surface methods for bi-randomisation structures. *J. Qual. Techn.*, 28(4), 381–97.
MacFie, H.J.H. (1986). Aspects of experimental design. Chapter 1 in J.R. Piggott (ed.), *Statistical Procedures in Food Research*. London and New York: Elsevier Applied Science.
MacFie, H.J.J., Bratchell, N., Greenhoff, K. and Vallis, L.V. (1989). Designs to balance the effect of order of presentation and first-order carry-over effects in Hall tests. *J. of Sens. Stud.* 4, 129–48.
Martens, H., Byrne, D.V., Dijksterhus, G. (2000) Power of experimental designs, estimated by Monte Carlo simulation. *J. Chemometrics* 14, 441–62.
Montgomery, D.C. (2005). *Design and Analysis of Experiments*, 6th edn. New York: John Wiley & Sons, Inc.
Næs, T., Aastveit, A., Sahni, N.S. (2007). Analysis of split-plot designs: an overview and comparison of methods. *Quality and Reliability Engineering International* 23, 801–20.
Sahni, N.S., Isaksson, T., Næs, T. (2001). The use of experimental design methodology and multivariate analysis to determine critical control points in a process. *Chemometrics and Intelligent Laboratory Systems* 56, 105–21.
Schlich, P. (1993b). Uses of change-over designs and repeated measurements in sensory and consumer studies, *Food Quality and Preference* 4, 223–35.
Zar, J.H. (1999). *Biostatistical Analysis*. 4th Edition. New Jersey: Prentice Hall.

13

ANOVA for Sensory and Consumer Data

Analysis of variance (ANOVA) is one of the most used and also most important methods for analysing sensory and consumer data. ANOVA comprises a large number of methods, but all of them have in common that the focus is on testing the importance of various factors on the outcome of an experiment. The absolutely simplest situation is when there is only one factor that is causing the variation, all other factors being kept constant. From this simplest situation we expand first to the situation with two factors and then to situations with several factors involved. The concept of random effect and mixed models is introduced in order to make the conclusions from ANOVA of sensory and consumer data more relevant for practical purposes. Models for special situations such as for split-plot designs and nested designs will also be discussed. Post hoc tests for detecting which levels of a factor that have different average values of the response will be covered at the end of the chapter.

The chapter is based on the material in Chapter 12 and is the basis for Chapters 5 and 8.

13.1 Introduction

Analysis of variance (ANOVA) is a useful methodology when focus is on investigating product differences in sensory and consumer studies (see Chapter 5 and 8). The main purpose of ANOVA methods is to identify and quantify the factors that are responsible for the variability of the response. In practice, this is most often achieved by first identifying which factors which are significant and then by investigating the most important factors in more detail by so-called post-hoc tests.

Statistics for Sensory and Consumer Science Tormod Næs, Per B. Brockhoff and Oliver Tomic

Even though the classical ANOVA is developed for the use with quantitative data, where random error variability can be described, at least approximately, by normal distributions, the ideas and principles are generic and carry over to other types of data and distributions (see Chapter 15). Although model residuals for sensory and consumer data are never 100 % normally distributes, it is usually a reasonable approximation, at least good enough to give useful results. The *p*-values etc. should of course, as is normal practice in all types of applied statistics, not be taken as exact values, only as good indications of the level of significance for the different factors. How to check validity of model assumptions is discussed in Chapter 15.

13.2 One-Way ANOVA

The one-way ANOVA is the simplest type of ANOVA and is suitable when only one factor is systematically varied in the experiment. Examples are the analysis of single assessor data in Chapter 3 and the model (5.9) for averaged descriptive sensory data. The one-way ANOVA model can be written as

$$y_{ir} = \mu + \alpha_i + \varepsilon_{ir} \tag{13.1}$$

where y_{ir} ($i = 1, \ldots, I$ and $r = 1, \ldots, R_i$) is the r'th observation (replicate) for the i'th level of the design variable. Here the μ is the overall mean and the parameters α_i measure the average effect of the different settings of the design factor. As usual, the ε is the random error term that for testing purposes is assumed to be normally distributed. For the model (13.1) to be unique, one needs to introduce the restriction that the sum of the $\alpha's$ is equal to 0 (or alternatively setting one of the values equal to 0). For further arguments leading up to this type of ANOVA model we refer to Lea *et al.* (1997).

For model (13.1), one is primarily interested in investigating the differences between the values of the α's. The formal hypothesis (see Chapter 11) of no level differences (all factor levels having zero deviations from the overall average) can be written as

$$H_0 : \alpha_1 = \alpha_2 = \cdots = \alpha_I = 0 \tag{13.2}$$

Conceptually, the test for this hypothesis is conducted by computing a statistic that measures the differences between the α's and by comparing this with a measure of the random variation in ε. In more technical terms, one-way ANOVA computes the total variation in the data and splits this in two according to the two sources of variation, the between level differences and the within level differences. These are called sums of squares (SS) for treatments (or between treatment levels) and errors respectively. In order to make the two quantities more easily comparable, one divides each of them by a so-called degrees of freedom (DF) that depends on the number of groups and the number of observations in total. The resulting quantities are called mean squares (MS) and are finally compared using the ratio, the so-called F-statistic :

$$F = \frac{MS(Treatment)}{MS(Error)} \tag{13.3}$$

Large values of the F indicate that the hypothesis should be rejected.

In more detail, the two quantities that are involved can be written as

$$MS(Treatment) = \sum_{i=1}^{I} R_i \, (\bar{y_i} - \bar{y})^2/(I-1)$$

$$MS(Error) = \sum_{i=1}^{I}\sum_{r=1}^{R_i} (y_{ir} - \bar{y_i})^2/(N-I) \tag{13.4}$$

where $N = R_1 + \cdots + R_I$, $\bar{y}$ is the mean of all observations and $\bar{y_i}$ is the average over all observations for level i. As can be seen, the first is essentially a sum of squares of differences between product means and overall mean while the latter is essentially a sum of squares of differences between observations and their corresponding level mean.

If there really are no level differences, statistical theory tells us that such an F-statistic will have an F-distribution (see Chapter 11) with numerator degrees of freedom *I-1* and denominator degrees of freedom *N-I*. This means that if the observed F, calculated from the data, is 'too large' compared to that distribution, we claim that the levels are significantly different (see Chapter 11). The definition of 'too large' depends on the significance level of the hypothesis test. The p-value, expressing the 'extremeness' of the observed F in this distribution, is given by most software packages as a part of the so-called ANOVA table. Such tables are very common in practice and contain the different effects as rows and the SS's, DF's, MS's, the F-values and the p-values as columns.

A simple example is given in Table 13.1 based on (fictitious) data from a small completely randomised consumer study. Thirty-two consumers were randomly allocated to evaluate one of 4 products. In this case the four observations for each product are independent of each other and also of the observations for the other products. This leads to a one-way ANOVA model. Note that this setup is not very common for consumer studies (see Chapter 8) and is here just used as an illustration of how to compute the statistics in one-way ANOVA. The basic ANOVA computations for these data are given by:

$$MS(Treatment) = \frac{1}{3} \cdot 8 \cdot \left[(3.50 - 4.375)^2 + (4.25 - 4.375)^2 + (5.75 - 4.375)^2 + (4.00 - 4.375)^2\right]$$

Table 13.1 *Thirty-two consumers evaluate four products: (one product for each consumer).*

	Prod.1	Prod.2	Prod.3	Prod.4
	4	4	5	4
	4	6	6	3
	3	4	5	4
	4	6	6	3
	3	2	7	5
	3	5	5	5
	4	2	5	4
	3	5	7	4
Average: 4.375	3.50	4.25	5.75	4.00

Table 13.2 One-way ANOVA table for data in Table 13.1. The symbol $p < 0.001$ means that the p-value is smaller than 0.001.

Source	DF	SS	MS	F	p-value
Treatment	3	22.5	7.5	7.2	<0.001
Error	28	29.0	1.0		

$$MS(Error) = \frac{1}{28}\left[\sum_{r=1}^{8}(y_{1r} - 3.50)^2 + \sum_{r=1}^{8}(y_{2r} - 4.25)^2 + \sum_{r=1}^{8}(y_{3r} - 5.75)^2 + \sum_{r=1}^{8}(y_{4r} - 4.00)^2\right]$$

leading to the ANOVA table in Table 13.2. In this case there is a clearly significant product difference ($p < 0.001$, see Chapter 11). As can be seen, in the table there is one row for each effect and one columns for each of the quantities mentioned above.

If a significant effect is found, one will normally be interested in comparing the means $\bar{y}_i$ of each of the levels for identifying where the actual effect is. The standard errors for the means can easily be computed and used together with the means for this type of comparison purposes, but usually it is better to use a more advanced strategy that takes the overall significance level into account when performing several tests at the same time. These tests are called post hoc tests and will be considered in Chapter 13.9.

13.3 Single Replicate Two-Way ANOVA

This is the type of analysis typically used for data from a randomised complete block design (see Chapter 12). The corresponding model accounts for systematic effects related to two independent factors (A and B), for instance the effect of an ingredient tested at a number of levels by different sensory assessors (blocks). The two-way main effects model can be written as

$$y_{ij} = \mu + \alpha_i + \beta_j + \varepsilon_{ij} \tag{13.5}$$

where y_{ij} ($i = 1, \ldots, I$ and $j = 1, \ldots, J$) is the observation for the ith level for factor A and jth level of factor B. The α's represent the different levels of factor A and the β's represent the different levels of factor B. In order to make the model unique, one needs to introduce some restrictions on the parameters. Usually, the sum of the α's and the sum of the β's are both set equal to 0. Note that this does not represent any loss of generality; it simply implies that the average level is equal to μ.

Again the purpose is to determine the effects of the two factors A and B on the response y, i.e. which of them is important and which levels that are different from the rest. As above, the total variance of y is decomposed into an SS measuring the effect of factor A (for instance assessor), an SS measuring the effect of factor B (for product or treatment) and an SS measuring the size of the random noise. Again the SSs are divided by their

Table 13.3 *Single-replicate two-way data example (randomised block design).*

	Prod1	Prod2	Prod3	Prod4	Average
Ass1	2	3	1	1	1.75
Ass2	7	6	7	8	7.00
Ass3	6	6	7	4	5.75
Ass4	5	4	5	1	3.75
Ass5	8	6	7	5	6.50
Ass6	2	3	3	1	2.25
Ass7	5	5	8	4	5.50
Ass8	4	3	5	2	3.50
Average	4.875	4.500	5.375	3.250	4.5

corresponding DFs before F—ratios are obtained for testing hypotheses. The sums of squares have a similar definition as above and details can be found in Lea *et al.* (1997).

The hypothesis of no assessor (factor A) differences

$$H_0 : \alpha_1 = \alpha_2 = \cdots = \alpha_I = 0 \tag{13.6}$$

is evaluated by the F-test statistic

$$F = \frac{MS(Assessor)}{MS(Error)} \tag{13.7}$$

using $(I - 1, (I - 1)(J - 1))$ degrees of freedom. Similarly the hypothesis of no product differences (for factor B)

$$H_0 : \beta_1 = \beta_2 = \cdots = \beta_J = 0 \tag{13.8}$$

can be evaluated by the F-test statistic

$$F = \frac{MS(Product)}{MS(Error)} \tag{13.9}$$

using $(J - 1, (I - 1)(J - 1))$ degrees of freedom.

An example based on two-way data can be found in Table 13.3. This is a situation with 8 sensory assessors each testing 4 different products. For the data set in Table 13.3 the ANOVA table is given in Table 13.4. This is again the typical way of representing the results, one row for each effect and one column for each of the quantities SS. DF, MS, F and p-value. As can be seen, both factors are significant (small p-values).

Table 13.4 *Two-way ANOVA table for data in Table 13.3. The symbol* p < 0.001 *means that the* p*-value is smaller than 0.001.*

Source	DF	SS	MS	F	*p*-value
Assessor	7	108.0	15.4	14.6	<0.001
Product	3	19.8	6.6	6.2	0.003
Error	21	22.3	1.1		

An illuminating exercise is to analyze the data as if they came from 32 assessors instead of 8, that is using a one-way ANOVA model. In this case, the *p*-value for the products is equal to 0.281, which means that the products are not longer significantly different. The extensive between individual variability expressed in the SS(assessor) of the two-way analysis now becomes a part of the random error and blurs the information about product differences. The blocking on individuals in the randomised block experiment hence makes it possible to account for this variability and obtain a more powerful experiment for (typically) lower cost.

13.4 Two-Way ANOVA with Randomised Replications

The next extension is to assume that there are more than one replicate for each combination of factors. This means that for each combination of factors, there are two or more observations which only differ in their random error. The full two-way ANOVA model with interactions and replicates (see also Chapter 5, model (5.1)) can be written as

$$y_{ijr} = \mu + \alpha_i + \beta_j + \alpha\beta_{ij} + e_{ijr} \tag{13.10}$$

where α, β are the so-called main effects for factors A and B and $\alpha\beta_{ij}$ is the term needed for accommodating a possible interaction (Chapter 12) between the two factors. The r here denotes replicate number ($r = 1\ldots,R$). Note that for the situation in Chapter 13.3, no such interaction was possible. The main effects are interpreted as average differences between the levels of the factors and the interactions represent the additional systematic effects of the factors not accounted for by the main effects.

The standard ANOVA analysis goes as above by defining SS, MS and constructing ratios. Each of the effects have their own series of SS., MS etc. In this case we end up with three different hypotheses with corresponding test statistics:

$$H_0 : \alpha_1 = \alpha_2 = \cdots = \alpha_I = 0 \text{ tested by } F = \frac{MS(A)}{MS(Error)} \tag{13.11}$$

$$H_0 : \beta_1 = \beta_2 = \cdots = \beta_J = 0 \text{ tested by } F = \frac{MS(B)}{MS(Error)} \tag{13.12}$$

$$H_0 : \text{ all } \alpha\beta_{ij} = 0 \text{ tested by } F = \frac{MS(A \times B)}{MS(Error)} \tag{13.13}$$

The general recommendation is to check for the interactions first. If the interactions are present, main effects tests must be interpreted as tests for average differences between the levels of the factors in addition to the differences already established to be present in the interactions. It is generally also recommended to study the quantitative sizes of the three mean squares (or equivalently the F-test statistics) to judge the relative importance of the effects – especially when all of them are statistically significant.

Alternatively one can, as proposed in for instance Næs and Langsrud (1998) consider the hypothesis that both the product main effect and interactions effects are equal to 0, corresponding to a kind of overall hypothesis of no product differences. This can easily be tested by pooling the SS and DF for products and interactions and dividing the obtained

Table 13.5 *Example of replicated two-way sensory data.*

	Prod1	Prod2	Prod3	Prod4	Average
Ass1	2, 2	3, 3	3, 1	1, 1	2.00
Ass2	8, 7	6, 6	7, 7	5, 8	6.75
Ass3	5, 6	4, 6	5, 7	1, 4	4.75
Ass4	4, 5	3, 5	5, 8	2, 4	4.50
Average	4.875	4.500	5.375	3.250	4.5

MS by the MS(Error) as usual. This may be a reasonable way to avoid possible problems with fixed effects tests when the ANOVA models are used for sensory data. These aspects will be discussed in Chapter 13.7.

Estimating the main effects and their standard errors is simple and is done by most computer packages. The estimates of the main effects for the different levels are simply defined as the averages for the actual levels minus the total average. The estimates of the interactions $\alpha\beta_{ij}$ can be written as the average values for the actual combinations of factors (i,j) minus the averages for the two factors (for indices i and j) plus the average of all observations (see Lea *et al.* (1997) for more detail, se also Chapter 5). Al these estimates can be considered as lease squares (LS) estimates of the corresponding parameter values (see Chapter 15).

For factors with two levels only, the main effects are usually defined slightly differently (but still representing the same phenomenon) as the average differences between the two levels of the factors. The interaction is in this case defined as the average difference between the effects of factor A for the two levels of factor B. This again represents the same phenomenon as in the more general case (see Box *et al.*, 1978).

In Table 13.5 is given a data set based on sensory analysis of 4 products with 4 sensory assessors. There are two replicates for each combination of assessor and product (separated by comma). The corresponding ANOVA table is given in Table 13.6. Here we see that the two main effects hypotheses must be rejected (small p-values). In Figure 13.1 is given a plot of all average values plotted for each assessor separately. This plot is called an interaction plot and shows clearly that assessor number 1 has a completely different pattern than the rest. He/she gives scores systematically lower than the rest and also has another interaction profile.

Table 13.6 *Fixed model ANOVA table for data in Table 13.5. The symbol* $p < 0.001$ *means that the* p*-value is smaller than 0.001.*

Source	DF	SS	MS	F	*p*-value
Assessor	3	91.0	30.3	19.4	<0.001
Product	3	19.8	6.6	4.2	0.022
Assessor*Product	9	14.3	1.6	1.0	0.469
Error	16	25.0	1.6		

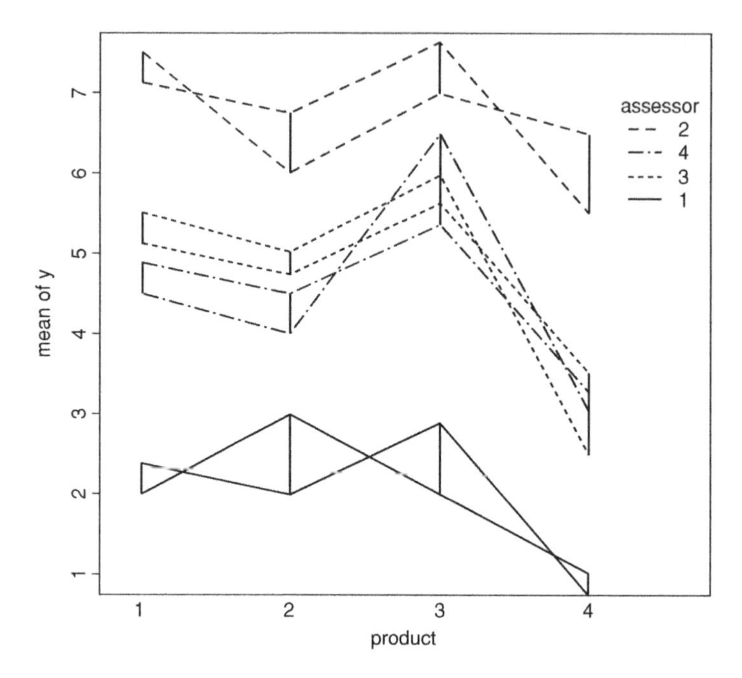

Figure 13.1 For each assessor there are two lines, one line which shows the average product profile for that assessor and one which is the closest possible parallel profile (parallel for each assessor).

13.5 Multi-Way ANOVA

In Chapters 5 and 8 full factorial designs with a number of factors A, B, C, D etc. are discussed. The principles for purely fixed ANOVA for such cases follow exactly the same rules and structure as for the two-way replicated situation. For instance using four factors (A, B, C and D and with R used as replicate as before) as an example, the data are represented as.

$$y_{ijklr,}i = 1, \ldots, I, j = 1, \ldots, J, k = 1, \ldots, K, l = 1, \ldots, L, r = 1, \ldots, R$$

1. Mean squares for main effects are simply scaled variances of the factor level averages, e.g.

$$SS(A) = \sum_{i=1}^{I} JKLR(\bar{y}_{i\bullet\bullet\bullet\bullet} - \bar{y}_{\bullet\bullet\bullet\bullet\bullet})^2 \tag{13.14}$$

   Here the dots indicate summation/averaging over the corresponding indices.
2. Mean squares for first-order (two-factor) interactions are scaled variances of the combined factor level averages corrected for the main effects, e.g.

$$SS(A \times B) = \sum_{i=1}^{I} \sum_{j=1}^{J} KLR(\vec{y}_{ij\bullet\bullet\bullet} - \bar{y}_{i\bullet\bullet\bullet\bullet} - \bar{y}_{\bullet j\bullet\bullet\bullet} + \bar{y}_{\bullet\bullet\bullet\bullet\bullet})^2 \tag{13.15}$$

3. Sum of squares for second-order (three-factor) interactions are scaled variances of the combined factor level averages corrected for the main effects and two-factor interactions, e.g. (where all the 'additional' summation dots were omitted from the indices)

$$SS(A \times B \times C) = \sum_{i=1}^{I}\sum_{j=1}^{J}\sum_{k=1}^{K} LR\left(\bar{y}_{ijk} - \bar{y}_{ij} - \bar{y}_{ik} - \bar{y}_{jk} + \bar{y}_{i} + \bar{y}_{j} + \bar{y}_{k} - \bar{y}\right)^2 \tag{13.16}$$

4. Etc.
5. The SS(Error) is computed as the total sum of squares minus the sum of the rest.

It should be noted that this computational procedure only holds for complete and balanced data. In all cases, however, one ends up with an ANOVA table with one row for each factor and one column for each of the quantities SS, MS, F and p-value. Usually, the p-value column is the most important since this best represents the effect of the factor. The recommendation for the analysis follows the same lines as for the two-way model: First look at the (higher order) interactions, then at the main effects and compare the relative sizes of the mean squares.

In many practical cases one will typically in such cases assume that higher order interactions, for instance from 3-way and upwards, are negligible as compared to the rest and thus left as a part of the error term. This is common practice and is based on the experience that higher order interactions are very often small or negligible as compared to the rest. A possible strategy is to first incorporate the effect of the higher order interactions and thus leave them out if insignificant.

Main effects and interactions plots are sometimes useful for visualising the size of the effects in this types of experiments. Examples of such plots are given in Figure 8.5 and Figure 13.1. One can also present the effects as columns with indication of standard errors of the effects on top as shown in Chapter 5.

If the data are not complete and balanced the ANOVA decomposition is not unique and there are different principles of constructing ANOVA tables for such more general cases, see e.g. Langsrud (2003). The most common tests are the so-called type III tests which, test the individual effects given that all the other model effects are in the model already. The most well know alternatives are the type I and type II tests. The type I tests test the effects in the sequence that they are represented to the computer program. This gives the user the possibility of building up the model from the most simple one containing a single factor to one which seems to be adequate for the data. An example is given in Table 5.5. For a description and discussion of the type II tests we refer to Langsrud (2003).

13.6 ANOVA for Fractional Factorial Designs

The analysis of fractional designs follows more or less the same structure as above; one decides the effects in the model and obtains the corresponding ANOVA table. Note, however, that if two factors are confounded, only one of them can be used as effect in the model (see e.g. Johansen *et al.* (2010a). Interpretation then has to be done according to the confounding pattern.

A possible complication with fractional designs, is that the DF's to use for error may be very small. This is a consequence of the fact that the number of experiments is per definition kept very low as compared to the number of factors. In such cases, one may improve significance testing by pooling the error SS with an error variance obtained by using replicated measurements of for instance a centre sample. Stepwise strategies for eliminating nonsignificant effects and pooling them with the random error have also been developed (see e.g. Langsrud and Næs, 1998). A third possibility is to use so-called *q-q* plots of the effects (see Chapter 15). These are plots of the estimated effects (see Chapter 13.4) vs. their expected values under the assumption of no significant effects. Points deviating from a straight line tendency are candidates to be denoted significant.

An example of a fractional factorial design used in product development can be found in Ellekjær *et al.* (1996). In that case, a 2^{7-2} design was used to study the effect of 7 factors on the sensory melting properties of processed cheese. The design was a Res IV design with some two-factor interactions confound with each other. A *q-q* plot of the effects is given in Figure 13.2. As can be seen, the main effects for factors B, C and D as well as the confounded interaction CD+EF (the two can not be distinguished) are clearly deviating

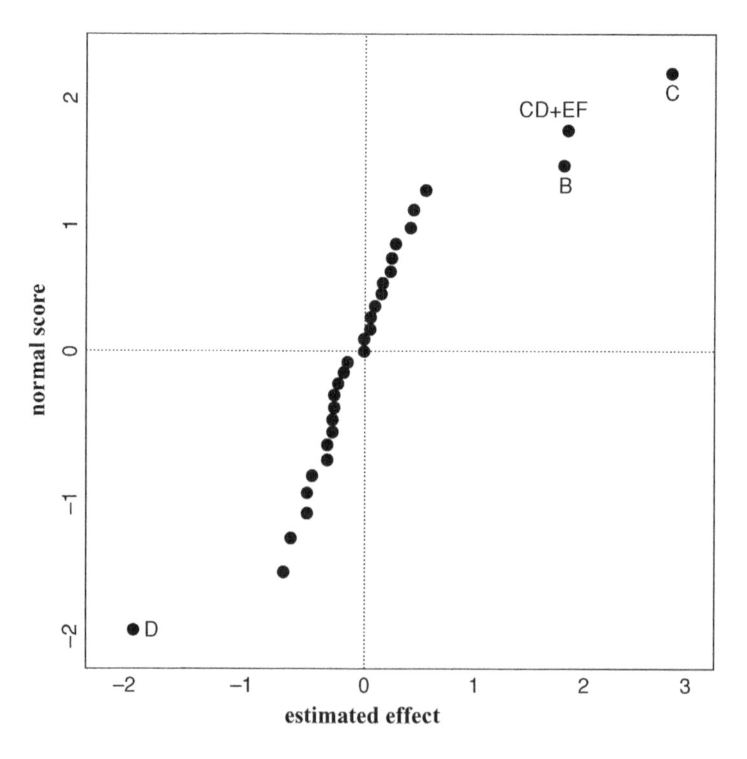

Figure 13.2 A q-q plot for data from a fractional factorial design. *The example in this case is taken from Ellekjær* et al. *(1996). Focus is on optimising sensory properties of processed cheese. Seven factors (process and ingredient factors) were investigated in a resolution IV design. Reprinted from Food Quality and Preference, 7, Ellekjær* et al.*, A case study of the use of experimental design and multivariate analysis in product improvement, 29–36, 1996, with permission from Elsevier.*

from the straight line tendency. They are therefore determined to be significant. Since both C and D have significant main effects, it is probable that the significant interaction is due to the contribution from CD. The results correspond well with the results obtained using regular ANOVA. Note that the DF's were here large enough for this experiment to allow for a proper significance testing, but that is not always the case.

13.7 Fixed and Random Effects in ANOVA: Mixed Models

In the analyses above, all effects were considered as so-called fixed effects, i.e. as unknown but fixed parameters. This means that specific focus is given to the actual levels of the factors involved. In sensory analysis (see Chapter 13.4), this means that focus is on both the actual products and assessors at hand. It has been shown that in situations with interactions between the two (Næs and Langsrud, 1998), this way of considering the assessors may lead to an exaggerated view on the differences between products. Therefore many researchers prefer to consider the assessor effect as a so-called random effect. This means that the assessors are considered as random representatives from a population. It can of course be argued that assessors are not really randomly chosen, but one can usually agree on the fact that the actual assessors are of little interest in themselves. Therefore, a random assumption is usually the best and also more robust. The hypothesis tests become more interesting and more representative for what one is really interested in. We refer to Næs and Langsrud (1998) for a broad discussion of this aspect and some of its consequences.

For consumer tests, treating the consumer effect as random is always the most natural since one is never interested in the actual consumers per se.

13.7.1 Mixed Version of the Replicated Two-Way Model

The mixed ANOVA model is the same as model (13.10), with the exception that the assessor effect and the interaction between assessor and product are now considered random normally distributed variables, i.e.

$$\begin{aligned} y_{ijr} &= \mu + a_i + \beta_j + \alpha\beta_{ij} + \varepsilon_{ijr}, \\ \alpha_i &\sim N(0, \sigma^2_{Ass}),\ \alpha\beta_{ij} \sim N(0, \sigma^2_{AssxProd}),\ \varepsilon_{ijr} \sim N(0, \sigma^2_{Error}) \end{aligned} \tag{13.17}$$

with all random contributions independent of each other.

In a mixed model all hypotheses refer to a population of assessors rather than to fixed ones. This implies that some of the hypothesis tests will be changed. The numerator in the F-statistics are the same as before, but now the denominator is sometimes different from the MS(Error). In most cases the denominator can be found as one of the other MSs in the ANOVA table, but in some cases, one may also need to combine the MSs in order to obtain the correct test. How to select the best MS is sometimes quite difficult and relies on theoretical calculations of the expectations of the MSs in the ANOVA table. In practice, however, the user does not need to know these rules, since most computer programs compute the correct tests without further manual calculations needed.

Table 13.7 Mixed model ANOVA table for data in Table 13.5. The difference in p-value for the products comes primarily because of different DF's for the test. The F's are very similar (identical when only one decimal is used). The symbol $p < 0.001$ means that the p-value is smaller than 0.001.

Source	DF	SS	MS	F-mixed	*p*-value
Assessor	3	91.0	30.3	19.2	<0.001
Product	3	19.8	6.6	4.2	0.042
Assessor*Product	9	14.3	1.6	1.0	0.469
Error	16	25.0	1.6		

For this particular model, it can be shown that the product effect (13.8) can be tested using the F-value

$$F = \frac{MS(Product)}{MS(Interaction)} \tag{13.18}$$

As can be seen, the denominator is now the MS for the interaction rather than the MS for the error term. The DF's for this test are the DF's for numerator and denominator respectively. Since the assessor effect and interaction effects are now random variables, the hypotheses of no assessor effect or interaction effect must be formulated in terms of their variances (their so-called variance components). For the assessor effect, the null hypothesis will be

$$H_0 : Var(\alpha_i) = \sigma^2_{Ass} = 0 \tag{13.19}$$

It can be shown that also for this test, the MS for the interaction must be used in the denominator. For the hypothesis of no interaction, however, the MS(Error) is as before used in the denominator.

In the two-way model without replicates, the interaction essentially coincides with the random noise and the model is reduced to a model with only main effects. It can be shown that in main effects model, the mixed model tests and the fixed model tests are identical, so in this case it does not matter whether the assessor is considered fixed or random.

The mixed model ANOVA table of the data in Table 13.5 is given in Table 13.7. As can be seen, the *p*-value for the products has now changed, but in this case, the differences are relatively small. This comes from the fact that interactions in this case are small. In many other cases, however, the difference between the two approaches may be substantial and must always be taken seriously.

13.7.2 Mixed Three-Way ANOVA

The situation with an additional systematic replicate structure can according to Chapter 5 be formulated with one fixed effect for products and two random effects for assessors and replicates. With all two-way interactions incorporated the model can thus be written as

$$y_{ijr} = \mu + a_i + \beta_j + \delta_r + \alpha\beta_{ij} + \alpha\delta_{ir} + \beta\delta_{jr} + \varepsilon_{ijr} \tag{13.20}$$

with the δ being the replicate effect. In this model all the effects except the product effect are random with mean 0 and a corresponding variance.

In this case, the three tests for the interactions can be conduced with the MS(Error) in the denominator. For the main effects, however, modifications have to be done. The most important hypothesis of no product differences:

$$H_0 : \beta_1 = \beta_2 = \cdots = \beta_J = 0 \tag{13.21}$$

can in this mixed model be validly tested by the following F-test statistic:

$$F = \frac{MS(\mathrm{Pr}oduct)}{MS(Assessor \times \mathrm{Pr}oduct) + MS(\mathrm{Re}plicate \times \mathrm{Pr}oduct) - MS(Error)} \tag{13.22}$$

The denominator degrees of freedom can here be computed using the so-called Satterthwaite's approximation (Satterthwaite, 1946). As can be seen, this is an example where it is not possible to find one single MS that matches the numerator. Similar F-tests can be used for testing the main effects of assessors and/or replicates. Again, both the test statistic and the DFs are computed by using a suitable software package.

A more comprehensive discussion of mixed model ANOVA can be found in for instance (Searle, 1971). The 'Proc mixed' procedure in SAS (Littel *et al.*, 1996) is an excellent tool for solving concrete problems.

For fractional factorial designs, the same type of methods can be applied just making sure that the appropriate effects are incorporated. If for instance two effects are confounded, only one of them can be used in the analysis and interpretation be done accordingly (Johansen *et al.*, 2010a).

13.8 Nested and Split-Plot Models

Two special cases of mixed models need special attention since they are used or referred to at various places in the book. In particular in connection with consumer conjoint studies these models are of special importance.

Nested models are characterised by the fact that all the levels of one factor are unique for each level of another. An important example is consumers who are always nested under gender, i.e. each consumer can only belong to one of the genders. A model with for instance one design factor, with gender and consumer within gender as systematic effects can be written as

$$y_{ijn} = \mu + \alpha_i + g_j + \alpha g_{ij} + c(g)_{jn} + \varepsilon_{ijn} \tag{13.23}$$

where the α is the design factor, the g is the gender effect and $c(g)_{in}$ represents consumer n nested within gender j. As can be seen, for each gender, the consumer within gender effect has two indices, one for the gender (j) and one for consumer (n, within gender). Note that there is no main effect for consumer since consumer n in one gender is another one than consumer n in the other. In this model the $c(g)_{jn}$ is considered random since one is never interested in the consumers per se (see Chapter 13.7).

Split-plot models will not play an important role in this book, but for completeness we will discuss them briefly; first of all since they are related to the different replicate structures in Chapter 5, but also since they can sometimes be of interest in consumer studies (see Hersleth *et al.*, 2003). A model with for instance 3 whole-plot effects and 2 sub-plot effects (see Chapter 12.6) and no interactions can be written as

$$y_{ijklm} = \mu + \alpha_i + \beta_j + \delta_k + e_{ijk} + \phi_l + \theta_m + \varepsilon_{ijklm} \quad (13.24)$$

In this model the ε_{ijklm} is the random sub-plot error term and the e_{ijk} is the whole-plot error term. The two error terms are independent and represent two different aspects of the experiment. It is generally not possible to merge the two and the analysis must therefore be adjusted accordingly. For most software packages the program is able to identify the right denominator, as was also mentioned above. An overview of split-plot models can be found in Næs *et al.* (2007).

13.9 Post Hoc Testing

Post hoc testing becomes important when one of the tests discussed above shows a significant effect (see e.g. Scheffe, 1959). If for instance one finds that the main product effect in model (13.10) is significant, one will be interested in knowing more about which products that are different from each other. Are all of them different or is it just a clear difference between two of the products? In such cases one can compute the averages for each of the products and compare them visually, but it is in general useful also to accompany this check with a statistical testing procedure. Comparing two and two products using t-tests as described in Chapter 11 is possible, but since there will be many tests involved, a possible risk is that too many tests are determined as significant. In order to control this type of error a number of methods have been developed that control the overall significance level. These procedures are called post hoc tests or multiple comparison tests and are generally based on adjusting the critical values of the individual tests in such a way that the overall significance level is controlled.

The most famous of these tests is the Tukey's test based on comparing all differences between means with an adjusted critical value. The critical values are for Tukey's method determined in such a way that there is a probability of no more than the significance level (for instance 0.05) to claim that at least one pair of means is significantly different when in fact the underlying means are identical. In practice, one does not need to know how to compute the adjustment factors since the software does this for us. The minimum value for being significant is called the least significant difference (LSD).

Other methods that can be used for post-hoc testing are the Bonferroni method, Newman-Keul's test and Duncan's test. These tests are based on different criteria for controlling the significance level. Some of these methods are discussed in some more detail in a sensory context in Lea *et al.* (1997). It should also be mentioned that the Bonferroni corrections can also be used in other situations when many tests are conducted, for instance when many factors in a model are tested simultaneously (see Chapter 8). In such cases one should always be aware of the over-optimism phenomenon described here and also consider this type of Bonferroni correction.

References

Box, G.E.P., Hunter, W., Hunter, S. (1978). *Statistics for Experimenter*. New York: John Wiley & Sons, Inc.

Ellekjær M.R., Ilseng M.R., Næs, T. (1996). A case study of the use of experimental design and multivariate analysis in product improvement. *Food Quality and Preference* 7(1), 29–36.

Hersleth, M, B-H. Mevik, Næs, T. Guinard, X. (2003). The use for robust design methods for analysing context effects. *Food Quality and Preference* 14, 615–22.

Johansen, S., Næs, T., Øyaas, J., Hersleth, M. (2010a). Acceptance of calorie-reduced yoghurt: Effects of sensory characteristics and product information. *Food Quality and Preference* 21, 13–21.

Langsrud, Ø. (2003). ANOVA for unbalanced data: Use Type II instead of Type III sums of squares. *Statistics and Computing* 13, 163–7.

Langsrud, Ø., Næs, T. (1998). A unified framework for significance testing in fractional factorials. *Computational Statistics and Data Analysis* 28, 413–31.

Lea, P. Næs, T., Rødbotten, M. (1997). *Analysis of Variance of Sensory Data*. New York: John Wiley & Sons, Inc.

Littell, R.C., Milliken, G.A., Stroup, W.W., Wolfinger, R.D. (1996). *SAS System for Mixed Models*, Cary, NC: SAS Institute Inc.

Næs, T., Langsrud, Ø. (1998). Fixed or random assessors in sensory profiling? *Food Quality and Preference* 9(3), 145–52.

Næs, T., Aastveit, A., Sahni, N.S. (2007). Analysis of split-plot designs: an overview and comparison of methods. *Quality and Reliability Engineering International* 23, 801–20.

Satterthwaite, F.E. (1946). An approximate distribution of estimates of variance components. *Biometrics Bulletin* 2, 110–14.

Scheffe, H. (1959). *The Analysis of Variance*. New York: John Wiley & Sons, Inc.

Searle, S.R. (1971). Linear Models. New York: John Wiley & Sons, Inc.

14

Principal Component Analysis

The principal components analysis (PCA) is a very versatile method that can be used for almost all types of data tables in order to obtain an overview. The method is particularly suitable in situations with lots of data and when little prior information is available. The PCA is based on the idea of finding the most interesting dimensions or directions of variability, called principal components. The idea behind the method as well as how to calculate the components will be discussed. The main results are presented in the scores plot, which describe the relation between the objects (rows in the data matrix) and the loadings plot, which describes the relations between the original variables (columns in the data matrix) and the principal components. These plots provide lots of information that can be used either for direct interpretation of the data set or as a preliminary step prior to further and more elaborate analyses. Rules for how to interpret the two plots will be given. Methods for validation and outlier detection will also be discussed.

The PCA is used throughout most of the book, but plays a particularly strong role in Chapters 3, 4, 5, 6, 9 and 10.

14.1 Interpretation of Complex Data Sets by PCA

A major challenge in both industry and research today is interpretation of large data sets. Typically, such sets can have several hundred and sometimes even thousands of columns and rows. In order to interpret such data, one needs statistical methods that can extract the most important information and present the results in such a way that they can be easily understood. The most useful methods are those that are based on few assumptions and that can be used for many types of applications.

Principal components analysis (PCA, see e.g. Mardia *et al.*, 1979) is an important method of this type. The PCA is based on the idea of finding the most important directions

Statistics for Sensory and Consumer Science Tormod Næs, Per B. Brockhoff and Oliver Tomic

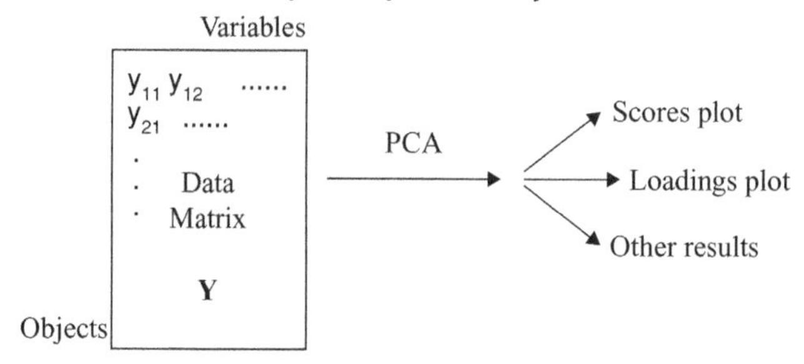

Figure 14.1 Illustration of information flow in PCA. *From data matrix with objects and variables to a small number of essential plots.*

of variability in the high-dimensional space of all the measured variables and presenting the results in plots that can be used for simple interpretation (see Figure 14.1). Other names for the PCA are singular value decomposition (SVD, Mardia *et al.*, 1979) and eigenvector decomposition.

The PCA can be looked upon as a purely descriptive mathematical method that simply extracts the main information in the data and presents the results graphically, but it is also possible to bring in statistical distribution theory and analyse the statistical distribution of the scores and loadings (see e.g. Mardia *et al.*, 1979).

Typically, the PCA is used in an early phase of an investigation as an explorative technique in order to provide an overview, but the method can also be used to generate hypotheses and ideas for further experimentation and for pre-processing prior to subsequent analyses. The latter is particularly important in regression and classification with many collinear variables. In such cases the classical solutions (MLR LDA, QDA, etc., see e.g. Ripley (1996) and Chapter 15) may be very unstable and give poor predictions (Næs and Indahl, 1998). PCA is then useful for first compressing the information down to a few dimensions or components and then using only these few components in further analyses.

14.2 Data Structures for the PCA

For the use of PCA it is necessary to organise the data in a so-called data matrix or data table where each row in the table corresponds to an 'object' and each column to a 'variable'. In this book the columns and rows can have different meaning dependent on application area (see for instance Chapters 3 and 9. In this chapter we will use the symbols N and K to denote the number of rows and columns respectively.

Using y_{nk} as the symbol for the observed value for the k'th variable for object n, the data matrix can be written as

$$\mathbf{Y} = \begin{pmatrix} y_{11} & \cdots & y_{1K} \\ \vdots & \cdots & \cdot \\ y_{N1} & \cdot\cdot & y_{NK} \end{pmatrix} \tag{14.1}$$

The first row corresponds to object 1, the next to object 2 etc. Likewise, the first column corresponds to variable 1, the next to variable 2 etc. Each row in the table is called a row vector and each column is a column vector.

In some cases a data table can have missing values. The PCA method can handle also such situations, by use of for instance the NIPALS algorithm (Wold, 1982), in particular if the number of missing values is not too large. If the data table contains very many missing values (for instance more than 20 % of the points), the result may in some cases become biased and more difficult to interpret, in particular if there is a systematic mechanism underlying the loss of data.

14.3 PCA: Description of the Method

The PCA is based on the idea of identifying the directions in the multivariate variable space with the highest possible variance. In the following we will describe the method step by step and we refer to Figure 14.2 for a graphical illustration of the idea. The data set in this case consists of 20 observations of 3 variables leading to a data matrix with 20 rows and 3 columns. In the figure, each row is represented as a point (vector) in the three-dimensional coordinate system. Note that this is not a typical situation for the use of PCA since there are only 3 variables in the data set, but it is simple and appropriate for illustration. For dimensions higher than 3, the method can not be visualised graphically, but the mathematics is identical.

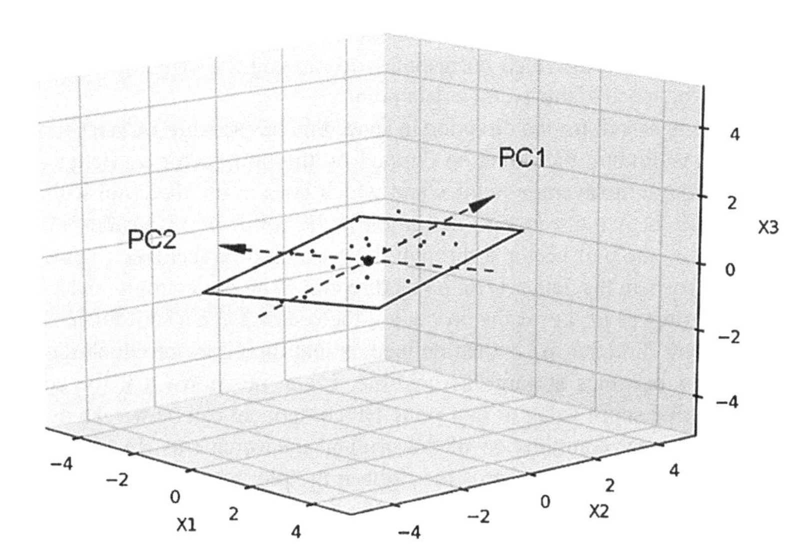

Figure 14.2 Illustration of the principle of PCA. *The illustration shows a three dimensional scatter plot of three measured variables. The points lie close to a two-dimensional plane in the three-dimensional space. The centre of the data is illustrated by the large point in the middle and the directions of the first two principal components are illustrated by arrows. The third principal component plays a minor role here since the data points lie close to a plane.*

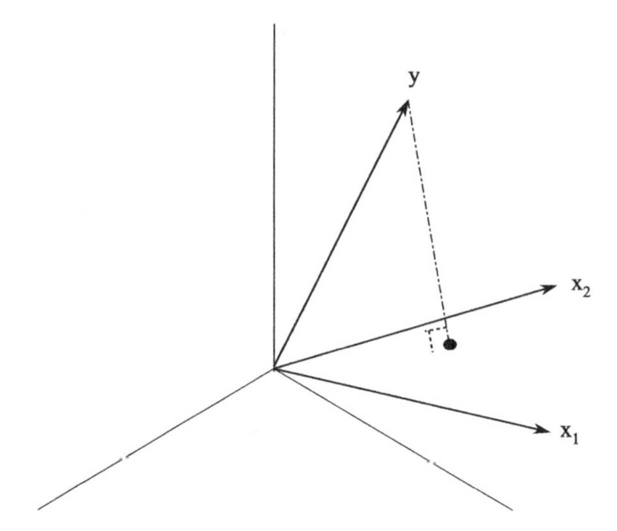

Figure 14.3 ***The idea of projection onto a plane illustrated in three dimensions***. *The vector* **y** *is projected onto the space spanned by the vectors* $\mathbf{x_1}$ *and* $\mathbf{x_2}$. *The vectors are indicated by solid arrows. The three axes represent the three axes in the three-dimensional coordinate system. The line connecting the* **y** *with the plane is perpendicular to the plane, i.e. this represents the shortest distance between the point and the plane.*

The first step of the PCA is to compute the average of all the variables. This average is plotted in Figure 14.2 as a larger point than the rest. Then the averages are subtracted from their corresponding variables which corresponds to moving the origin of the coordinate system or the vector space to the average data point $\bar{\mathbf{x}}$.

The next step is to search for the direction in space that has as large variance as possible. This corresponds to finding the direction defined by the unit vector $\hat{\mathbf{p}}_1$ (length equal to 1) which goes through the average point $\bar{\mathbf{x}}$ and which goes in the direction with as much variability as possible. A more precise definition is the following: Consider an arbitrary direction denoted by the unit vector $\hat{\mathbf{p}}$ through the mean and project (see Figure 14.3) all the data points **y** (rows in the data set) onto this direction. The projection of each such data vector **y** can be written as $\hat{\mathbf{p}}\hat{t}$, i.e. as the product of the vector $\hat{\mathbf{p}}$ and a coordinate $\hat{t}$. The first principal component direction $\hat{\mathbf{p}}_1$ is then defined as that direction for which the variance of the coordinate $\hat{t}$ becomes as large as possible. It can be shown that the coordinate, usually called the first **score** value or first principal component can be written as $\hat{t} = \hat{\mathbf{p}}^{\mathrm{T}}\mathbf{y}$, which is simply a linear combination of the original measurements. In Figure 14.2 the arrow that goes in the direction towards the right in the plot represents the first principal component.

The elements of $\hat{\mathbf{p}}_1$ are called the **loading** values for components 1 and as can be seen they define the contribution of each of the original variables in the calculation of the first score. A high loading value for instance for variable 1, means that the first principal components is strongly related to variable 1. In Figure 14.2 it can be seen that the first component goes in the negative direction for x_1 and in the positive direction for x_2. The contribution along the direction x_3 is negligible since the arrow is more or less parallel to the x_1/x_2 plane.

The next step is to search for the second principal component. This is done by the same procedure as for component number 1, but under the restriction that the direction of the second component is orthogonal to the first, i.e. that the $\hat{\mathbf{p}}_2$ is orthogonal to $\hat{\mathbf{p}}_1$. It can be shown that this is identical to subtracting the contribution of the first component from the data matrix and then doing the same as for component 1 without using a restriction. It can also be shown that the vector of scores for component 2 ($\hat{\mathbf{t}}_2$) is orthogonal to the vector of scores for component 1 ($\hat{\mathbf{t}}_1$). Since the original variables are mean centred, this means that component 1 and 2 are uncorrelated. In Figure 14.2 the arrow that goes towards the left represents the second principal component. As can be seen it goes in the negative direction for both x_1 and x_2. Again the contribution along the third axis is negligible.

Generally, the procedure continues by extracting new components that describe as much variance as possible under the restriction that each new principal component direction is orthogonal to the previous. This is done until the desired number of components, here called A, have been extracted. In our case, however, 2 components were enough to explain most of the variation. The number of components is limited by $\min(K,N-1)$.

It can be shown that the sum of the variances of the principal components is equal to the sum of the variances of the original variables. This result implies that it is meaningful to talk about explained variance in percent for each of the components. If for instance the total variance is 100 and the variances for the first and second components are 50 and 20 respectively, we will say that the two first components explain 70 % of the variance in the data set. In the example in Figure 14.2, the explained variance for the two first components is 87 % + 9 % = 96 %. Note that the explained variance is largest for the first component, next largest for the second etc.

If we extract all the possible components, the PCA can be considered simply as a full rotation of the original variables (or as a change of co-ordinate system). Often, however, the relationship among the variables is so strong that only a few components, for instance 2 or 3, are needed in order to explain a substantial amount of the total variance in the data set. In practice, it is also difficult, unless one is very experienced, to interpret more than 3 components. For the present example, the third component describes a very small percentage of the variation (4 %).

14.4 Projections and Linear Combinations

As described above, the score for the first principal component for one of the objects with observation vector $\mathbf{y}$ can be written as

$$\hat{t}_1 = \mathbf{y}^{\mathrm{T}}\hat{\mathbf{p}}_1 = \sum_{a=1}^{K} y_a \hat{p}_{a1} \tag{14.2}$$

i.e. as a linear combination of the elements of $\mathbf{y}$ (scalar product of two vectors). This can also be interpreted as the projection of the vector $\mathbf{y}$ onto the space defined by the vector $\hat{\mathbf{p}}_1$ (see Figure 14.3). The same holds for $\hat{t}_2$, $\hat{t}_3$ etc. .... This implies, using standard matrix algebra, that the full set of principal component scores $\hat{\mathbf{T}}$ can be written as linear combinations of the original data table $\mathbf{Y}$ represented by the matrix product

$$\hat{\mathbf{T}} = \mathbf{Y}\hat{\mathbf{P}} \tag{14.3}$$

where $\hat{\mathbf{P}}$ is the matrix of loadings (with rows corresponding to the variables and columns to the components). When all possible components have been extracted, this can be inverted to yield

$$\mathbf{Y} = \hat{\mathbf{T}}\hat{\mathbf{P}}^{\mathrm{T}} = \sum_{a=1}^{K} \hat{\mathbf{t}}_a \hat{\mathbf{p}}_a^{\mathrm{T}} \tag{14.4}$$

where the $\hat{\mathbf{t}}$'s are the columns of $\hat{\mathbf{T}}$ and the $\hat{\mathbf{p}}$'s are the columns of $\hat{\mathbf{P}}$. This equation shows how the matrix $\mathbf{Y}$ can also be written as a linear function of the principal components $\hat{\mathbf{T}}$.

It is quite common to decompose the Equation (14.4) into two independent contributions

$$\mathbf{Y} = \sum_{a=1}^{A} \hat{\mathbf{t}}_a \hat{\mathbf{p}}_a{}^{\mathrm{T}} + \sum_{a=A+1}^{K} \hat{\mathbf{t}}_a \hat{\mathbf{p}}_a^{\mathrm{T}} \tag{14.5}$$

or

$$\mathbf{Y} = \sum_{a=1}^{A} \hat{\mathbf{t}}_a \hat{\mathbf{p}}_a^{\mathrm{T}} + \hat{\mathbf{E}} = \hat{\mathbf{T}}_1 \hat{\mathbf{P}}_1 + \hat{\mathbf{E}} \tag{14.6}$$

where the first part corresponds to those components that describe most of the variation in $\mathbf{Y}$ and which usually are the most important ones for plotting and the last part is the rest of the components, often considered as noise and denoted by $\hat{\mathbf{E}}$. In other words, one can consider the matrix $\mathbf{Y}$ as composed into a contribution from a few major principal components plus noise.

If the first few components extracted contain a large portion of the information in $\mathbf{Y}$, the elements of $\hat{\mathbf{E}}$ will be small. In this case the first few components scores multiplied by the corresponding loading vectors approximate the information in the matrix $\mathbf{Y}$.

Note that for the principal components and also the loadings, we have used a hat on top of the symbols. The reason for this is that they are computed from the data. Another argument for this notation is that sometimes component models are thought of as more regular statistical models (see for instance Chapter 14.11) and in that case it is natural to present the symbols without any hats on top.

14.5 The Scores and Loadings Plots

The scores plot is a scatter plot of the columns of $\hat{\mathbf{T}}$ and the loadings plot is a scatter plot of the columns of $\hat{\mathbf{P}}$. Usually, two-dimensional plots based on the two first components are used, but three-dimensional plots of the first three components are also possible. They are, however, often, more difficult to interpret. For interpreting three components, it is also possible to plot component 1 vs. component 2 in one plot and components 2 vs. component 3 in another.

In Figure 14.4.a is presented a view of the data in Figure 14.2 from the top. This plot represents the projection of all data points onto the plane spanned by the two arrows in Figure 14.2. The score plot is then simply based on presenting this projection directly onto the computer screen (Figure 14.4.b).

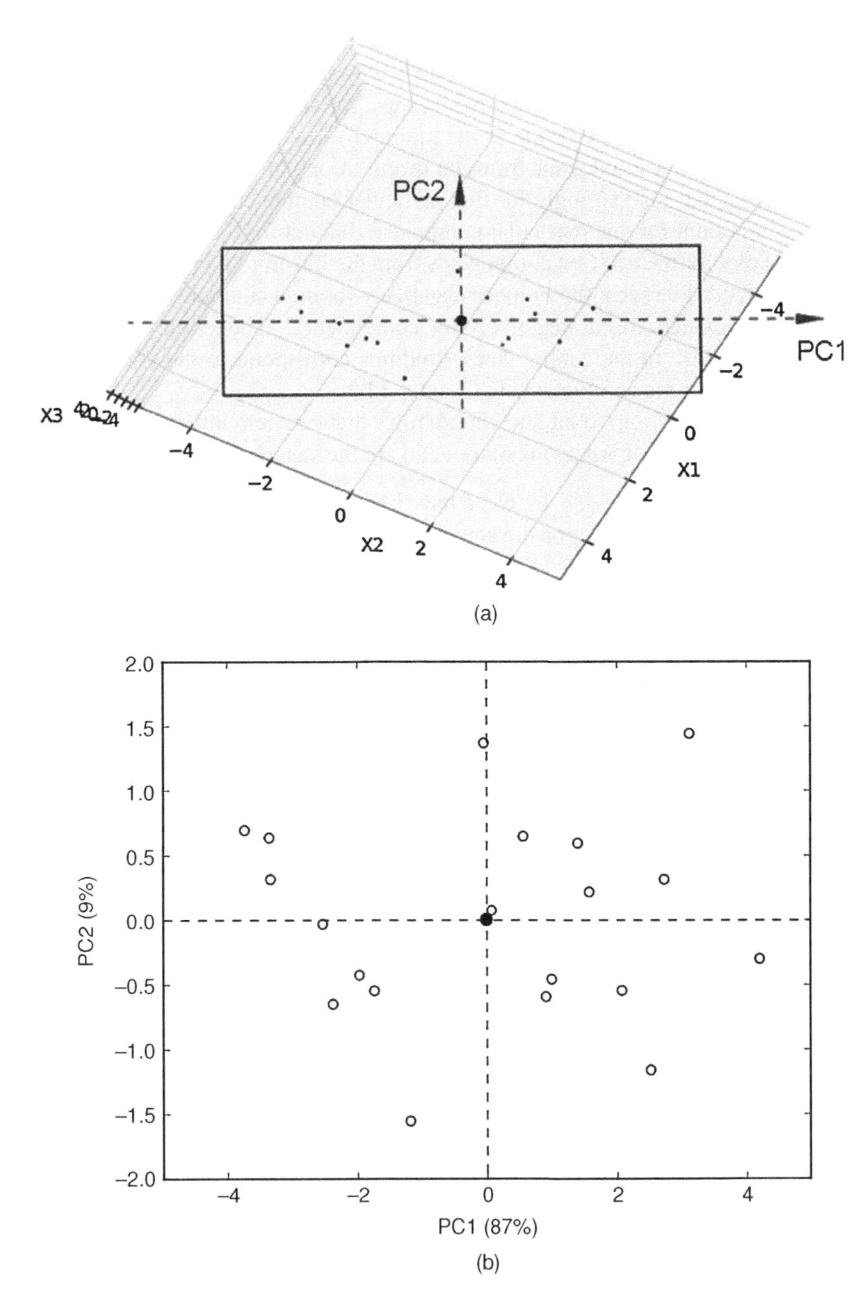

Figure 14.4 Score plot. *This plot shows the scatter plot of the scores for the two first components for the data used in the illustration in Figure 14.2. In a) is presented the plot in Figure 14.2 seen from the top and in b) is presented the regular score plot. The two plots contain essentially the same information.*

The scores plot is therefore simply a plot of the relation between the objects in the projected space. The loadings plot is a plot which shows the relation between the original variables and the principal components themselves. Variables with a large loading value will show up far away from the origin in the plot and variables with a small loading will fall close to the origin. Variables far from the origin in one of the directions/components only, are variables that load strongly for that particular component. This means that these variables are important for that particular component, but not for the others.

The loading plot for the two first components from the illustration in Figure 14.2 is given in Figure 14.5. As can be seen, the x_1 loads negatively for both components, while x_2 loads positively for component 1 and negatively for component 2. The x_3 plays a neutral role since it is placed close to the centre. These findings correspond to the comments given about the directions of the components in Chapter 14.3.

There are a number of important and simple rules that can help interpreting these plots. The higher the explained variance, the more valid are the statements:

- **Loadings plot:** Variables which are close have high correlation.
- **Loadings plot:** Variables on opposite side of the origin have negative correlation.
- **Scores plot:** Samples which are close to each other have similar overall properties and samples which are far apart are very different.
- **Scores plot and loadings plot together:** Samples to the right in the scores plot are dominated by (have large values of) variables to the right in the loadings plot, objects at the top of the scores plot are dominated by variables at the top in the loadings plot etc.

Some users of PCA prefer to use the scores and loadings separately, while others like to plot them on top of each other, in so-called bi-plot. This is to a large extent a matter of

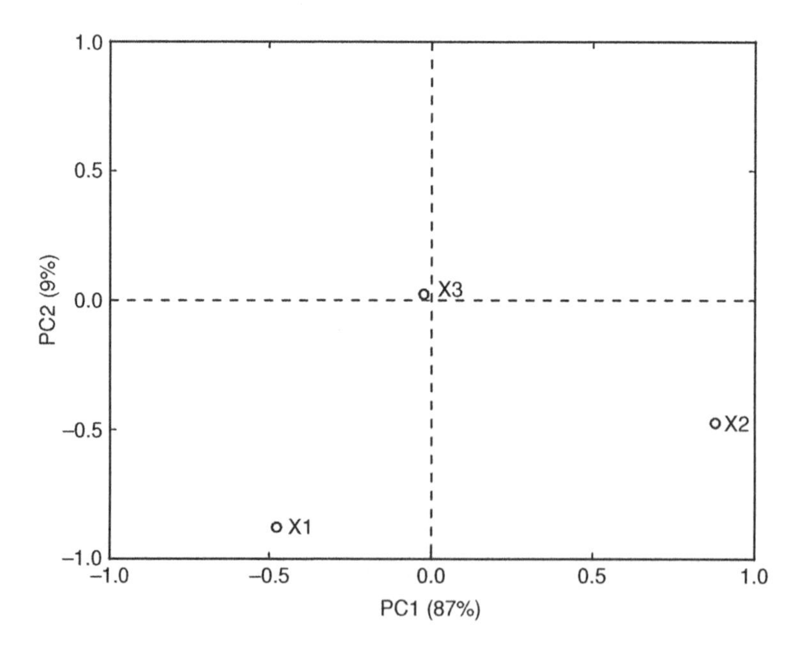

Figure 14.5 ***Loading plot.*** *This plot shows the scatter plot of the loadings for the two first components for the data used in the illustration in Figure 14.2.*

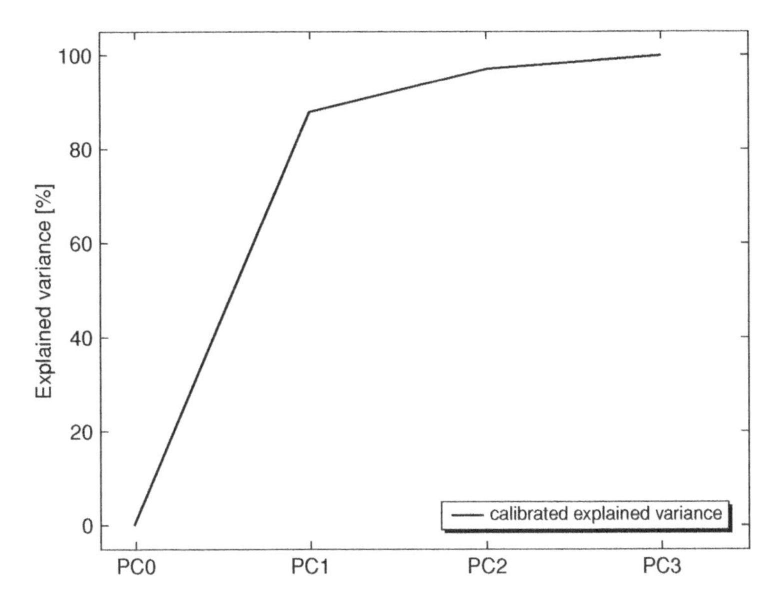

Figure 14.6 Explained variance plot. *This plot visualises the overall explained variance for each of the components for the data used in Figure 14.2.*

taste. The concept of bi-plot can be extended in various ways as was described in Gower and Hand (1995) in order to provide additional information.

In addition to the scores and loadings plot one will normally also look at a plot of the explained variances of the different components. An example of this is given in Figure 14.6 based on the data in Figure 14.2. One can see that the numbers here correspond to the numbers mentioned above and also to the numbers given along the two axes in Figure 14.4b and Figure 14.5.

Another way of plotting the explained variances was proposed in Dahl *et al.* (2008) for situations where there are many explained variances to look at simultaneously. This plot is called the Manhattan plot and may for instance be useful when one is interested in the explained variance for each of the attributes or for each of the assessors in a sensory panel. The plot presents the results by the use of shades of grey, with black corresponding to 100 % explained and white to 0 % explained variance. An example of a Manhattan plot is given in Figure 3.10.

In some cases, one may be interested in considering the objects as randomly selected representatives from a population. In such cases it is possible to compute and incorporate confidence ellipses in the plot indicating the variability around samples or groups of samples, for instance by using statistical resampling techniques. This aspect will, however, not be pursued here and we refer to Pages and Husson (2005) for examples of similar type in sensory analysis.

14.6 Correlation Loadings Plot

A modification of the above regular loadings plots, called the correlation loadings plot, was proposed in Martens and Martens (2001) (see also Mardia *et al.*, 1979). It is a two-dimensional scatter plot based on the idea of plotting the correlations between the components and the variables themselves instead of the loadings as they are (See Figure 14.7). The advantage of this plot is that one can get direct information about how much the different variables are correlated with or explained by the different components. In particular when the units of the variables are different, this may give additional and useful information. When variables are already standardised, the differences between the loadings and the correlation loadings plots will generally be smaller.

Note that the correlation loadings plot also opens the possibility of drawing circles in the plot corresponding to various degrees of explained variances. These circles are based on the concept that the explained variance of a variable y considered as a linear function of two uncorrelated components $\hat{t}_1$ and $\hat{t}_2$, can be written as a sum of the explained variance for each of the two variables separately, i.e.

$$R^2_{\hat{t}1,\hat{t}2} = R^2_{\hat{t}1} + R^2_{\hat{t}2} \tag{14.7}$$

(see Chapter 15). Using the Pythagorean theorem for triangles, the square length from the centre to a correlation loading point can be considered a sum of the explained variances of the two components separately and can thus according to Equation (14.7) be interpreted as the explained variance from a linear combination of the two t's. Typically one will present a circle for 100 % explained and for 50 % explained variance by the two components.

An interesting application of the method of correlation loadings is to incorporate a dummy matrix representing the objects in the **Y**-matrix. The object variables are thus given a very low weight in the PCA, meaning that they have no influence on the PCA solution

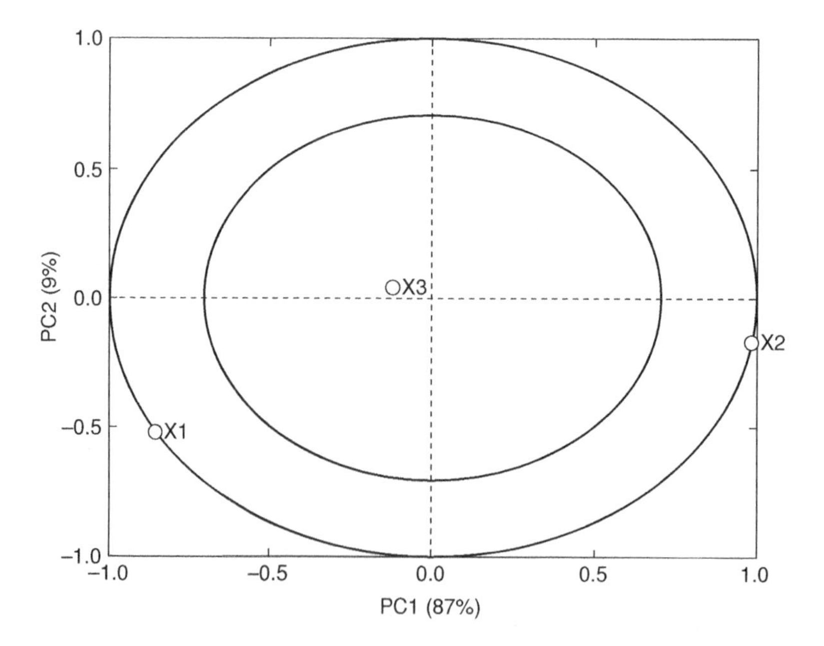

***Figure 14.7* Correlation loadings.** *This plot is the correlation loadings analogue for the loadings plot in Figure 14.5.*

(Martens and Martens, 2001). When it comes to correlation loadings, however, the 'object variables' will find their place in the plot corresponding to the scores in the scores plot. In other words, this is an alternative to producing the bi-plots mentioned above.

14.7 Standardisation

The only purpose of PCA is to look for directions with high variance. This means that if there are some variables that have a larger variance than others, they will be given the most attention and dominate the components extracted and then also the plots. This may be natural in some cases, but generally one is interested in letting all variables play a role in the estimation of components. A way to obtain this is to divide all variables by their own standard deviation ending up with variables that all of them have the same standard deviation equal to 1.

For sensory and consumer data, the standardisation may have a certain effect, but since variables are measured on more or less the same scale, the impact of standardisation will often have only a moderate effect on conclusions drawn from the study. The plots may be slightly different, but the overall conclusions will often be similar. Some more specific aspects for sensory and consumer data are given next. Here we will also discuss briefly centring and not centring before PCA which may possibly be an issue in particular for consumer data.

14.7.1 Sensory Analysis

The comments given here hold regardless of whether one uses PCA on the average sensory profiles or on the unfolded matrix of individual sensory profiles (see Tucker-1 in Section 14.11 below).

For sensory analysis, mean centring is always done for each attribute. Standardisation is also often done, but as was indicated above this has usually only a moderate effect on the conclusions drawn. One should, however, be aware of the fact that if some of the variables have a small variance and this is due to the fact that there is no or very little significant information in them, automatic standardising will 'blow up' the noise in these variables and make the noise equally important as the more informative variables in the rest of the sensory profile. Such variables may have an unfortunate effect on the PCA solution. Therefore, it is a sensible practice to eliminate the variables that do not significantly distinguish between the products before standardisation. This can always be done by the use of two-way ANOVA as is discussed in Chapters 5 and 13 if two or more replicates are used.

It is important to emphasise that the variables after standardisation have a slightly different interpretation than before standardisation. Performing a standardisation implicitly assumes that the original length of the scale used for the different attributes have no direct interpretation and meaning. In other words, differences in range for the different attributes are considered as uninteresting and only due to the calibration of the panel.

14.7.2 Consumer Preference Data

For consumer data, one sometimes uses consumers as rows (see above, Chapter 8, Section 8.5.2) and sometimes as columns (Chapter 9). In both cases mean centring for each consumer will usually be done. Note that for the situation with consumers as columns, this will be done automatically within standard PCA. In the other case, it must be done prior to PCA (see above, Chapter 8, Section 8.5).

Mean centring eliminates information about where on the scale the assessors give their scores. For instance, an assessor which gives values 1, 3 and 5 for three products will end up with the same centred values as one which gives the values 3, 5 and 7. In other, words, if mean centring is done, one ends up with an analysis of relative differences and forgets about whether a person generally scores high or low. If differences in the use of the scale can be considered a nuisance factor, this is a reasonably strategy. If not, one can avoid it by using PCA without mean centring, which is an option in some computer packages.

If the data are standardised, information about the range of the scale vanishes. Standardisation by setting standard deviation equal to 1 is usually not done for consumer data. If in doubt, the reader is recommended to or try both and compare the results.

14.8 Calculations and Missing Values

It can be shown that the formula (14.4) is essentially the same as the singular value decomposition (SVD) of the matrix **Y** (Mardia *et al.*, 1979). This mans that principal components scores and the loadings vectors can be found as eigenvector of the $\mathbf{YY}^T$ and $\mathbf{Y}^T\mathbf{Y}$ matrices respectively. The explained variances of the components correspond to the eigenvalues of the matrices.

A method with attractive numerical properties for situations with missing values is the NIPALS method (Wold, 1982). This is based on a series of simple regression models, and in each step one simply skips those observations or cells that are missing. If the number of missing values is large, the method will still give results, but the results may be unreliable and misleading. This is in particular the case if there is a systematic structure related to why values are missing. For more information and comparison of methods we refer to Grung and Manne (1998).

Another advantage of the NIPALS method is that one does not need to compute all components at the same time. The NIPALS method estimates the most important component first and then the next important etc. This may be important for large data sets, for which the calculations of all components may take some time. Usually, however, with modern computers this aspect is less important unless the data set is extremely large.

14.9 Validation

In regression analysis (Chapter 15) empirical validation of models is very important for the purpose of getting information about how good the model is for prediction. In the following we will discuss briefly how these methods can be used for determining how many PCA components that can be looked at safely. In most cases one will use only 2 or 3 components, but even for this small number of components one is interested in how reliable the components are.

The most fortunate situation to be in from an empirical validation point of view is when the samples are selected at random from a population of samples and the number of samples is reasonably large. In this case one can use cross-validation (CV) or prediction testing (see

Chapter 15) to get an idea of the importance or significance of the different components extracted. Both of these methods are based on the idea of modelling some of the samples in the data set (training data) and testing out on some others (test data). Prediction testing is the best and simplest, but requires that a number of objects (for instance 30 %) are set aside for testing only. Cross-validation, based on leaving one or a few samples out in sequence, is more appropriate when the number of samples is small.

For both methods the testing is done by calculating the residuals after fitting of the test vectors $\mathbf{y}$ to the loadings matrix $\hat{\mathbf{P}}_1$ estimated by the training data using the model

$$\mathbf{y} = \hat{\mathbf{P}}_1\mathbf{t} + \mathbf{e} \tag{14.8}$$

The residuals are computed using standard regression procedures (see Chapter 15) before the sum of squares of the residuals is computed. Usually one will do a correction for the degrees of freedom involved following the formulae (15.7). The procedure is repeated for $A = 1$, $A = 2$ and so on until the maximum wanted number (for instance 6–7) has been extracted. The residual sum of squares is then plotted as a function of the number of components. Note that the residual sum of squares will decrease (at last when the number of degrees of freedom is not corrected for) when new components are incorporated, which means that when using either prediction testing or cross-validation for PCA, one must be more careful than when using the same two methods for regression. Instead of looking for a minimum value, one must rather look for a place where the residual curve flattens out and becomes stable. This is the point where incorporating new components does not improve prediction and where one should stop interpreting components. A better alternative than plotting the residual sum of squares, is to plot the explained validation variance instead. This means that one looks at the variance explained by the various components as relative to the total sum of squares (see Figure 14.3).

In situations with a small set of samples or samples based on an experimental design, the prediction testing and CV are more problematic to use. The reason for this is that each sample is unique and it is not necessarily possible that each sample fits into the model determined by the rest of the samples. In such cases other methods may be more useful for obtaining information about validity of the components for interpretation purposes.

One possibility is to look at the explained variance for different number of components when fitting the model. If this explained variance curve goes up in a steep way before coming to a point where it clearly flattens out, this point is an indication of where one should stop interpreting components. Another important 'tool' is interpretation itself. If the interpretation makes sense in terms of previous knowledge or the design behind the study, this is a good argument for validity of the results. In some cases, one may also have prior knowledge regarding for instance correlation structure among variables and/or similarity between samples, which can be useful.

An alternative type of CV for PCA was proposed in Wold (1978) where observations were removed according to a certain pattern in the data table. One should, however, always remember that in practice it is difficult to interpret more than 2–3 components.

14.10 Outlier Diagnostics

As for all other statistical methods, outlier detection is important also for PCA. This may be important for obtaining additional information about the data set, but in most cases it

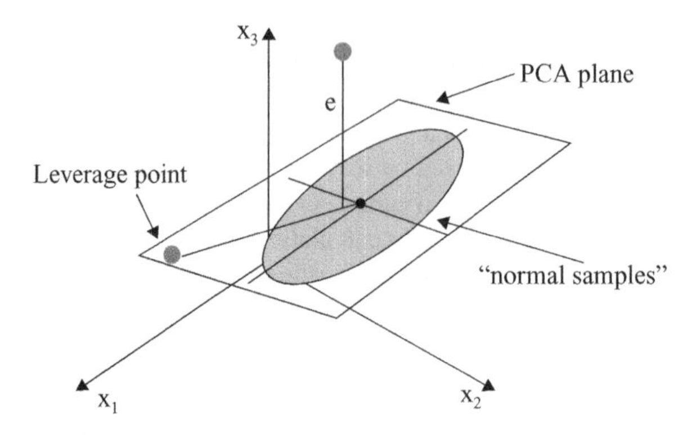

Figure 14.8 The leverage and X-residuals illustrated in three dimension. *The ellipse in the plane represented by the two first principal components represents the region where the observations are located. The two filled circles represent a high leverage observation and an observation with large residuals. An observation with high leverage is an observation lying far from the centre within the PCA space and an observation with large residuals is an observation which is lying far from, measured orthogonally to, the space spanned by the first (in this case 2) principal components.*

is important for detecting samples or variables that have an extraordinary influence on the data model. In extreme cases, the direction of one or several of the first few components in a PCA may be driven by a few samples only.

The simplest tool for detecting outliers and also for detecting the most influential observations is the scores plot. If there are samples which are positioned far from the centre as compared to the rest, these are obvious candidates for further study. If there are many components to look at, it may be helpful to support this search by the so-called leverage which is a measure of the distance from a sample to the centre (see Figure 14.8). The leverage for sample n is defined as

$$h_n = y_\mathbf{n}^\mathbf{T}(\mathbf{Y}^\mathbf{T}\mathbf{Y})^{-1}\mathbf{y}_n = \sum_{a=1}^{K} \hat{t}_{na}^2/\lambda_a \tag{14.9}$$

where λ_a is the eigenvalue of component a. These values can be plotted as a function of the sample number in order to reveal which samples that have an extreme position in the score space. Often 2 or 3 times the average value is used as an indication of outlyingness. The leverage will, however, usually be split according to Equation (14.5) and one will only look at the first part of it.

Another tool which may be of some interest to look at is the residual plot or the squared residual plot. The residual plot is a plot of the values in the residual matrix obtained by regressing the different observation vectors onto the PCA model (see Figure 14.8). The residuals matrix can be written as

$$\hat{\mathbf{E}} = \mathbf{Y} - \hat{\mathbf{Y}} = \mathbf{Y} - \hat{\mathbf{T}}_1\hat{\mathbf{P}}_1^\mathbf{T} \tag{14.10}$$

i.e. as the differences between the true observations and the predicted values of the observations from the model based on the dominant eigenvectors (defined by $\hat{\mathbf{P}}_1$, see Equation

(14.6)). These values can be plotted for each sample separately or for several samples overlaid in the same plot (see Martens and Næs, 1989). The residuals can also be squared and averaged for each sample and thus plotted as a function of the sample number in order to get an overview. Again, 2 or 3 times the average can be used as a quick guide as to when to determine a sample to be an outlier.

The leverage and the squared residual can be plotted against each other in a so-called influence plot (see Martens and Næs, 1989). For further discussion of geometrical aspects of outlier detection and also about how to detect outliers in PCA we refer to Næs *et al.*, 2002).

When a serious outlier is detected, one will in most cases end up with eliminating the sample from further analysis, but at the same time reporting and being aware of the fact that one is then analysing only the main part of the data set. In some cases, one may, however, be able to understand why a sample is detected as an outliers and take proper actions based on the actual explanation. In some cases it may even be possible to correct a possible error, for instance a misprint which has been detected in this way. Note, however, that an outlier is not necessarily an erroneous observation.

It is always a good practice to compare the analyses before and after outlier detection in order to see what effect an outlier has on the solution. If the influence is small, it matters little what is done with the sample.

14.11 Tucker-1

The PCA method is most often used on a well defined two-dimensional matrix. It can, however also be of interest in situations where the original data set has a so-called three-way structure as is the case for descriptive sensory data (see Chapter 2). But in order to use PCA on this type of data, one needs to unfold the data first, which means that the slices for the different assessors are either set beside each other or under each other. One of these unfoldings ends up with a matrix having J rows corresponding to the samples and I^*K columns corresponding to the different assessor and attribute combinations. The other one results in a matrix with K columns corresponding to the different attributes and I^*J rows corresponding to the different assessor and sample combinations.

The former of these unfolding structures gives rise to PCA models (see above, end of Section 14.4) of the type

$$\mathbf{Y}_\mathrm{i} = \mathbf{T}\mathbf{P}_\mathrm{i}^\mathrm{T} + \mathbf{E}_\mathrm{i} \tag{14.11}$$

and the other one to

$$\mathbf{Y}_\mathrm{i} = \mathbf{T}_\mathrm{i}\mathbf{P}^\mathrm{T} + \mathbf{E}_\mathrm{i} \tag{14.12}$$

In both models, the $\mathbf{TP}^\mathrm{T}$ part describes the most important components while $\mathbf{E}$ represents the rest, usually thought of as noise. The first one is a common scores model allowing for different loadings for the different assessors while the second one is a common loadings model allowing for different scores. The first model is the one used for analysing individual sensory data for the purpose of detecting individual differences in Chapter 3. The other one is the one used in the PC-ANOVA (Chapter 5) where the purpose is to relate the scores to external data. We refer top Figure 3.8 for an illustration of the principle of unfolding.

Both of these models are obtained by using regular PCA directly on the corresponding unfolded matrix. Using PCA on this type of data set is often called Tucker-1 according to the author of the first publications in the area (Tucker, 1964, 1966). The method is also known under the name of consensus PCA (CPCA. Westerhuis *et al.*, 1998).

The results from a Tucker-1 method are presented in scores plot and loadings plot. For assessor evaluation purpose, the correlation loadings are often used. The explained variances can be plotted for each combination of assessor and variables, which is best done using the so-called Manhattan plot proposed in Dahl *et al.*, 2008) (see Chapter 3).

14.12 The Relation between PCA and Factor Analysis (FA)

A closely related methodology called factor analysis (FA) has been developed and also found important applications in many disciplines (see e.g. Mardia *et al.*, 1979). The method is similar to PCA in the sense that it searches for a few factors or components that describe the data structure in an adequate way. The difference lies in the fact that factor analysis is based on a classical statistical model for the data structure, not only defined by a variance criterion. The FA model for an observation vector can be written as:

$$\mathbf{y} = \Lambda\mathbf{t} + \mathbf{e} \tag{14.13}$$

with the assumption that the elements of the vector $\mathbf{e}$ are uncorrelated and that all $\mathbf{y}$ are independent and identically distributed. The vector $\mathbf{t}$ is a vector of random score variables and Λ is a matrix of unknown parameters. This means that the vector $\mathbf{y}$ can be described as a linear function of a small number of components $\mathbf{t}$ plus uncorrelated random errors $\mathbf{e}$. Each vector of the data set is assumed to be independent from the other vectors.

The model parameters are usually estimated using maximum likelihood (ML, see Mardia *et al.*, 1978), assuming that all variables have a normal distribution. A restriction is needed in order for the elements of Λ to be estimated uniquely. There is no requirement in FA that the first factors in $\mathbf{t}$ describe as much information as possible. Therefore, one will usually need to look for a meaningful interpretation after the elements have been estimated. This is usually done by some type of rotation, or another linear transformation of the Λ. Such transformations are possible since one can always multiply Λ by a matrix and $\mathbf{t}$ with the inverse without loosing fit or changing any of the assumptions. In other words,

$$\Lambda\mathbf{t} = \Lambda\mathbf{H}\mathbf{H}^{-1}\mathbf{t} = \Lambda^{*}\mathbf{t}^{*} \tag{14.14}$$

Rotating Λ boils down to trying to find the most useful rotation matrix $\mathbf{H}$ that can give a meaningful interpretation of the Λ. A number of different methods have been developed based on different criteria. Most of them, however, seek as many large contrasts as possible between large absolute values and 0 in the loadings matrix Λ. In principle the factors of PCA can also be rotated, but this is seldom done in practice.

Note that model (14.13) is very similar to the model (14.6) above. The main differences lie in the fact that here the residuals are explicitly assumed to be uncorrelated. In practice, if the error terms are small as compared to the variance of the systematic structure, the two methods PCA and FA will give very similar results in terms of the space spanned by the components extracted. As noted above, however, for FA there is no natural ordering of the factors so a rotation is always needed. If the residual errors are large and the model is adequate, the differences between the two methods can be substantial if the error variances are different. If they are identical, the solutions will again be similar.

Calculations for FA are somewhat more complex than for PCA, Generally one needs more experience in order to use FA in practice. Therefore PCA is used as the method of choice in this book.

References

Dahl, T., Tomic, O. Wold, J.P., Næs, T (2008). Some new tools for visualising multi-way sensory data. *Food Quality and Preference* 19, 103–13.

Gower, J.C, Hand, D.J. (1995). *Biplots*. London: Chapman & Hall.

Grung, B., Manne, R. (1998). Missing values in principal components andlaysis. *Chemometrics and Intelligent Laboratory Systems* 42, 125–39.

Mardia, K.V., Kent, J.T., Bibby, J.M. (1979). *Multivariate Analysis*, London: Academic Press.

Martens, H., Næs, T. (1989). *Multivariate Calibration*. Chichester: John Wiley & Sons, Ltd.

Martens, H., Martens, M. (2001). *Multivariate Analysis of Quality: An Introduction*. Chichester: John Wiley & Sons, Ltd.

Næs, T., Indahl, U.G. (1998). A unified description of classical classification methods for multicollinear data. *Journal of Chemometrics* 12, 205–20.

Næs, T., Iskasson, T., Fearn, T., Davis, T. (2002). *A User-Friendly Guide to Multivariate Calibration and Classification*. Chichester: NIR Publications.

Pages, J., Husson, F. (2005). Multiple factors analysis with confidence ellipses: a methodology to study the relationships between sensory and instrumental data. *J. Chemometrics* 19, 138–44.

Ripley, B.D. (1996). Pattern Recognition and Neural Networks. Cambridge: Cambridge University Press.

Tucker, L.R. (1964). The extension of factor analysis to three-dimensional matrices. In Frederiksen, N., Gulliksen, H. (eds). *Contributions to Mathematical Psychology*. New York: Holt, Rinehart & Winston, 110–82.

Tucker, L.R. (1966). Some mathematical notes on three-mode factor analysis, *Psykometrica* (31) 279–311.

Westerhuis. J.A., Kourti, T., MacGregor, J.F. (1998). Analysis of multiblock and hierarchical PCA and PLS models. *Journal of Chemometrics* 12, 301–21.

Wold, H. (1982). Soft modelling: The basics and some extensions. In *Systems under Indirect Observation* (eds Jøreskog, K.G., Wold, H). Amsterdam: North Holland.

Wold, S. (1978). Cross-validatory estimation of the number of components in factor analysis and principal components models. *Technometrics* 20, 397–406.

15

Multiple Regression, Principal Components Regression and Partial Least Squares Regression

Regression methods relate two sets of data to each other and are among the most useful and most used statistical methods. They are particularly useful in sensory science where one of the main focus points is to find relations between data sets, for instance between sensory panel data and consumer liking data. This chapter starts by pointing out the importance of regression methodology before outlining the most useful methods with focus on the situation with several explanatory variables. The classical method of least squares (LS) estimation will be discussed together with some of its shortcomings when the explanatory variables are collinear. Methods developed for solving the collinearity problems will be considered with focus on why these methods represent an improvement. Some properties of the methods as well as techniques for validation and outlier detection will be discussed. Main focus will be on linear methods, but some attention will also be given to some simple generalisations which can be used to handle nonlinear relations and categorical response data. Relations between regression analysis and ANOVA will also be discussed briefly.

This chapter is based on concepts discussed in Chapter 11 and will be the basis for most of the methodology discussed in Chapters 4, 6 and 9.

15.1 Introduction

Simple linear regression analysis with only one X-variable (explanatory variable) and one Y-variable (response variable) was covered in Chapter 11. In the present chapter we will focus on the more general and realistic situation with several X-variables, and also discuss

Statistics for Sensory and Consumer Science Tormod Næs, Per B. Brockhoff and Oliver Tomic

methodology for situations with several Y-variables which among others is important for the preference mapping methodology in Chapter 9. We will give main attention to linear regression methods, but we will also show how these methods can easily be extended to handle nonlinear polynomials. Regression methodology for binary output will also be covered briefly since these methods are the basic building blocks for analysing choice based experiments (Chapter 8).

The importance of having regression methodology for several explanatory variables has its origin in what is sometimes referred to as the nonselectivity problem (Martens and Næs, 1989). This means that more than one explanatory variable are important to predicting and understanding the relationship to the response variable(s). In our context this means for instance that several sensory attributes are important for the purpose of predicting and understanding the liking of a product.

The linear regression situation with only one x and one y can easily be handled by calculating just variances and covariances of the variables as was shown in Chapter 11. The present situation with several x's and also possibly several y's is much more complex and in order to be able to describe this type of methodology properly one needs matrix algebra. The reader who is not familiar with this area can skip the paragraphs or sections below where matrices are presented without loosing the conceptual understanding of the methodologies.

For an illustration of the regression problem on a conceptual level we refer to Figure 15.1. Here we see that both a model and a data set containing both response values (y) and explanatory variables (x) are needed. The model is fitted to the data, possibly after a pre-processing (see e.g. Næs *et al.*, 2002). The model is evaluated and an outlier check is made. The model is then interpreted and possibly also used for prediction of new response values based on measurements of explanatory variables only.

We will here, as opposed to the treatment of ANOVA models in Chapter 13, mainly consider the situation with continuous explanatory variables. The two methodologies can both be considered as special cases of a general theory (Searle, 1971), but in this book, they will usually be treated separately. In Chapter 15.3 we will give a brief description of the relation between them, but this is not necessary for understanding the rest of the book. In the present chapter, we will briefly also discuss how to combine categorical and continuous explanatory variables, so-called analysis of covariance (ANCOVA, see e.g.

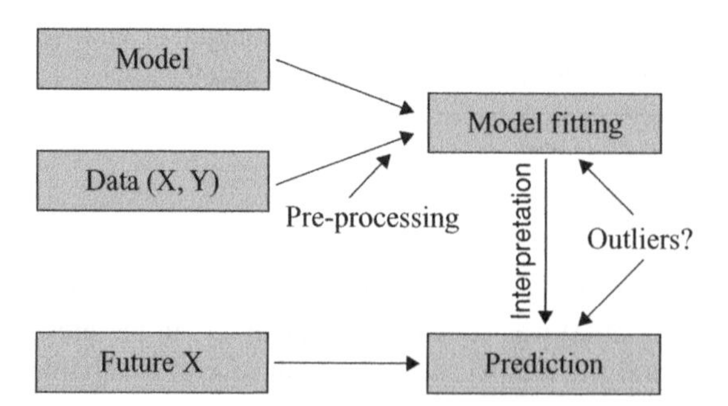

Figure 15.1 *Relating X data to Y-data for the purpose of prediction and improved understanding.*

Weisberg, 1985). When concerns model diagnostic tools, the two types of methods will be treated together in Chapter 15.10.

The main advantage of regression methods over ANOVA based methods in situations where one has a choice (see e.g. Chapter 8) is the smaller number of parameters in the models. Instead of letting each level of a variable be represented by its own parameter, the linear regression model uses only one parameter for each variable. This means that the estimates will be more precise if the model is adequate. The drawback is that more assumptions are made about the functional form of the relation and therefore unreliable estimates may be the result if the model gives a poor fit to the data.

15.2 Multivariate Linear Regression (MLR)

The multiple linear regression model for a single observation (n) can be written in the following way:

$$y_n = \beta_0 + \sum_{k=1}^{K} x_{nk}\beta_k + \varepsilon_n \tag{15.1}$$

Here y and the x's are the measured variables for object n, the β_0 is the so-called intercept, the other β-values are the regression coefficients for the X-variables and ε is the random error term. A regression coefficient equal to for instance 2 means that the Y-value on average increases by an amount of 2 for each increase of 1 unit for the corresponding X-variable. Note that if the X-variables are centred before analysis, the regression coefficients do not change, only the intercept. If the x's are centred, the intercept is equal to the average of y. If both x and y are centred, the need for an intercept vanishes completely. This fact will be used for PCR and PLS regression as discussed below in Sections 15.7 and 15.8.

As can be seen, this is a direct generalisation of the simple linear model (11.10). It is important to note that linear models will seldom fit exactly to real underlying relations, but experience has shown that they are often reasonable and useful for many different real world applications.

In matrix notation, the same formula can be written as

$$\mathbf{y} = \mathbf{X}\boldsymbol{\beta} + \boldsymbol{\varepsilon} \tag{15.2}$$

where $\mathbf{y}$ is the column of Y-values for the N objects, $\mathbf{X}$ is the matrix of X-variables

$$\mathbf{X} = \begin{pmatrix} 1 & x_{11} & x_{12} & & x_{1K} \\ 1 & & & & \\ . & & & . & . \\ . & & & . & . \\ 1 & x_{N1} & & & x_{NK} \end{pmatrix} \tag{15.3}$$

with a column of *1*'s incorporated to accommodate the intercept β_0. The $\boldsymbol{\beta}$ is the corresponding vector of regression parameters, now containing the intercept, and ε is the vector of residual random errors. Usually, the error terms are assumed to be independent and identically distributed with mean equal to 0 and variance equal to σ^2. Sometimes, in particular for testing and confidence interval purposes one also needs the assumption of normality of the error terms. Other possibilities also exist, but these are more difficult to handle mathematically. Tools are also available for checking these assumptions as will be discussed below. Many of the tests presented here are reasonably robust to small deviations from normality.

The main purpose of regression analysis is to find good estimates for the regression coefficients to be used either for prediction of new Y-values from measurements of new X-values or for direct interpretation. The classical solution to the estimation of β is to minimise the least squares (LS) criterion defined by

$$\sum_{n=1}^{N}\left(y_n - \beta_0 - \sum_{k=1}^{K} x_{nk}\beta_k\right)^2 \tag{15.4}$$

The solution can in matrix notation be written as

$$\hat{\beta} = (\mathbf{X}^{\mathrm{T}}\mathbf{X})^{-1}\mathbf{X}^{\mathrm{T}}\mathbf{y} \tag{15.5}$$

The methodology is also sometimes referred to as multiple linear regression (MLR). It can be shown that the optimal solution can also be found by maximising the correlation between y and linear functions of the x's.

For situations where the number of variables is moderate and the correlation between them is not too large, the formula in (15.5) represents a useful method, but in modern science these requirements are not always fulfilled. Quite often, also in sensory science, the number of variables may be large and the correlations among some of them may be very high. This can create stability problems of the regression coefficients (Gunst and Mason, 1977, 1979). This is often called the collinearity problem and will be given more attention below. In some cases, the number of variables can also be larger than the number of objects leading to mathematical rank problems when computing Equation (15.5).

An illustration of the collinearity problem in the situation with two X-variables and one y is given in Figure 15.2. In this case, the observations of the two X-variables lie close to a straight line which means that the correlation is high. Fitting a linear equation to the observations in this case is identical to fitting a plane to the data points. The plane will be stable along the direction of the first principal component in the $\mathbf{X}$-space (symbolised by

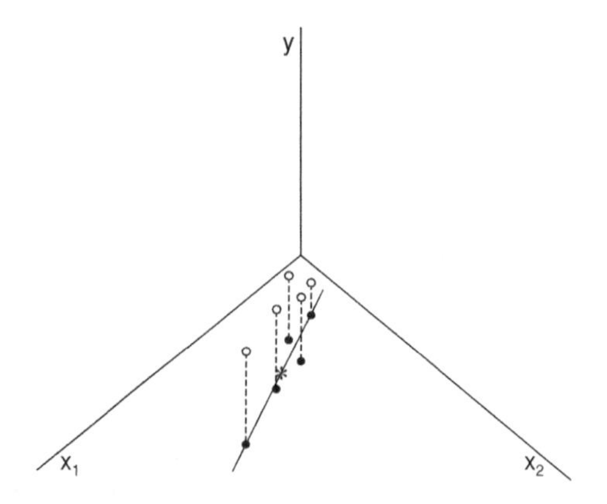

Figure 15.2 *Illustration of the collinearity problem in a situation with two x-variables. Martens and Næs (1989) Multivariate Calibration with permission from John Wiley and Sons.*

a line, i.e. the first principal component direction) since this direction represents a large variability. In the orthogonal direction, however, a small change in one or a few of the observations may create a large change in the fitted plane. This has a detrimental effect on the stability of the regression coefficients and it is known that it can also influence future prediction results strongly (Næs and Martens, 1989). Therefore, other regression solutions are needed, both for improving stability and for improving interpretations.

Possible ways of solving the collinearity problem will be discussed below in Sections 15.6, 15.7 and 15.8. First of all we will discuss the solution based on eliminating unimportant variables from the set of X-variables. With respect to the illustration in Figure 15.2, this would mean selecting only one of the variables for the regression: This is reasonable since the two contain more or less the same information. This is generally a sensible strategy, but in many cases, one will end up with very different variable sets depending on which variable selection method is used. This is particularly true when the number of objects is low. Therefore most of the focus in the present book will be on another strategy, so-called full-spectrum data compression regression, in particular the methods principal components regression (PCR) and partial least squares (PLS) regression. These are methods which compress the information down onto a few important components before regression, and then use only the variability along these directions in the regression.

The collinearity problem is most often diagnosed by looking at the eigenvalue structure (see Chapter 14) of the matrix $(\mathbf{X}^T\mathbf{X})$, which is the same as investigating the explained variances of the different components in a PCA of the matrix $\mathbf{X}$. If the largest eigenvalue is much larger than the smaller ones, there is collinearity in the data, i.e. the variance in the directions with the largest variability is much larger than the variance in the directions with small or moderate variability.

Once the regression coefficients are estimated, the prediction equation can immediately be established using Equation (15.1) with the β's substituted by $\hat{\beta}$. In other words, for new measurements of the x's, the corresponding Y-value can be predicted by

$$\hat{y} = \hat{\beta}_0 + \sum_{k=1}^{K} x_k \hat{\beta}_k \tag{15.6}$$

The variance of the error term is normally estimated by

$$\hat{\sigma}^2 = \sum_{n=1}^{N} (y_n - \hat{\beta}_0 - \sum_{k=1}^{K} x_{nk}\hat{\beta}_k)^2/(N - K - 1) \tag{15.7}$$

i.e. using essentially the same formula as can be found in Equation (11.12). This estimate plays an important role in the estimates of the variances of the regression coefficients. The covariance matrix of the regression coefficient estimators can be written as

$$\text{Cov}(\hat{\boldsymbol{\beta}}) = (\mathbf{X}^T\mathbf{X})^{-1}\sigma^2 \tag{15.8}$$

with the variance estimates on the diagonal. Substituting the σ with its estimate in (15.7) gives unbiased estimates of all variances and covariance. The variances of the predicted values can be written as

$$var(\hat{y}) = \mathbf{x}^T(\mathbf{X}^T\mathbf{X})^{-1}\mathbf{x}^t\sigma^2 \tag{15.9}$$

where the $\mathbf{x}$ is the measurement vector for a new sample.

The significance test of the different coefficients can be obtained by dividing the regression coefficient by the corresponding standard error estimates obtained from (15.8) and by comparing the ratio to a percentile in the t-distribution. Unfortunately, the standard errors of the regression coefficients do not have closed form solutions, and can only be computed by statistical software packages. An alternative to t-tests for the regression coefficients is to use an F-test obtained by comparing sums of squares, as was done in the ANOVA chapter, Chapter 13. Note that the ANOVA method also opens the possibility of testing for more general hypotheses, for instance for several coefficients at the same time. For testing single regression coefficients, the two tests are equivalent. In fact it can be shown that $t^2 = F$. Note that for the distribution results to hold exactly, one needs all the assumptions listed above for the error terms ε.

A much used method for model validation is the R^2 obtained as the square of the correlation between y and the predicted values of y for the data set. This can also be interpreted as the explained variance in percentage. If for instance R^2 is equal to 0.8, one possible interpretation is that the model can explain 80 % of the variability of the response y in the data set. The problem with R^2 is that it always increases as the number of explanatory variables increases. If the number of variables is high as compared to the number of objects, the R^2 will always be large. This means that R^2 is only reliable when the number of samples is large and the number of variables is small, for instance in the ratio of at least 4/1. A possible solution to this problem is to use the corrected version of R^2, adjusted R^2 or the C_p (see e.g. Weisberg, 1985), which is an estimate of the prediction ability taking this aspect into account. Good models have a C_p value close to the number of variables K (Mallows, 1973). A similar measure is the Akaike's information criterion (see e.g. Weisberg, 1985). An even better strategy is to use cross-validation or the method of predictions testing that will be discussed below.

15.3 The Relation between ANOVA and Regression Analysis

Both regression analysis and ANOVA can be formulated as special cases of a general linear model theory. The focus and the tools for interpretation and visualisation of results are different, but the basic theory is the same. In regular regression analysis, the **X**-matrix is based on measurements of continuous variables, while for ANOVA models, the **X**-matrix has a special structure based on so-called dummy variables (0/1 variables) corresponding to the different levels of the design variable(s). As an example, consider the one-way ANOVA model with one design factor at 2 levels. In this case one defines 2 dummy 0/1 variables, each variable corresponding to one of the levels of the design factor. The first dummy variable takes on a value of 1 when the observation comes from level one of the design and 0 elsewhere. The second dummy variable is constructed the same way. The **X** matrix in (15.3) for such a situation with three replicates for each factor level can then be written as

$$\mathbf{X} = \begin{pmatrix} 1 & 1 & 0 \\ 1 & 1 & 0 \\ 1 & 1 & 0 \\ 1 & 0 & 1 \\ 1 & 0 & 1 \\ 1 & 0 & 1 \end{pmatrix} \qquad (15.10)$$

where the first column corresponds to the intercept as before, the second column corresponds to the first level of the design factor and the third column to the second level of the design factor.

Denoting the regression vector $\boldsymbol{\beta}$ by

$$\boldsymbol{\beta} = (\mu, \alpha_1, \alpha_2)^T \tag{15.11}$$

and multiplying this with the **X** matrix (Equation (15.2)), the resulting model can be written as

$$y_{ir} = \mu + \alpha_i + \varepsilon_{ir}, \quad i = 1, \quad 2\ r = 1, 2, 3 \tag{15.12}$$

As can be seen, this is the same as the ANOVA model described in Chapter 13, Section 13.1. The parameters can as usual be estimated by LS, but in this model one needs a restriction on the parameter estimates in order to obtain a unique solution. The restriction usually applied is to assume that the sum of the coefficients is equal to 0. Another possibility is to omit one of the levels of the design factor in the model and consider the intercept as representing the omitted level. The classical ANOVA test for no effect of the design factor can be written as

$$H_0 : \alpha_1 = \alpha_2 \tag{15.13}$$

As described in Chapter 13, this hypothesis can be tested by an F-test.

15.4 Linear Regression Used for Estimating Polynomial Models

A few places in this book there is a need for modelling nonlinear relations. The simplest way of doing this is to use polynomial models. The main reason for this is that polynomial models can be estimated by the same linear methods as presented above, They are also quite flexible and easy to use.

An example of a polynomial model of order two in two X-variables is

$$y = \beta_0 + \beta_1 x_1 + \beta_2 x_2 + \beta_{11} x_1^2 + \beta_{22} x_2^2 + \beta_{12} x_1 x_2 + \varepsilon \tag{15.14}$$

As can be seen, there are, in addition to the linear terms, both quadratic terms and a product term in the model. As can also be seen, if we denote the three nonlinear terms by x_3, x_4 and x_5, the resulting model is essentially a linear model with 5 X-variables. The extension to several X-variables and to higher polynomial degree is obvious. It is strongly recommended to center the variables before forming quadratic and product terms. This makes both interpretation and estimation easier.

For situations with many collinear X-variables, the simplest way of incorporating non-linearities is to use polynomial functions of the principal components. In the case of two principal components, the only change needed to model (15.14) is that the X-variables are substituted by principal components scores. Note that mean centring is already done for the principal components. These types of models are extensively discussed in for instance Næs *et al.* (2002).

Polynomial models can be useful both for prediction and for interpretation. An important way of interpreting polynomial models is by the use of contour plots. These are plots that present the surface in terms of level curves or contours in the same way as level curves on a regular map of a forest or a mountain area. These are useful for identifying ‘peaks’, ‘valleys’ and other model characteristics. They are, however, only applicable for two-dimensional situations. Maximum and minimum points can be found by visual inspection of contour plots or by using tailor made formulae for the purpose (Myers and Montgomery, 1995). An example of a contour plot of a polynomial model is given in Figure 9.8. As can be seen, the relation between the response and the two X-variables is clearly nonlinear with maximum points at $x_1 = 0$, and $x_2 = 0$.

15.5 Combining Continuous and Categorical Variables

The simplest example of combining continuous and categorical variables is the linear model with one variable of each type, i.e.

$$y_{ni} = \beta_0 + \alpha_i + \beta_i x_n + \varepsilon_{ni} \tag{15.15}$$

where $i = 1{,}2$ denotes the two levels of the categorical variable and n denotes observation number. As can be noted, this corresponds to two different regression models combined in one; one model for each level of the categorical variable. The intercepts in the two models, $\beta_0 + \alpha_1$ and $\beta_0 + \alpha_2$, are different and so are also the regression coefficients β_1 and β_2. A graphical illustration of this model is given in Figure 15.3. If a common regression slope can be assumed for the different groups, the index i from the regression coefficient can be eliminated.

If we define the **X**-matrix as

$$\mathbf{X} = \begin{pmatrix} 1 & 1 & 0 & x_1 & 0 \\ 1 & 1 & 0 & x_2 & 0 \\ 1 & 1 & 0 & x_3 & 0 \\ 1 & 0 & 1 & 0 & x_4 \\ 1 & 0 & 1 & 0 & x_5 \\ 1 & 0 & 1 & 0 & x_6 \end{pmatrix} \tag{15.16}$$

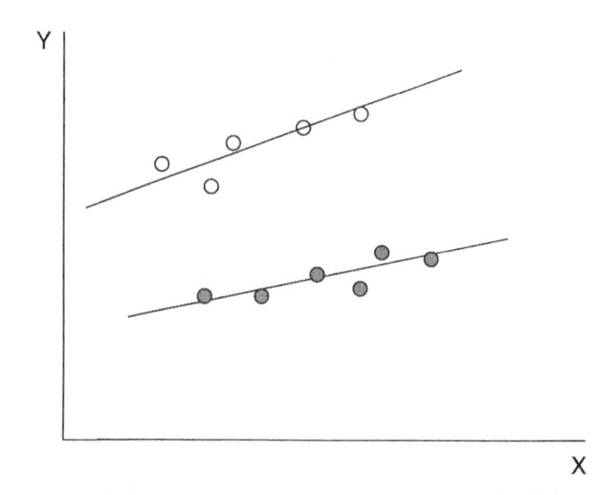

Figure 15.3 *Combining continuous and categorical variables. The plot shows two groups of data with different regression slope.*

and the regression vector as $\beta = (\beta_0, \alpha_1, \alpha_2, \beta_1, \beta_2)^T$, the model (15,.15) can be formulated as (15.2). As above we advocate centring the X-variables before modelling. The reason for this is that if not done, the test for equality of intercepts becomes quite meaningless. More interesting hypotheses are obtained by comparing the groups at the average X-value and this is exactly what is obtained by centring.

In some more complex cases, the number of X-variables is quite large. In such cases one can reduce the full set of variables before the regression by principal components and use the principal components instead of the original variables in the regression (Jørgensen and Næs, 2004). A more sophisticated method is to use the method LS-PLS (Jørgensen *et al.*, 2004) based on first fitting the categorical variables by LS and then using PLS on the residuals. The method has been shown to be better than the PCA approach from an interpretation point of view.

15.6 Variable Selection for Multiple Linear Regression

There exist a large number of variable selection methods in statistics. The main focus of these methods is to obtain improved interpretation and prediction in situations with many collinear variables, but in some cases it may also be important to reduce the number of variables to make future analyses simpler. If a smaller set of variables contains more or less the same information as a larger set, it is often advantageous to use the smaller set in future studies in order to save time and resources.

Important techniques are the forward and backward selection methods. The forward methods are methods for which the first step is to find the variable which has the strongest significant effect on y. The next step is to search for the variable which is most significant given that the first one is already in the model. This procedure continues until there are no more significant variables to be incorporated. The backward methods start with all (if that is possible) variables in the model. The first step is then to eliminate the variable which is least significant. This process continues until all variables left are significant at a pre-specified level. The two strategies will in most cases lead to different models. The explanation of this is that in cases with several collinear variables, there exist many models with more or less the same fit. It is a good practice to compare the various candidate models towards the end of the selection process by using criteria like for instance the R^2, C_p or cross-validation (CV see below, Section 15.9). In order to make the final choice, it is important to use these criteria together with prior knowledge. If for instance two models have more or less the same R^2, one will usually select the one which makes most sense from other points of view. For even better validation strategies we refer to the double CV methods discussed in Hjort (1993).

An alternative to the stepwise methods is to use best subset regression, which means that all possible models up to a certain size are compared using a criterion such as for instance R^2 or C_p. Typically, one will look at the best 5–10 solutions based on for instance up to 10 regressors in the model. Usually there will be several models with similar R^2 or C_p and the choice is usually not obvious. One will typically prefer models with a small number of variables with comparable fit or prediction ability. Cross-validation can also be useful for choosing the number of variables.

15.7 Principal Components Regression (PCR)

The variable selection methods have many nice properties, but it is generally easier to obtain good and interpretable solutions using one of the 'data compression' methods to be discussed next. The most basic method within this class of techniques is the principal component regression (PCR, see e.g. Næs and Martens, 1988). The idea behind this method is to base the regression analysis on the first few (A) principal components of $\mathbf{X}$ only (those with the largest variance). This means that instead of eliminating variables, PCR compresses the information down to a few linear combinations or components and uses only these components in the regression. All variability along the minor principal component axes are thus disregarded in the regression analysis. The method is based on the idea that it is often better to replace the information along the minor unstable components by 0 instead of using them directly in the model. Other strategies for selecting components can be found in Joliffe (1982).

The two models defining the PCR method are

$$\mathbf{X} = \mathbf{T}\mathbf{P}^{\mathrm{T}} + \mathbf{E} \tag{15.17}$$

$$\mathbf{y} = \mathbf{T}\mathbf{q} + \mathbf{f} \tag{15.18}$$

where the $\mathbf{X}$ and $\mathbf{y}$ are mean-centred input (explanatory) and output (response) matrices respectively. As can be seen, the index 1 is (because of mean centring) omitted from $\mathbf{T}$ and $\mathbf{P}$ as opposed to in Equation (14.6). The first model corresponding to the principal components model itself and the second corresponds to the regression step with the $\mathbf{q}$ vector being the vector of regression coefficients. The principal components corresponding to the dominating eigenvalues are computed using model (15.17) and these components are then used directly in the regression model (15.18). The regression coefficients are estimated using the LS criterion. The score matrix $\mathbf{T}$ is the only carrier of information from $\mathbf{X}$ over to $\mathbf{y}$, not the $\mathbf{X}$ matrix itself. Figure 15.4 illustrates the PCR estimation process.

Note that after the principal component scores $\hat{\mathbf{T}}$ have been computed the two equations (15.17) and (15.18) are very similar in the sense that the $\mathbf{P}$ and $\mathbf{q}$ can be computed by regressing $\mathbf{X}$ and $\mathbf{y}$ onto $\hat{\mathbf{T}}$ respectively. The same model can be used for several Y-variables, but as for MLR, the regression equations for the different Y-values are computed independently. How to determine the number of components to be used in Equations (15.17) and (15.18) will be discussed below in Section 15.9.

As soon as the model parameters, i.e. the loadings $\hat{\mathbf{P}}$ and the regression coefficients $\hat{\mathbf{q}}$, are computed, they can be combined to give the regression equation

$$\hat{y} = \bar{y} + \mathbf{x}^{\mathrm{T}}\hat{\mathbf{P}}\hat{\mathbf{q}} \tag{15.19}$$

The regression coefficients can then be interpreted as for the MLR. Most often, one will, however, be more interested in interpreting the PCR model estimates $\hat{\mathbf{P}}$, $\hat{\mathbf{q}}$ and $\hat{\mathbf{T}}$, using scatter plots. The estimates of the residuals $\mathbf{E}$ and $\mathbf{f}$ obtained by comparing the true values with the estimated model values are sometimes also used for detection of outliers (see below, Section 15.10).

In most cases when PCR is used in this book, the focus is mainly on the interpretation of the first few components in the equation. Typical situations where this is the case can be

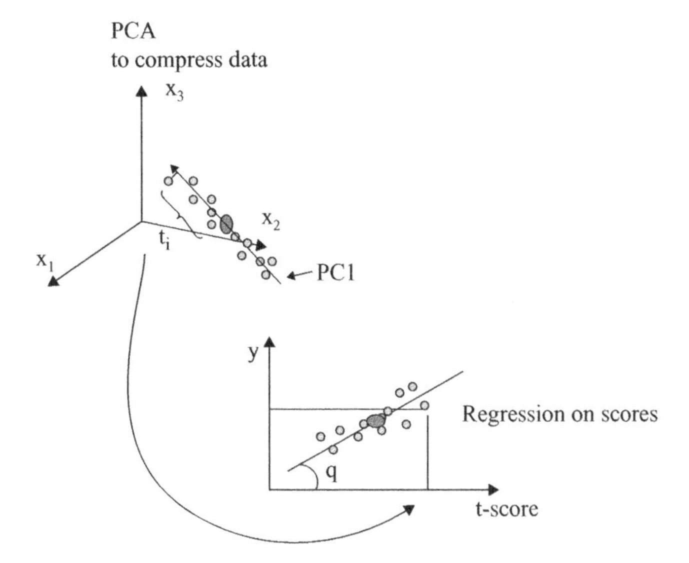

Figure 15.4 PCR for data compression and regression. First, the PCA is used on the X-data (upper left part) and only the information along the first few components (here only the first) is used for regression vs. the response y (lower right part).

found in Chapter 9. Again referring to practical experience, these few components often carry a large portion of the most important information in the data set.

The PCR method solves the collinearity problem in an elegant way and provides a number of good tools for interpretation. It is also well understood theoretically (see e.g. Næs and Martens, 1988; Gunst and Mason, 1977, 1979) and has shown to perform very well both for prediction and interpretation in many practical applications also within sensory science. It is also easy to compute and easy to modify for more complex relations. An example of the latter is given in Chapter 9 when discussing ideal point preference mapping. In that case, the principal components are combined using a polynomial model.

A possible problem with the PCR is that all the components in model (15.17) are extracted based on **X** only. This may be a drawback in situation where the first few components of **X** have less relation to y than the components with minor variability. This is quite common in for instance spectroscopy, but in sensory science this effect is usually less important. Therefore for most purpose, the PCR will be sufficient for solving the collinearity problem and for providing useful tools for interpretation. A possible improvement over the PCR that has been proposed is the PLS regression method to be discussed in the next section.

15.8 Partial Least Squares (PLS) Regression

The PLS regression (Martens and Næs, 1989; Wold *et al.*, 1983) is based on the same general model structure as PCR (models (15.17) and (15.18)), the only difference is that the components $\hat{\mathbf{T}}$ are computed in another way.

The PLS components are obtained by maximising the covariance between linear functions of **X** and the response **y** (both centred as for PCR). Then the effect of the first factor is subtracted from the **X** and **y** and the residuals are used for computing the second component using the same covariance criterion. The procedure continues until the desired number of components, A, has been extracted. The components $\hat{\mathbf{T}}$ obtained by the PLS regression are orthogonal.

This way of computing components ensures that the components extracted are more relevant for the prediction of **y** than the principal components. This may sometimes lead to models with a smaller number of components, which may possibly be easier to interpret, but from a prediction point of view the difference between the two methods is usually small. In sensory and consumer science the difference between the two methods is also often quite small from an interpretation point of view.

The covariance criterion is a compromise between the variance criterion used for PCR and the correlation criterion used for MLR (see above, Section 15.2). This means that PLS is a compromise between the very stable and conservative PCR and the MLR which uses the Y-information as actively as possible.

For the multivariate case, PLS regression maximises the covariance between linear functions of **X** and linear functions of **Y**, i.e. it maximises

$$\text{cov}(\mathbf{Xa}, \mathbf{Yb}) \tag{15.20}$$

over both the vectors **a** and **b**. Note that the PLS solution for several Y-values is not obtained by separate fitting of each individual Y-variable. For interpretation purposes it is usually advantageous to use all Y-variables simultaneously, but for obtaining good predictions, the best choice is often to treat each variable separately. For several Y-values, the PLS method is often referred to as PLS2.

PLS regression is easy to compute, in particular for the univariate case, where no iteration is needed, it solves the collinearity problem and offers the same interpretation tools as for PCR. The tools are used the same way, so from a user's point of view the two methods look very similar. The PLS algorithm also works for missing data.

Also for PLS the use of bi-plots and correlation loadings is possible. The PLS method can be extended to several blocks of variables in the same way as the PCA (see Tucker-1, Chapter 14), but this is not considered further here (Westerhuis *et al.*, 1998).

In Figure 6.2 is given an example of score and loadings obtained by the PLS method.

15.9 Model Validation: Prediction Performance

For all the methods above it is important to be able to assess the prediction ability, i.e. to determine how well the X-data can predict the Y-data. For the PCR and PLS methods it is also important to determine the best choice of the number of components for prediction purposes. In many cases in sensory and consumer science, however, the goal is less ambitious with a main focus on the validity of the first few components (see also Chapter 14).

The simplest method for obtaining this information is the so-called prediction testing, which is based on splitting the data in two sets, building the model on one of them and

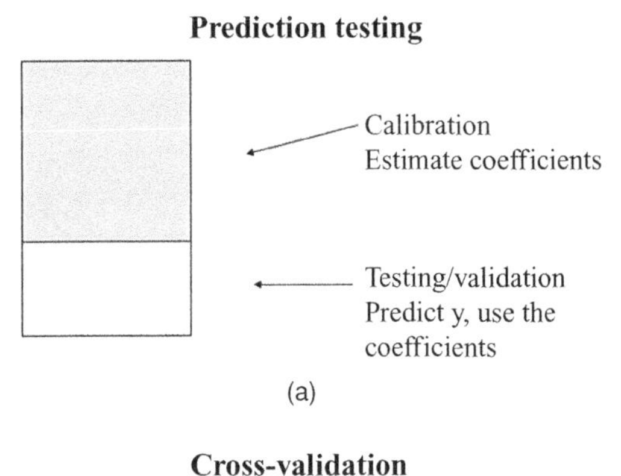

Cross-validation

Model fitting, find y=f(x)
estimate coefficients

Predict y, use the coefficients

(b)

Figure 15.5 *Validation, prediction testing vs. cross-validation. The illustration shows conceptually the differences between a) prediction testing and b) cross-validation. In cross-validation, all test samples are taken from the same set, while for prediction testing the equation is tested out on a totally unused data set.*

testing it out on the other (see Figure 15.5). If the data set is too small for splitting, cross-validation can be used instead. This is a method which does prediction testing several times, each time keeping one sample out of the data set, calibrating on the rest and testing it out on the single sample kept out (see Figure 15.5, Stone, 1974). This is repeated with replacement until all objects have been tested. Sometimes one can take out blocks of samples, so-called segmented cross-validation. A commonly used criterion for model quality is the root mean square error of prediction (RMSEP) defined by

$$RMSEP = \sqrt{1/N \sum_{n=1}^{N} (y_n - \hat{y}_n)^2} \tag{15.21}$$

i.e. the square root of the averaged squared differences between the measured and predicted response value. Small values of the *RMSEP* are favoured since these indicate small differences between measured and predicted *Y*-value. Sometimes one uses the square value *MSEP*.

For practical purposes it is often better to report the explained variance of an equation obtained by comparing the M*SEP*'s with the *MSEP* obtained by using the averages of the *Y*-values for prediction. This is typically done by computing the difference between the *MSEP* for the mean minus the *MSEP* for the actual predictor and dividing this differences

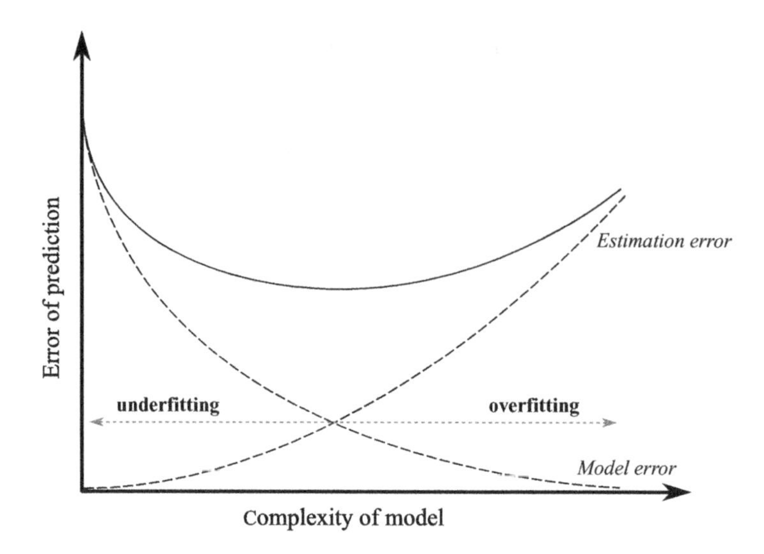

Figure 15.6 *Example of prediction ability vs. number of components. When building a model, the estimation error increases and the model error decreases as the model size increases. The prediction error is a sum of the two types of error. Finding the optimal complexity is a trade-off between the two effects is important and is usually done by some sort of validation. The optimal value for prediction purposes usually lies between the two extremes. Adapted from Martens and Næs (1989) John Wiley and Sons.*

by the *MSEP* for the mean. The higher the explained variance is, the better the equation is at predicting y. A closely related alternative is to use the correlation coefficient between measured and predicted y-values (from prediction testing or cross-validation), which is a prediction analogue of the standard R^2 discussed above.

The principle behind these methods is that it is only when the equation is tested out on new data that the prediction ability can be properly assessed. It can be shown by using theory and examples that testing the equation out on the same data that were used for model estimation will give overoptimistic estimates of the prediction ability.

When concerns determining the best number of components for prediction purposes, one will usually repeat the prediction testing or cross-validation for several values of A and select the best solution. The values are often presented in a plot of the prediction error (RMSE) vs. the component number (Figure 15.6). If two values of A give similar results one will normally select the model with the smallest number of components. The general tendency is that when using either too few (underfitting) or too many (overfitting) components, the prediction ability is poor. The optimal is usually in the middle representing a good compromise between the contributions from the estimation bias and estimation variance. A statistical test has been developed for checking for significant difference between the prediction ability of two models (CVANOVA, Indahl and Næs(1998)). When focus is on the first few components for interpretation only, one is primarily interested in how much the two components explain in a cross-validation sense of the total variability.

It should be mentioned that in some cases, even the use of cross-validation may become complicated and even questionable. For very small data sets, and in particular when the samples are obtained using a designed experiment, each sample is unique and it is often difficult to predict this unique information from the rest of the samples. The method of cross-validation is therefore best suited for situations where the samples span a natural space of variability and when the number of samples is not too small. Note that these considerations become more serious as the number of components increases.

Even in situations where CV is not fully appropriate (see Chapter 9), the PCA and PLS scores and loadings can still be meaningful if one simply considers them as graphical representations of the data without any further generalisation property.

15.10 Model Diagnostics and Outlier Detection

15.10.1 Multiple Regression and ANOVA Models

In this chapter we will discuss both multiple linear regression and ANOVA models, since they are essentially treated the same way when concern model diagnostics and outlier detection. For both types of models there are two important aspects; the first is to check the model assumptions and the second has to do with the detection of outlying and influential observations. Both are important for the validity of the conclusions drawn from an analysis. Usually, one will check for outliers first since these can potentially have a very harmful effect on all the results obtained. If serious outliers are detected, proper actions must be taken as will be discussed below. If, not, one will normally proceed to check the assumptions behind the model used.

15.10.1.1 Model Assumptions Made

For regression models the most important assumption to check is the functional form assumed. This is important for all purposes and all possible applications. For full ANOVA models, with all interactions present, this aspect is of less interest since each level of a factor has its own average without any assumed relation to the other levels.

The following assumptions are usually made for the random errors in regression and ANOVA models (see above, Section 15.2).

1. normality
2. homogeneity of the variances
3. independence

For coefficient estimation and prediction purposes in regression, these are usually of minor importance from a practical point of view, but when computing standard errors, significance tests etc. they become important. The tests discussed in this book are reasonably robust to small deviations from the assumptions, but it is recommended that this aspect is always given some attention in order to detect large deviations.

15.10.1.2 Residuals and Leverage

There are basically two different tools that are used for checking model assumptions and for detecting influential observations; the residuals (or variants thereof) and the leverage. The residuals are obtained by simply calculating the difference between the measured and predicted response values

$$\hat{\boldsymbol{\varepsilon}} = \mathbf{y} - \hat{\mathbf{y}} \tag{15.22}$$

These residuals are estimates of the true errors and are very useful for investigating both general model assumptions and for detecting outliers and are computed in most modern software packages.

It turns out, however, that the (theoretical) variance of these residuals is generally not homogeneous. It can be shown that the variance of the residual $\hat{\varepsilon}_n$ is equal to:

$$Var(\hat{\varepsilon}_n) = \sigma^2(1 - h_n) \tag{15.23}$$

where σ^2 is the model error variance and h_n is the so-called *leverage* (see Chapter 14) for observation n. A possible modification of the standard residuals in order to correct for the differences in variance is to use the studentised residuals

$$\hat{\varepsilon}_n^{Stud} = \hat{\varepsilon}_n / \hat{\sigma}\sqrt{1 - h_n} \tag{15.24}$$

which has a stable variance over all observations. For situations without outliers or serious model problems, approximately 5 % of the studentised residuals should be outside the interval between −2 and 2.

The leverage is itself a useful tool for detecting outliers. It is a measure of the distances in X-space from the n'th observation to the mean of all observations. So-called X-outliers, i.e. samples which deviate from the rest in the X-space are the samples with very high leverage value. Note that the leverage is close to 0 for points near the centre and close to 1 for the points furthest away from the centre. High leverage (atypical) observations are potentially highly influential on the results of the analysis (Cook and Weisberg, 1982). For regular ANOVA models applied to designed data, the leverage values are all moderate and seldom used.

15.10.1.3 Different Ways of Using Residuals and Leverages

The residuals (or studentised residuals) are usually presented in various types of plots for detecting model problems. The most frequent choices are histograms, *q-q* plots (see Chapter 11), plots of residuals versus predicted values and plots of residuals versus the values of the factors in the model.

For detecting lack of model fit, the most useful tool is the plot of the predicted values vs. the residuals (see Figure 15.7). This plot is also very useful for detecting outliers. Another possibility for the same purposes is to plot the residual vs. observation number or residuals vs. one of the input variables.

For detecting lack of normality, a quick and useful check is the *q-q* plot as discussed in Chapter 13. A deviation from normality will result in points deviating from a straight line tendency. This type of plot is also useful for detecting outliers. A histogram can also be used. For more formal tests of normality one can use Kolmogorov Smirnov tests (Weisberg, 1985). Figure 15.8 shows an example of a *q-q* plot of regression residuals which indicates an error distribution close to a normal one.

For checking homogeneity of variance it is useful to plot predicted values versus the values/levels of the quantitative/qualitative factors in the model. If a systematic tendency is observed regarding larger residual in one end as compared to the other, this is an indication of heterogeneous random error (so-called heteroscedasticity).

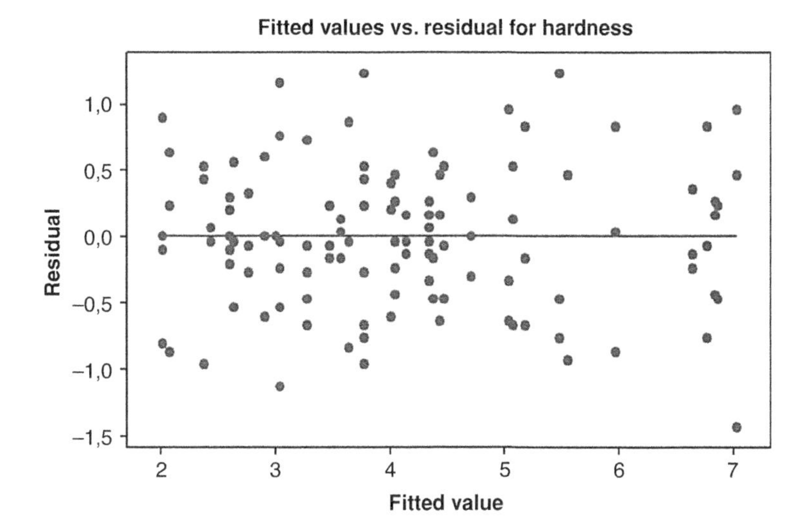

Figure 15.7 *Example of a plot of residuals vs. predicted value for a sensory analysis.*

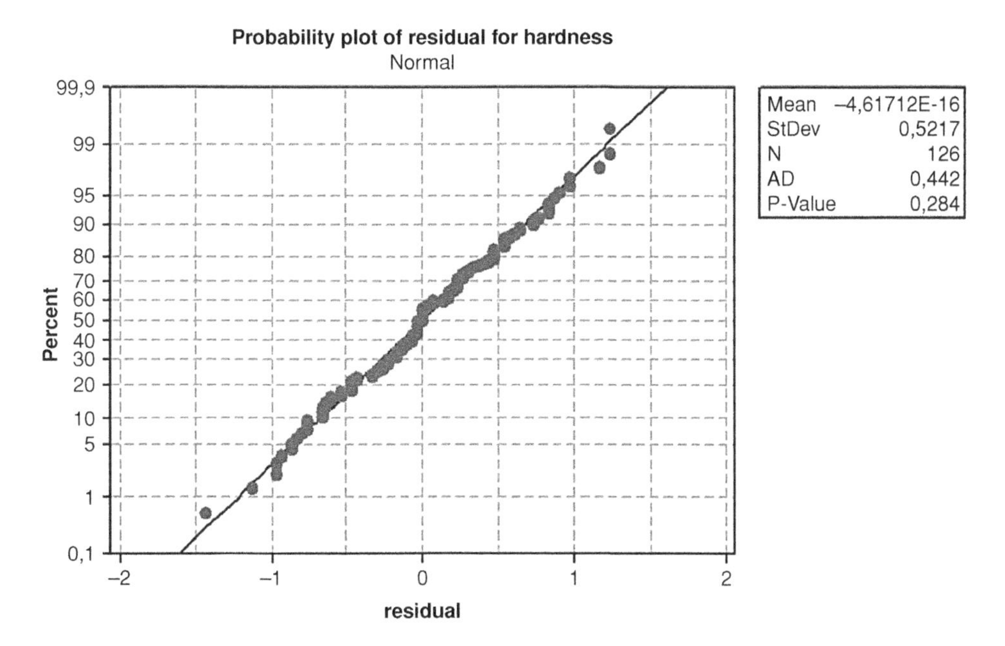

Figure 15.8 *Example of a q-q plot of residuals from a sensory analysis. The straight line indicates the line corresponding to exact normal distribution. As can be seen, for this data set all points lie relatively close to the straight line and the distribution can safely be determined as very close to the normal one.*

For checking independence of the residuals, the Durbin-Watson statistic can sometimes be useful (Durbin and Watson, 1951, 1971).

For an observation to be really dangerous for the model, one needs a large leverage in addition to a large residual. One can plot leverage and residuals separately or against each other in influence plots (Martens and Næs, 1989) or one can simply combine them in the regular Cooks' influence measure (Cook and Weisberg, 1982).

If a possible outlier is found, the most natural thing to do first is to try to understand the reason and correct errors if possible. If no reason is found, one will typically check for influence using for instance the Cook's influence measure. If the observation has little influence it matters little what is done, so the most natural thing to do is to leave the point in the model. If the observation is highly influential, one has to make a decision. In most cases, the best is to eliminate the observation, and report that one presents the results based on the main part of the data set leaving the rest as outliers. This is usually better than having an analysis depending strongly on only one or a few observations.

For more information about model diagnostics and how to detect outliers, we refer to Weisberg (1985). See the same reference for transformations that can sometimes be useful to remedy model problems.

15.10.2 PCR and PLS Models

For PCR and PLS models, which are here mainly used for prediction and interpretation, main focus is on model shape and detection of outliers and not so much on random error distribution.

Since data compression for both these methods is done for the X-data according to model (15.17), both the X-residuals, scores plot and the leverage for different number of components are available in the same way as for PCA described in Chapter 14. Note that the main difference from MLR lies in the availability of the X-residuals and the scores.

When relating scores to y in the regression step (model (15.18)), regression residuals are obtained as

$$\hat{\varepsilon} = \mathbf{y} - \hat{\mathbf{y}} = \mathbf{y} - \mathbf{X}\hat{\mathbf{b}} \tag{15.25}$$

As for X-residuals, the Y-residuals can be computed for different number of components.

Usually, one will concentrate on the residuals and leverage for the optimal number of components as determined by CV or prediction testing, but components in vicinity are also useful to look at. These tools are all easy to compute and are easily available in software. Plotting is done as for MLR, either vs. the observation number or vs. the predicted values.

15.11 Discriminant Analysis

Discriminant analysis is a methodology used to develop classification rules that can be used for interpreting differences between classes/clusters of observations and for allocating new objects to the best fitting class. In sensory analysis, this type of analysis is not as important as in many other branches of food science (for instance in spectroscopy), but for some cases within consumer segmentation it may have some relevance (see Chapter 10).

Discriminant analysis is sometimes also referred to as supervised classification. The name supervised is used because all the techniques assume that there exist a number of pre-defined classes of observations that are used for training purposes. The scope of discriminant analysis is then to establish the relation between the vectors of measurements **x** and the class membership. In this way, discriminant analysis can be viewed as a kind of categorical regression analysis. Note that the major difference between discriminant analysis and cluster analysis (see Chapter 16) is that for the latter no prior information about classes is available prior to analysis.

A large number of different methods have been developed for discriminant analysis (Ripley, 1996). The most frequently used are probably the linear discriminant analysis (LDA) and quadratic discriminant analysis (QDA). These are both based on modelling each of the classes using multivariate normal distributions and then developing a classifier that selects the class with the largest a posteriori probability given the actual measurement vector.

Another type of discriminant analysis is based on direct use of regression analysis, either standard MLR or PLS regression. Theoretically they may be less attractive than the LDA, QDA and logistic regression, as will be discussed below, but from a user's point of view they are very simple and useful. For these methods, one simply defines the Y-variables as dummy variables representing the different classes. For instance, in the situation with three classes and three objects in each class, the Y-matrix can be written as

$$\mathbf{Y} = \begin{pmatrix} 1 & 0 & 0 \\ 1 & 0 & 0 \\ 1 & 0 & 0 \\ 0 & 1 & 0 \\ 0 & 1 & 0 \\ 0 & 1 & 0 \\ 0 & 0 & 1 \\ 0 & 0 & 1 \\ 0 & 0 & 1 \end{pmatrix} \tag{15.26}$$

The relation between **Y** and **X** is then analysed for instance by regular PLS2 regression. The scores and also the loadings can be plotted the usual way in order to understand the relation between the segments and the different X-variables in the data set. If this type of method is used for allocating new samples to the different classes, one will typically predict all Y-values and put the new sample in the class with the largest predicted Y-value. A better possibility even is to use a regular discriminant analysis method, like LDA or QDA, using the PLS scores as input (Indahl *et al.*, 1999).

15.12 Generalised Linear Models, Logistic Regression and Multinomial Regression

Generalised linear models (GLM, see e.g. MacCullagh and Nelder, 1983, 1989; Olsson, 2003; Fahrmeir and Tutz, 2001) is a useful class of methods for analysing the relation between a categorical response and a number of explanatory variables (either continuous or categorical). This is a large and well developed area within modern statistics and comprises a number of methods for different types of applications, one of them being discriminant analysis as discussed above.

The simplest GLM method is logistic regression which is appropriate when the outcome is a dichotomous variable with values 0 and 1. An approach often taken is to simply model the probability of the outcome 1 as a function of the explanatory variables **x**. A linear function is not fully satisfactory since this may give values outside the actual region between 0 and 1. A number of other solutions are therefore proposed, but the most frequently used is the so-called logit transform. Using this function, the regression equation for the probability becomes

$$P(Y = 1) = \frac{exp(\beta_0 + \sum_{k=1}^{K} x_k \beta_k)}{1 + exp(\beta_0 + \sum_{k=1}^{K} x_k \beta_k)} \tag{15.27}$$

As usual, the most important aspect of a regression model is to compute estimates for the regression coefficients, compute standard errors and to perform significance tests. This is here done by setting up the likelihood (i.e. joint probability distribution of all variables) for the data and then maximising this likelihood in terms of the unknown parameter vector $\boldsymbol{\beta}$. The solution, which is called the maximum likelihood (ML) estimate, can be found in standard software packages. Statistical tests are obtained by using the Wald test (see e.g. Olsson, 2003), where each coefficient is divided by the corresponding estimate of the standard error, or by using the so-called deviance. A full set of diagnostic tools and goodness of fit measures are also developed for this method. Details can be found in McCullagh and Nelder (1989).

When several categorical outcomes are possible, it is natural to use a multinomial model. It is possible to extend the above methodology to such situations. Sometimes one distinguishes between two types of such models, models with no ordering of the outcomes (nominal variable) and one in which there is an ordering (ordinal variable). Common for these situations is that one models the probability for each of the outcomes using a similar transform of the external variables as used in Equation (15.27). Regression coefficients are found the same was as for the standard logistic regression. This type of model will not play an important role in this book, but will be touched upon in the chapter about consumer choice experiments (Chapter 8). For these situations the consumer is asked to select one of several possible outcomes and one is interested in how the conjoint factors (here **x**) are related to the different outcomes of the experiment, i.e. how they are related to the probability of choosing the different products.

This type of models can also be useful for modelling the relation between consumer attributes (demographics, attitudes etc.) and consumer segments (Chapter 10). If a number of segments have been identified, the relation between the segments can easily be studied by the use of these methods.

If there are several and collinear regression variables, a possibility is to use a generalisation of PLS regression developed for the generalised linear models. A weakness with these methods is they are not yet fully developed for significance testing of the regression coefficients.

References

Cook, R.D., Weisberg, S. (1982). *Residuals and Influence in Regression*. New York: Chapman & Hall.

Durbin, J., Watson, G.S. (1951). Testing for serial correlation in least squares regression II, *Biometrika* 38, 159–78.

Durbin, J., Watson, G.S. (1971). Testing for serial correlation in least squares regression, III, *Biometrika* 58, 1–19.

Fahrmeir, L., Tutz, G. (2001). *Multivariate Statistical Modelling Based on Generalized Linear Models*, Springer Series in Statistics. New York: Springer.

Gunst, R.F., Mason, R.L. (1977). Biased estimation in regression. An evaluation using mean squared error. *Journal of American Statistical Association* 72, 616–28.

Gunst, R.F., Mason, R.L. (1979). Some considerations in the evaluation of alternate prediction equations. *Technometrics* 21, 55–63.

Hjort, J.S.U. (1993). *Computer Intensive Statistical Methods: Validation, Model Selection, and Bootstrap*. London: Chapman & Hall/CRC.

Joliffe, I.T. (1982). *Principal Components Analysis*. New York: Springer Verlag.

Jørgensen, K., Næs, T. (2004). A design and analysis strategy for situations with uncontrolled raw material variation. *J. Chemometrics* 18, 45–52.

Jørgensen, K., Segtnan, V., Thyholt, K., Næs, T. (2004). A comparison of methods for analysing regression models with both spectral and designed variables. *J. Chemometrics* 18, 451–64.

McCullagh, P., Nelder, J.A. (1983). *Generalized Linear Models*. London: Chapman & Hall.

McCullagh, P., Nelder, J.A. (1989). *Generalized Linear Models* (2nd edn). London: Chapman & Hall/CRC.

Mallows, C.L. (1973). Some comments on Cp. *Technometrics* 15, 661–76.

Martens, H., Næs, T. (1989). *Multivariate Calibration*. Chichester: John Wiley & Sons, Ltd.

Næs, T., Martens, H. (1988). Principal components regression in NIR analysis. *J. Chemometrics* 2, 155–67.

Næs, T., Iskasson, T., Fearn, T., Davis, T. (2002). *A User-Friendly Guide to Multivariate Calibration and Classification*. Chichester: NIR Publications.

Olsson, U. (2003). *Generalised Linear Models: An Applied Approach*. Lund, Sweden: Studentlitteratur AB.

Ripley, B.D. (1996). *Pattern Recognition and Neural Networks*. Cambridge: Cambridge University Press.

Searle, S.R. (1971). *Linear models*. John Wiley and Sons. New York.

Stone, M. (1974). Cross-validatory choice and assessment of statistical precision. *J. Roy. Stat. Soc*, B. 39, 111–33.

Weisberg, S. (1985). *Applied Linear Regression*. New York: John Wiley & Sons, Inc.

Westerhuis, J.A., Kourti, T., MacGregor, J.F. (1998). Analysis of multiblock and hierarchical PCA and PLS models. *Journal of Chemometrics* 12, 301–21.

Wold., S., Martens, H., Wold, H. (1983). The multivariate calibration problem in chemistry solved by the PLS method. *Proc. Conf., Matrix Pencils* (A. Ruhe, B. Kågström, eds). March 1982. Lecture Notes in Mathematics. 286–93. Heidelberg: Springer Verlag.

16

Cluster Analysis: Unsupervised Classification

In sensory and consumer science it is often of interest to detect similarities between individuals in their response pattern. A number of such examples and challenges are discussed in Chapter 10. In this chapter some of the most basic techniques used for cluster analysis and segmentation will be presented. We will distinguish between the hierarchical methods and the partitioning methods and discuss how the results from the two can be validated and interpreted. The hierarchical methods are usually based on Euclidean or Mahalanobis distances between objects while the partitioning methods can also be used for some alternative distances with special importance in consumer science. The use of the residual distance in fuzzy clustering is such an example. How to detect outliers and how to determine the number of clusters will be given some attention. The last section of the chapter is devoted to cluster analysis of matrices.

The Chapter 10 depends strongly on the theory of this chapter. The present chapter is also strongly linked to Chapter 14.

16.1 Introduction

An important problem area within multivariate statistics is to identify clusters in the data set, either clusters of objects or clusters of variables, without using prior information. This is usually called cluster analysis or unsupervised classification (Mardia *et al.*, 1979; Ripley, 1996). Within the framework of this book, the most typical example of cluster analysis is segmentation of consumers for improved understanding of individual differences (see Chapter 10), but the problem is also relevant within the area of quality control of sensory analysis as discussed in Chapter 3. An illustration of the concept of cluster analysis is given in Figure 16.1.

Statistics for Sensory and Consumer Science Tormod Næs, Per B. Brockhoff and Oliver Tomic

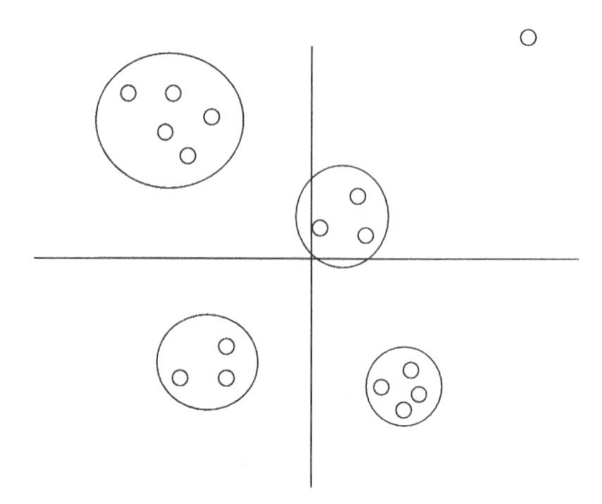

Figure 16.1 The idea of cluster analysis illustrated by four different clusters and an obvious outlier. *As can be seen, there are 4 quite clear clusters in this data set and one single outlier in the upper part of quadrant 1. The goal of cluster analysis is to identify this type of tendencies in data sets with several more variables than the two used for illustration here.*

The area of cluster analysis consists of a large number of methods and techniques, which can be divided into two main categories, the so-called hierarchical methods and the partitioning methods. In the following we will discuss the two subclasses separately.

An important aspect of cluster analysis within the area of consumer and sensory science is that one can seldom expect to find clearly separated clusters. This is particularly true in consumer studies where one typically finds overlap regions where no clear cluster memberships can be identified. Therefore the main aim of cluster analysis is here merely to identify those individuals that are most similar to each other, not necessarily to find clearly separated clusters.

Also for cluster analysis, principal components analysis (PCA) is an important technique (Figure 10.1 in this book and Næs *et al.*, 1998). First of all it can be used for visual clustering of objects, using the scores and loadings plots, without the use of more formal statistical procedures. An important example of this is within preference mapping (Chapter 9) where a visual splitting into subgroups based on the first few principal components is often sufficient. Another use of the PCA in cluster analysis is for improved interpretation of clusters already identified by another algorithm. This is typically done by the use of different symbols for the different clusters in the scores plot and comparison of scores plot with the loadings plot (Næs *et al.*, 1998).

Usually, cluster analysis is performed directly for the multivariate observation vectors, but some clustering methods can also be used for other types of data, for instance matrices and residuals from various types of regression methods. Below we will first concentrate on the former of these situations describing the important class of hierarchical clustering based on distances defined prior to analysis. After that we will discuss some important partitioning methods which can both be used for vector valued observations and more complex implicitly defined distances, for instance regression residuals. As is discussed in Chapter 10, the latter is very important for applications with segmentation of consumers both in conjoint analysis and in preference mapping. According to Wajrock *et al.* (2008), partitioning methods are superior to hierarchical methods in his context, but in our opinion it is too early to draw such a clear conclusion and therefore both approaches will be discussed here.

We will concentrate on clustering of objects, not of variables and we will use the generic symbols N and K for the rows (objects) and columns (variables) respectively. The observation vector for object n is denoted $\mathbf{y}_{\mathrm{n}} = (y_{n1}, y_{n2}, \ldots, y_{nk})^T$. For methodology developed particularly for clustering of variables we refer to Qannari *et al.* (1997) and to Berget *et al.* (2005).

16.2 Hierarchical Clustering

The starting point for the hierarchical clustering methods is always a matrix of distances between the objects to be clustered. A number of different distances are possible, but the most common is the Euclidean distance d_{ij} defined by

$$d_{ij} = sqrt\left(\sum_{k=1}^{K}(y_{ik} - y_{jk})^2\right) \tag{16.1}$$

where i and j represent two different objects. In two- and three-dimensional space, this distance corresponds to the usual distance as can be measured by a ruler. Another important distance is the Mahalanobis distance defined by

$$d_{ij} = sqrt((\mathbf{y}_{\mathrm{i}} - \mathbf{y}_{\mathrm{j}})^{\mathrm{T}} \hat{\boldsymbol{\Sigma}}^{-1} (\mathbf{y}_{\mathrm{i}} - \mathbf{y}_{\mathrm{j}})) \tag{16.2}$$

where the $\hat{\boldsymbol{\Sigma}}$ is the estimated covariance matrix of the vectors $\mathbf{y}$ (see Mardia *et al.*, 1979). The difference between the two distances is that the latter weighs the contribution in the different directions relative to the variability in the corresponding directions as measured by the covariance matrix. Usually, the distances are collected in a distance matrix

$$\mathbf{D} = \{d_{ij}\} \tag{16.3}$$

where i,j again denotes different objects.

The difference between the various hierarchical methods is the way they use these distances, in particular how they define the distances between an object and a cluster and between two clusters. Below we will discuss two such methods.

Regardless of which distance measure is used, the first step is always to identify those objects that are closest. These two objects are then put in one cluster, ending up with N-1 clusters, one consisting of two objects and the others consisting of only one object each. In the next step the goal is again to find the objects (or clusters) which are the closest using the same criterion. The process thus continues until all objects are collected within the same cluster or until the desired number of clusters is reached.

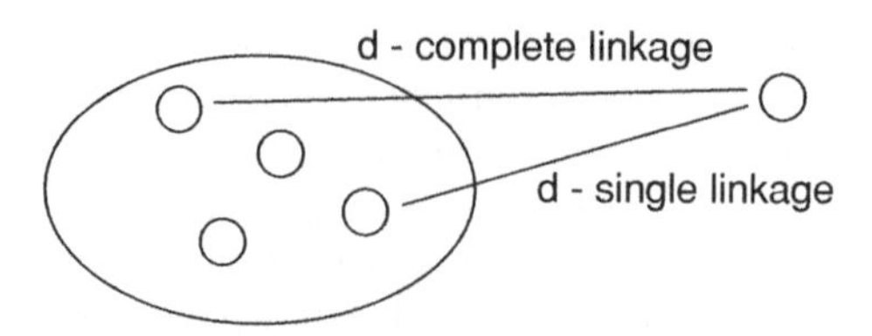

Figure 16.2 ***Complete and single linkage illustrated.*** *The distance from a sample to a group is for compete linkage defined as the distance between the actual sample and the sample in the cluster which is furthest apart. For single linkage the corresponding distance is defined as the distance between the actual sample and the sample in the cluster which is closest.*

16.2.1 Complete Linkage

This is probably the most common of the hierarchical methods and defines the distance from an object to a cluster as the distance between the actual object and the object in the cluster which has the largest distance to it (Mardia *et al.* (1979) and Figure 16.2). When comes to distances between two clusters, the definition is the same; the distance between the individual objects within the two clusters that are farthest apart. An important property of this method is that it generally provides compact and rather equally sized clusters. The reason for this is that it is easier to join two objects than joining an object with a cluster. The method is sometimes also referred to as furthest distance clustering.

16.2.2 Single Linkage

For this method the distance from an object to a cluster is defined as the distance between an object and the object within the cluster which is closest to it (Mardia *et al.* (1979) and Figure 16.2). The distance between two clusters is the distance between the two objects in the two clusters which are closest to each other. This sometimes gives rise to the so-called chaining effects, which means that clusters are just extended instead of joining two clusters.

For well separated clusters, the two approaches will give similar results, but for other situations the results may be quite different. When in doubt, we recommend the complete linkage as the method of choice.

16.2.3 Dendrogram for Visual Inspection of the Results

The results from hierarchical clustering are usually presented in so-called dendrograms. These are illustrations of the results which follow the process of identifying similar objects and putting them in clusters. An example of a dendrograms is given in Figure 16.3 for an analysis of a sensory data set. In this case the objects are the assessors and the distance between the assessors is the Procrustes distance discussed in Chapter 17. As can be seen, at the bottom of the plot all objects are considered as separate clusters. Then after a certain distance (along the vertical axis), the objects 1 and 4 are identified as the two closest to each

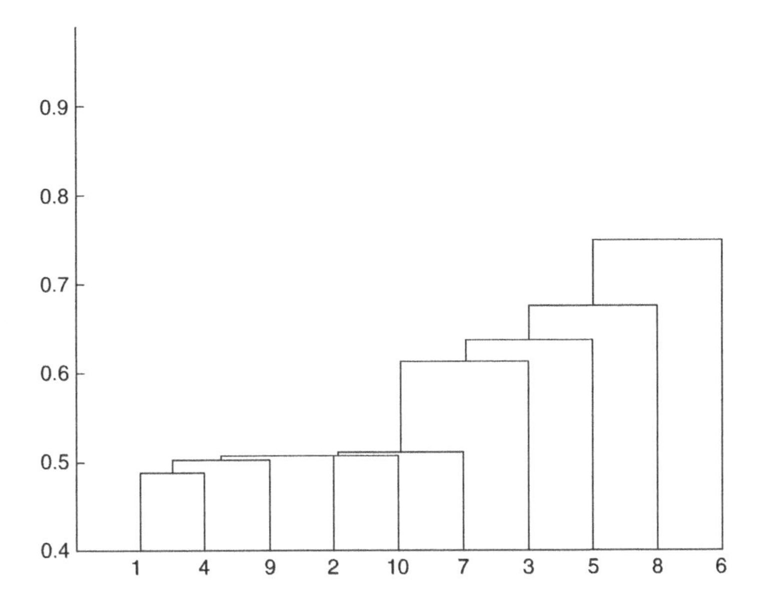

Figure 16.3 Dendrogram illustrated for the purpose of clustering of assessors in sensory profiling. *In this case, the Procrustes distance described in Chapter 16.4 is used in connection with hierarchical clustering. As can be seen, 6 of the samples form a clear cluster with the other assessors being quite different. Reprinted from Food Quality and Preference, 15, Dahl and Næs, Outlier and group detection in sensory panels using hierarchial cluster analysis with the Procrustes distance, 195–208, 2004 with permission from Elsevier.*

other and then put in the same cluster. Then the process continues by putting also objects 2, 7, 9 and 10 into the same cluster. As can be seen, there is a clear difference between the objects in this first cluster and the rest of the objects since one needs to go a quite long distance along the vertical axis before object 3 is merged with the same cluster. This means that the cluster consisting of the objects 1, 2, 4, 7, 9 and 10 form a cluster of very similar objects which are quite different from the rest. As can be seen, natural cluster structures will manifest themselves in the dendrogram by long vertical line segments before a new object is merged with the cluster or before two clusters are merged.

In some cases, one stops the process at a fixed number of clusters if one is interested in identifying exactly this number of segments or subgroups. This is the case for instance for the sample selection methodology discussed in Chapter 9. Note that if consumer segments are very small, little can be said about them. Usually, one will therefore stop segmentation at 3 or 4 segments.

An important feature of the hierarchical methods is that they can be useful for identifying outliers in the data set. An object which is an outlier is very different from the rest and will then have a large distance to all other object. This object will show up as an object which is put in a cluster at the top level of the dendrograms, i.e. in a cluster with a very large distance to the rest.

16.3 Partitioning Methods

16.3.1 K-Means Clustering

The K-means clustering (MacQueen, 1967) aims at partitioning the N observations into C clusters $\mathbf{S}=\{S_1, S_2, \ldots, S_C\}$ in such a way that the criterion

$$\underset{\mathrm{S}}{argmin} \sum_{\mathbf{c}=\mathbf{1}}^{\mathbf{C}} \sum_{y_n \in S_c} (\mathbf{y_n} - \hat{\mathbf{v}}_\mathbf{c})^\mathrm{T} (\mathbf{y_n} - \hat{\mathbf{v}}_\mathbf{c}) \tag{16.4}$$

is minimised. In this formula the $\hat{\mathbf{v}}_\mathrm{c}$ is the mean of cluster S_c

Note that the number of clusters C has to be decided before the optimisation starts. A common way of solving the problem of unknown C is to try different C values and consider how natural the different splittings are. This can be done by computing an index for cluster validity, for instance by comparing the between cluster variability with the within cluster variability. Alternatively, one can simply look at the PCA plots of the objects as was indicated above. In this book, no further consideration will be given to this method. Instead we will concentrate on the following closely related technique which we think has a number of advantages.

16.3.2 Fuzzy Clustering (FCM)

Instead of allocating a membership value of either 0 or 1 (0 – 'does not belong to' and 1– 'belongs to') for all objects and cluster combinations, fuzzy clustering introduces membership values that lie in the interval between 0 and 1, allowing for degree of membership. The sum of the membership values for each object to all clusters is defined to be equal to 1. The membership values can thus be interpreted as probabilities of membership to the different groups (Krishnapuram and Keller, 1993).

The principle of fuzzy clustering is illustrated in Figure 16.4. In the upper plot, the two clusters are identified by two different symbols. When FCM (Bezdek, 1981) is run on these data, one ends up with the plot in the lower panel with membership values indicated by different symbols.

The fuzzy c-means method (FCM) is based on minimising the criterion

$$J = \sum_{c=1}^{C} \sum_{n=1}^{N} u_{nc}^m d_{nc}^2, \qquad m \geq 1 \tag{16.5}$$

over membership values u_{nc} and distances d_{nc} between the objects (n) and clusters (c). Minimising J with respect to $\mathbf{U}=\{u_{nc}\}$ and $\mathbf{D}=\{d_{nc}\}$ will favour combinations of large values of u and small values of d and vice versa, corresponding to as clearly separated clusters as possible.

The parameter m is often called the fuzzifier and determines the fuzziness of the clustering, see e.g. (Krishnapuram and Keller (1993, 1996). The larger the value of m, the more fuzzy is the clustering. Usually m is set equal to 2, as this value has proven to give good results with the FCM. If $m = 1$, it has been shown that the FCM criterion is essentially equivalent to the K-means method. In Johansen *et al.* (2010b) is illustrated how the m can

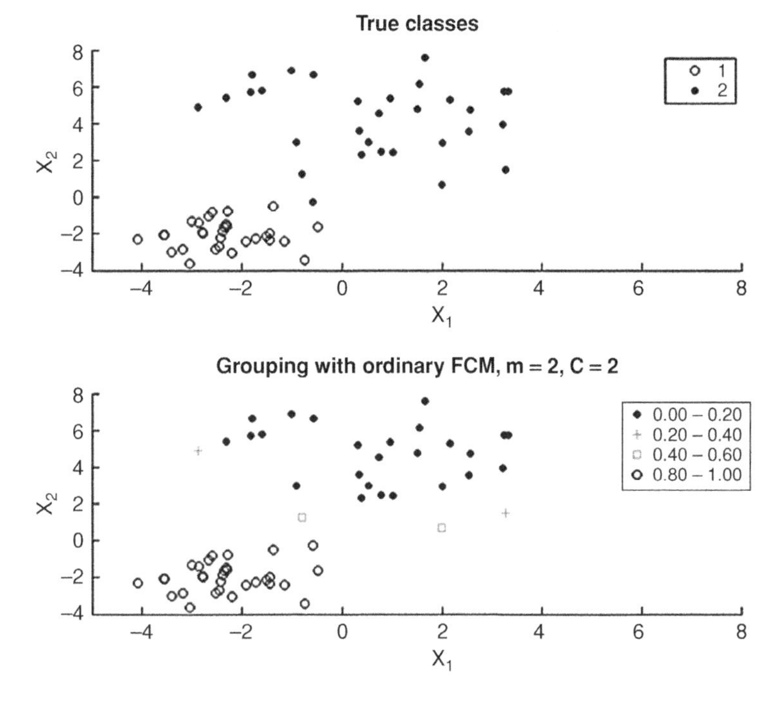

Figure 16.4 Illustration of FCM for bivariate normally distributed data. *Data are generated from two different distrbutions,* $X_1 \sim N(\mu_1, \Sigma_1)$ *and* $X_2 \sim N(\mu_2, \Sigma_2)$ *where* $u_1 = [-2\ -2]$, $u_2 = [1\ 3.5]$*and* $\Sigma_1 = [0.98\ 0.2;\ 0.2\ 0.7]$, $\Sigma_2 = 3^*[1\ -0.5;\ -0.5\ 1]$. *Top: Scatter plot of data labelled with true classes, Bottom: Scatter plot (as obtained by FCM) of simulated data where the different symbols mark membership values to cluster 1 according to the legend. Reprinted from Computational Statistics & Data Analysis, 52, New modifications and applications of fuzzy C-means methodology. 2403–2418, 2008 with permission from Elsevier.*

be optimised for a particular case by comparing prediction results within each group for different values of m.

As can be noted, also for this method, the C has to be decided on before calculations can start. The same comments as for the K-means hold here, but here inspection of the u-values is an additional option. If for instance all membership values fall close to the extremes 0 and 1, the clustering is a very natural splitting. A discussion of various criteria for assessing cluster validity is given in Bezdec (1981). A technique based on PCA of the membership matrix was used in Næs and Hildrum (1997) in the special case of three clusters.

Minimisation of J is achieved by iteratively optimising for **U** and **D**. When J ceases to change, the procedure is stopped. Initial values of **U** are usually selected at random, or chosen according to prior knowledge of group structure. In the case of Euclidean distances (denoted by $\|\,\|$) the algorithm for minimising J can be summarised by the following steps 1–4:

1. Initialise **U**

2. Update distances for given **U**. This is obtained by first computing the weighted averages $\mathbf{v_c}$ for each cluster and then computing the Euclidean distances directly:

$$d_{nc}^2 = \|\mathbf{y_n} - \hat{\mathbf{v}}_\mathbf{c}\|^2, \qquad \hat{\mathbf{v}}_\mathbf{c} = \frac{\sum_{n=1}^{N} u_{nc}^m \mathbf{y_n}}{\sum_{n=1}^{N} u_{nc}^m} \tag{16.6}$$

3. Update membership values for given **D**

$$\begin{aligned} u_{nc} &= \left(\sum_{d=1}^{C}\left(\frac{d_{nc}}{d_{nd}}\right)^{\frac{2}{m-1}}\right)^{-1} & m > 1 \\ u_{nc} &= \begin{cases} 1 & if\, d_{nc} = min_d(d_{nd}) \\ 0 & else \end{cases} & m = 1 \end{aligned} \tag{16.7}$$

In the special case that $d_{nc} = 0$, i.e., $\mathbf{y_n} = \hat{\mathbf{v}}_\mathbf{c}$, u_{nc} is set to 1. If two or more centres happen to coincide, i.e. $\hat{\mathbf{v}}_{\mathbf{c}'} = \hat{\mathbf{v}}_\mathbf{c}$ for some $c' \neq c$, u_{nc} is set to $1/r$, where r is the number of coinciding centres.

4. Calculate the criterion J and check for convergence. If $|J_{old} - J_{new}| < \varepsilon$ stop, else go to step 2. Other stopping criteria may also be used.

Even though an improvement is obtained for each step in the iteration, there is no guarantee of convergence to the optimal value. Therefore, different starting points should always be tested. Experience has, however, shown that the general structure of the algorithm seems to work well in many types of applications, also for other types of distance measures (see e.g. Berget *et al.*, 2008; Berget and Næs, 2002; Næs and Mevik, 1999). It has been shown that the FCM algorithm with Euclidean distance converges in the sense that given a data set where the number of distinct samples is greater than C, the iteration sequence will converge to a saddle point or local minimum of J (Gröll and Jäkel, 2005). According to Rousseeuw (1995), convergence properties of the FCM are better than for K-means clustering.

16.3.3 Using the Residual Distance

Note that only step 2 is dependent on the distance between an object and a cluster. The most common is the Euclidean distance, but other methods based on PCA-based distances and the Mahalanobis distance are also well established (see Bezdec, 1981; Bezdec *et al.*, 1981a,b; Gustafson and Kessel, 1979). Another criterion that was put forward in Wedel and Steenkamp (1989, 1991) and in Næs *et al.* (2001) is the use of the residual distance, giving the criterion

$$J = \sum_{c=1}^{C}\sum_{n=1}^{N} u_{nc}^m (y_n - \mathbf{x}_\mathbf{n}^\mathbf{T} \boldsymbol{\beta}_\mathbf{c})^2. \tag{16.8}$$

As can be seen, this is identical to J in Equation (16.5) except that the Euclidean distance is replaced by a residual distance between a response variable y_n and a linear function of some explanatory variables $\mathbf{x}_n$. Note that different regression coefficients $\boldsymbol{\beta}_c$ are used for each cluster. In consumer segmentation, the y will typically correspond to consumer liking for a product and the $\mathbf{x}$ to either the sensory properties or the design of the study (Næs *et al.*, 2001). Using this criterion focuses directly on identifying groups of individuals that have the same relation between the stimuli and liking.

One of the advantages of the present method for consumer segmentation purposes is that different products can be tested by the different consumers and the number of objects per consumer can be very small (Wedel and Steenkamp, 1989, 1991; Johansen *et al.*, 2010b). From a practical point of view this is important since a single consumer can only test a relatively small number of possible products while the number of interesting combinations may be large (Næs *et al.*, 2001). We refer to Chapter 8.3 for further discussion of advantages and limitations of such strategies.

Since each consumer usually tests several products and will therefore be involved in more than one term in the sum of (16.8), all u-values for all the terms corresponding to one single consumer must be identical. This requires a slight modification of the standard optimisation of the u-values (see Næs *et al.*, 2001); the u-values are now updated using sums of the squared residuals instead of squared residuals themselves, the sums being taken over all observations within the consumers.

The optimisation for the regression distance is simple and goes along the same lines as described above, iteration between computing the membership values and updating the distances. The u-step based on given values of the distances is identical while the optimisation of the distances for given u-values can be solved by using regular weighted least squares (WLS, see Næs and Isaksson, 1991) regression. The latter is obvious if we for given u-values interpret the criterion in (16.8) as a weighted LS criterion with the u's as the weights. Note that the method provides both a proposed splitting and separate linear regression coefficients (β-values) within each group.

16.3.4 Sequential Clustering and the Use of the Noise Cluster

The FCM method has a number of additional features which can also be important in sensory science. One of them is the noise cluster method for handling outliers and the use of the sequential clustering for determining the number of clusters. Both these features are used in for instance the method discussed below in Section 16.4 and in Chapter 3, Section 3.9.4.

The idea of noise clustering (Dave, 1991) is to allocate outlying points/observations into a noise cluster, such that they do not affect the clustering of the other points. The noise cluster is defined as a 'universal entity' that has the same distance from every point in the data set. For all point $n = 1,2,\ldots,N$ it is assumed that

$$d_{n,C+1}^2 = \delta^2 \tag{16.9}$$

where $C+1$ corresponds to the noise cluster. As the algorithm proceeds, the good points increase their chance of belonging to the good cluster.

Dave (1991) suggested to recalculate the noise prototype distance for every iteration as a multiple of the simple average of all distances using

$$\delta^2 = \frac{\lambda}{NC}\sum_{c=1}^{C}\sum_{n=1}^{N} d_{nc}^2 \tag{16.10}$$

It is, however, not always easy to know what value of the parameter λ is the optimal one and some trial and error may be necessary (for guidelines, see Dave, 1991). Large values of λ will lead to all observations being members of the good cluster while small values will give the opposite result.

The sequential clustering proposed in Berget *et al.* (2005) was based on the use of the concept of noise cluster. The method starts by identifying the most clear or natural cluster treating the rest of the observations as a noise cluster. This means that C is set equal to 1 in (16.9). When the first cluster is identified, this is removed from the data set and the procedure starts over again with the rest of the data. The sequential procedure stops when there is little cluster tendency left. The method has the advantage that it identifies a sequence of clusters with the most obvious cluster first, the next most obvious after that etc. It is also useful for identifying the number of clusters in the data set.

16.3.5 Finite Mixture Model Clustering

A closely related, but somewhat more sophisticated method is the finite mixture model approach (Raftery, 1998; Banfield and Raftery, 1993). This is based on the likelihood for the data given an assumption on the distribution function within each group. The likelihood function for such data sets can be written as

$$L = \prod_{n=1}^{N} \sum_{c=1}^{C} \pi_c f_c(\mathbf{y_n}, \psi_c) \tag{16.11}$$

where the f's are the conditional distributions within each of the segments or clusters. The π_c is the probability that an object belongs to groups c and ψ_c is the set of model parameters for the cluster c. For multinormal distributions, each f in Equation (16.11) can be written as

$$\mathrm{f_c}(\mathbf{y_n}, \psi_c) = \frac{1}{\sqrt{2\pi \boldsymbol{\Sigma}_c}} e^{-(\mathbf{y_n} - \mathbf{x_n^T} \boldsymbol{\beta}_c)^T \boldsymbol{\Sigma}_c^{-1} (\mathbf{y_n} - \mathbf{x_n^T} \boldsymbol{\beta}_c)} \tag{16.12}$$

As can be seen, the f distributions are essentially dependent only on the distribution of the residuals. Note that if the different covariance matrices are assumed to be equal to the identity, this approach is essentially an optimisation of the regression vectors $\boldsymbol{\beta}$ only, showing the close relation to the FCM discussed above.

The L is usually optimised using the EM algorithm (see Poulsen *et al.*, 1997)). As for the FCM, the method provides β-values for each of the groups.

The same approach can be extended to random coefficient models within each of the groups. This opens up for individual differences and correlation structure within the different groups. In addition to the model requirements above, it is assumed that individual regression coefficient vectors $\mathbf{b_{nc}}$ (individual n member of group c) within each group have a random distribution representing the distribution of the individual differences within the groups, i.e.

$$\mathbf{b_{nc}} \approx \mathrm{N}(\boldsymbol{\beta}_c, \boldsymbol{\Sigma}_c) \tag{16.13}$$

The average within each group, the $\boldsymbol{\beta}$, is still the most interesting part of the model. Note that this approach is closely related to Bayesian statistics. Optimisation is again done by ML and the EM algorithm. We refer to Erichsen and Brockhoff (2004) and Poulsen *et al.* (1997) for further applications and further discussion of the approach within consumer science (see also Gustafsson *et al.*, 2003).

For testing the adequacy of the model and for comparing different solutions a number of methods for estimation of prediction ability of the model have been proposed. The most famous is the generally applicable AIC (Akaike, 1973), but other and modified versions like the BIC and other have been proposed. Alternatively, one can use chi square tests for the hypothesis that the model is correct.

The standard segmentation methodology used for choice data is usually based on the multinomial logit model. One starts by assuming that the observed choices y_{jkm} *(I correspond to ., J)* are independent and follow a mixture of multinomial distributions. Again one ends up with a so-called mixture model similar to Equation (16.11). A further refinement of the model is also possible here by assuming random regression coefficients within each cluster to allow for correlations among the responses given by one single assessor. The same model can also allow for a correlation structure between alternatives within choice sets. (More details can be found in Gustafsson *et al.*, 2003, p. 393; see also Wedel and Kamakura, 1998.)

16.4 Cluster Analysis for Matrices

In Sections 16.1–16.3 main attention has been given to clustering of observation vectors or residuals. It is, however, easy to generalise both hierarchical and partitioning methods to clustering of matrices as well. A couple of such alternatives, which have previously been applied for detecting outlying assessors in descriptive sensory analysis (see Chapter 3, Dahl and Næs, 2004, 2009), will be discussed next. Assume for the rest of this chapter that there are I different (mean centred) Y-matrices (for instance corresponding to I different assessors) available for clustering.

One of these methods is called Proclustrees since it combines Procrustes analysis with hierarchical clustering. The distance between two matrices $\mathbf{Y}_1$ and $\mathbf{Y}_2$ is assumed to be the Procrustes distance given by

$$\min_{\mathbf{H}} \left\| \mathbf{H}\left(\frac{\mathbf{Y}_1}{\|\mathbf{Y}_1\|}\right) - \frac{\mathbf{Y}_2}{\|\mathbf{Y}_2\|} \right\| \tag{16.14}$$

where $\mathbf{H}$ is an orthogonal matrix and the norm (given by $\| \|$) is the Frobenius norm, i.e the squared root of the squared sum of all matrix elements. This means that similarities are considered after correction for optimal rotations (Gower, 1975; Dijksterhuis, 1996; Gower and Dijksterhuis, 2004; see also Chapter 17). The normalisation of each matrix within the norm ensures that the need for isotropic scaling vanishes.

As soon as the Procrustes distances have been computed between all matrices, the regular hierarchical methods discussed above can be used directly using the corresponding distance matrix. In Dahl and Næs (2004) the method was used to identify and eliminate outlying assessors before computing the average sensory profiles. It was illustrated how this could be used to improve the relation between the sensory average and external spectroscopic data (see Figure 6.3 based on the same data).

A slightly more general method was published in Dahl and Næs (2009). This method is based on using FCM and was tested out on distances based on Tucker-1 and Tucker-2

(Chapter 17) models. This means that the method looks for cluster tendencies relative to degree of fit to models defined by either the Tucker-1 or the Tucer-2 equations, which are both important and flexible models within sensory analysis (Brockhoff *et al.,* 1996). The method can, however, also easily be modified to handle other distances such as the Procrustes distance. The distances (to the good cluster) considered can be written as

$$d_{i1}^2 = \|\mathbf{Y}_i - \mathbf{T}\mathbf{P}_i^{\mathrm{T}}\|^2 \tag{16.15}$$

$$d_{i1}^2 = \|\mathbf{Y}_i - \mathbf{T}_i\mathbf{P}^{\mathrm{T}}\|^2 \tag{16.16}$$

$$d_{i1}^2 = \|\mathbf{Y}_i - \mathbf{T}\mathbf{W}_i\mathbf{P}^{\mathrm{T}}\|^2 \tag{16.17}$$

for two different Tucker-1 models (16.15 and 16.16) and a Tucker-2 model (16.17). Again the norm is the Frobenius norm. The first of the Tucker-1 distances corresponds to the common scores model discussed in Chapter 3. The other one corresponds to the common loadings model, which is obtained by unfolding the three-way sensory matrix the other way (see Chapter 5).

Dahl and Næs (2009) uses these distances in combination with the first step of the sequential FCM technique described in Chapter 16.3. The distances in (16.15)–(16.17) are used directly in the FCM criterion and the solution is found by using the general algorithmic setup in Equations (16.6) and (16.7). The assessors allocated to the noise cluster were identified as outliers. A special plot based on the membership values to the noise cluster (outlier cluster) which can help determining which assessors are the most outlying, was also developed.

References

Akaike, H. (1973). Information theory and an extension to the maximum likelihood principle. In B.N. Petrov, F. Csaki (eds). *Second International Symposium on Inference Theory*, Budapest: Akademiai Kiado, 267–81.

Banfield, J.D., Raftery, A.E. (1993). Model-based gaussian and non-gaussian clustering clustering. *Biometrics* 49, 803–21.

Berget, I., Næs, T (2002). Optimal sorting of raw materials based on predicted end product quality, *Quality Engineering* 14, 459–78.

Berget, I., Mevik, B.-H., Vebø, H., Næs, T. (2005). A strategy for finding biological relevant clusters in microarray data, *Journal of Chemometrics* 19, 482–91.

Berget, I., Mevik, B-H., Næs. T. (2008). New modifications and applications of fuzzy C- means methodology. *Computational Statistics and Data Analysis* 52, 2403–18.

Bezdek, J.C. (1981) *Pattern Recognition with Fuzzy Objective Function Algorithms.* New York: Plenum Press.

Bezdek, J.C., Coray, C., Gunderson, R., Watson, J. (1981a). Detection and characterization of cluster substructure .1. Linear structure – Fuzzy C – Lines, *SIAM J. Appl. Math.* 40, 339–57.

Bezdek, J.C., Coray, C., Gunderson, R., Watson, J., (1981b). Detection and characterization of cluster substructure. 2. Fuzzy C – Varieties and convex combinations thereof, *SIAM J. Appl. Math.* 40, 358–72.

Brockoff, P, Hirst, D., Næs, T. (1996). In T. Næs, E. Risvik (eds), *Multivariate Analysis of Data in Sensory Science*. Amsterdam: Elsevier.

Dahl, T., Næs, T. (2004). Outlier and groups detection in sensory panels using hierarchical cluster analysis with the Procrustes distance. *Food Quality and Preference* 15, 195–208.

Dahl, T., Næs, T. (2009). Identifying outlying assessors in sensory profiling using fuzzy clustering and multi-block methodology, *Food Quality and Preference* 20(4), 287–94.

Dave, R.N. (1991). Characterization and detection of noise in clustering, *Pattern Recognition Letters* 12, 657–64.

Dijksterhuis, G. (1996). Procrustes analysis in sensory research: In T. Næs, E. Risvik (eds), *Multivariate Analysis of Data in Sensory Science*, pp. 185–217. Amsterdam: Elsevier Science Publishers.

Erichsen, L., Brockhoff, P.B. (2004). An application of latent class random regression coefficient regression. *Journal of Applied Mathematical and Decision Sciences* 8(4), 247–60.

Gower, J.C. (1975), Generalized Procrustes analysis, *Psychometrica* 45(1), 3–24.

Gower, J.C., Dijksterhuis, G. (2004). *Procrustes Problems*. Oxford: Oxford University Press.

Gröll, L., Jäkel, J. (2005). A new convergence proof of fuzzy c-means. *IEEE Trans. Fuzzy Systems* 13, 717–20.

Gustafson, D.E., Kessel, W.C. (1979). Fuzzy clustering with a fuzzy covariance matrix. *Proceedings of the 1978 IEEE Conference in Decision and Control*. IEEE Control Systems Society, San Diego, California, USA, 761–6.

Gustafsson, A., Herrmann, A., Huber, F (2003). *Conjoint Measurement: Methods and Applications*. Berlin: Springer.

Johansen, S., Hersleth, M., Næs, T.M. (2010b). The use of fuzzy clustering for segmentation in linear and ideal point preference models. *Food Quality and Preference* 21, 188–96.

Krishnapuram, R., Keller, J.M. (1993). A possibilistic approach to clustering, *IEEE Transactions on Fuzzy Systems* 1, 98–110.

Krishnapuram, R., Keller, J.M. (1996). The possibilistic C-means algorithm: Insights and recommendations, *IEEE Transactions on Fuzzy Systems* 4, 385–93.

MacQueen. J.B. (1967). Some methods for classification and analysis of multivariate observations. In *Proceedings of 5th Berkeley Symposium on Mathematical Statistics and Probability*. 1: 281–97, University of California Press.

Mardia, K.V., Kent, J.T., Bibby, J.M. (1979). *Multivariate Analysis*, London: Academic Press.

Næs, T., Isaksson, T. (1991). Splitting of calibration data by cluster-analysis. *Journal of Chemometrics* 5(1), 49–65.

Næs, T., Hildrum, K.I. (1997). A comparison of multivariate calibration and discriminant analysis for determining tenderness of meat by NIR spectroscopy. *Applied Spec.* 51(3), 350–7.

Næs, T., Lynch, H., Drava, G (1998). The use of chemometrics in food authenticity studies. In M. Lees (ed.), *Food Authenticity, Issues and Methodologies*. Nantes, France: Eurofins Scientific.

Næs, T., Mevik, B.H. (1999). The flexibility of fuzzy clustering illustrated by examples, *Journal of Chemometrics* 13, 435–44.

Næs, T., Kubberød, E., Sivertsen, H. (2001). Identifying and interpreting market segments using conjoint analysis. *Food Quality and Preference* 12(2), 133–43.

Poulsen, C.S., Brockhoff, P.M.B, Erichsen, L. (1997). Heterogeneity in consumer preference data – a combined approach. *Food Quality and Preference* 8(5/6), 409–17.

Raftery, C. (1998). Algorithms for model based Gaussian hierarchical clustering. *SIAM J. Sci. Comput.* 20(1), 270–81.

Ripley, B.D. (1996). *Pattern Recognition and Neural Networks*. Cambridge: Cambridge University Press.

Rousseeuw, P.J. (1995). Discussion: Fuzzy clustering at the intersection, *Technometrics* 37, 283–6.

Wajrock, S., Antille, N., Rytz, A, Pineau, N., Hager, C. (2008). Partitioning methods outperform hierarchical methods for clustering in preference mapping. *Food Quality and Preference* 19, 662–9.

Wedel, M., Kamakura, W.A. (1998). *Market Segmentation. Conceptual and Methodological Foundation*. New York: Kluwer Academic Publishers.

Wedel, M., Steenkamp, J.-B.E.M. (1989). A fuzzy clusterwise regression approach to benefit segmentation. *International Journal of Research in Marketing* 6(4), 241–58.

Wedel, M., Steenkamp, J.-B.E.M. (1991). A clusterwise regression method for simultaneous fuzzy market structuring and benefit segmentation. *Journal of Marketing Research* 28(4), 385–96

17

Miscellaneous Methodologies

This chapter describes some alternative methods that have been mentioned briefly elsewhere as extensions, generalisations or modifications of more basic methodology. The methods will be described at a rather general level with reference to some of their most important applications within sensory and consumer science. The first part of the chapter covers methods which are useful for analysing the full set of individual sensory profile data. These are sometimes referred to as three-way methods since there are three 'ways' in the data table; assessors, attributes and products as discussed in Chapter 3. We will in particular discuss generalised procrustes analysis (GPA) and three-way generalisations of principal components analysis, such as for instance PARAFAC and the different Tucker methods. Path modelling is another important area covered in this chapter. This is a type of methodology which is useful when one is interested in relations between several data sets, for instance attitudes, habits, demographic variables and liking. Multidimensional scaling (MDS) is described as a method which uses data given as distances between objects. The chapter ends with a discussion of how to estimate missing values in a data set.

This chapter is based on methodology presented in Chapter 14 and 15. The methods treated here are of particular interest in Chapter 3, 5 and 6, but the last section is relevant for all situations where missing values are present.

17.1 Three-Way Analysis of Sensory Data

Sensory panel data can always be looked upon as three-way data tables with assessors, products and attributes as the three 'ways' (see Figure 2.1). In order to analyse both objects, assessors and the correlation structure among the attributes, the full three-way structure of the data has to be taken into account. This can be done in various ways using different

Statistics for Sensory and Consumer Science Tormod Næs, Per B. Brockhoff and Oliver Tomic

underlying ideas and philosophies as has been demonstrated in for instance Chapter 3 and Chapter 6. In the present chapter we will discuss briefly a number of related methods that can be used as alternatives.

17.1.1 Generalised Procrustes Analysis (GPA)

The GPA (Gower, 1975; Gower and Dijksterhuis, 2004) is a method which has been used for a long time in sensory analysis (Dijksterhuis, 1996; Langron *et al.*, 1984; Arnold and Williams, 1987) both for detecting and eliminating individual differences. The method is based on a model assuming that the different assessor matrices, the slices in Figure 2.1, differ with respect to translation, isotropic scaling (the same scaling for all attributes) and rotation. This means that an assessor's data matrix can be transformed to another assessor's matrix (except noise) by changing the mean, the scale and by 'rotating' the data table. In mathematical terms, the three transforms can be described together by the formula

$$\mathbf{T_i} + c_i \mathbf{Y_i} \mathbf{H_i} \tag{17.1}$$

where $\mathbf{T_i}$ is the matrix of translation constants, c_i is the scalar that represents the isotropic scaling and $\mathbf{H_i}$ is a rotation matrix (orthogonal matrix, $\mathbf{H}^{\mathbf{T}}\mathbf{H} = \mathbf{H}\mathbf{H}^{\mathbf{T}} = \mathbf{I}$). The rotation is the most complex and difficult to relate to specific sensory aspects, but one important example is switching of two related attributes (Arnold and Williams, 1987). As before $\mathbf{Y_i}$ is the data matrix for assessor i (with products as rows and attributes columns).

The so-called generalised procrustes analysis (GPA) estimates all three model elements in such a way that the different assessors become as similar to each other as possible (see Figure 17.1). The translation is first simply eliminated by mean centring of each attribute

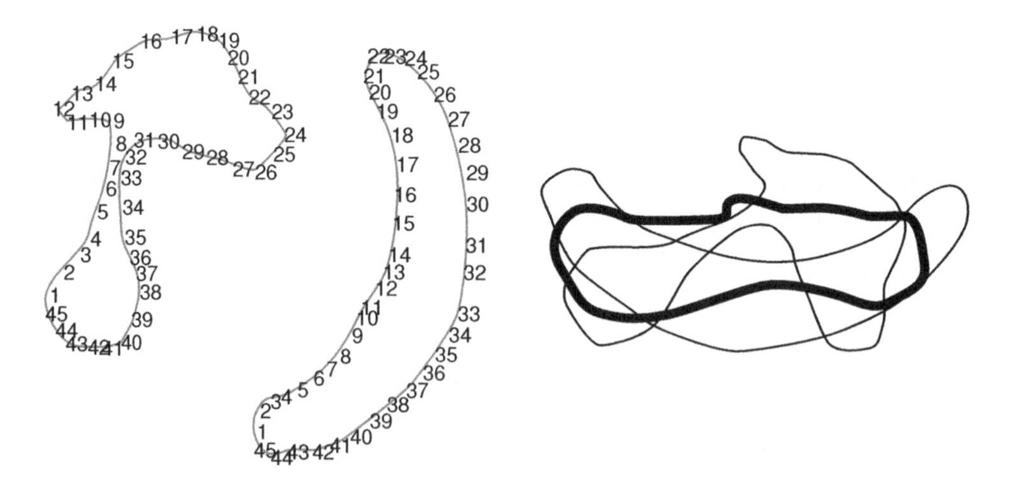

Figure 17.1 An illustration Procrustes rotation. *In a) is presented two different configurations before tranformation and in b) after transformation. The solid line in (b) is the consensus. From Dahl and Næs(2004). Reprinted from Food Quality and Preference, 15, Dahl and Næs, Outlier and group detection in sensory panels using hierarchial cluster analysis with the Procrustes distance, 195–208, 1995, with permission from Elsevier.*

for each assessor. Then the mean centred data are used to minimise the criterion

$$\sum_{i=1}^{I} \|c_i \mathbf{Y}_i \mathbf{H}_i - \mathbf{V}\|^2 \tag{17.2}$$

over all possible choices of c_i and $\mathbf{H}_i$. In this case, the norm is the Frobenius norm which is obtained by first squaring all elements of the matrix, then taking the sum of all squares before the square root is taken of the sum. In order to avoid a trivial solution, a restriction is put on the variances of **V**. The **V** matrix is called the consensus matrix and is estimated as a by-product of the optimisation. The consensus can be thought of as an average after the elimination or reduction of the individual differences.

For interpretation of the results, one usually applies principal component analysis on the consensus and looks at the differences between the products. The components can then be related to the original variables by regression and the results presented in the same plot. In order to understand the individual differences in the original data, one can look at the c-values and also look at the projections of the transformed individual data onto the principal component space obtained from the consensus **V**.

If there are reasons to believe that some assessors do not fit to this model, one can look at the size of the contributions in formula (17.2) for each of the individuals separately. A better alternative is, however, to use a clustering method which takes the Procrustes distance into account (see Chapter 16 and Dahl and Næs, 2004). This is a method which looks at the differences between assessors after elimination of the Procrustes model differences. These distances are then submitted to hierarchical clustering. Assessors which are different from the others with respect to this criterion will then show up as outliers in the dendrogram.

17.1.2 Three-Way Generalisations of PCA

A completely different way of handling three-way data is to use data compression methodology inspired by the PCA method. A number of different methods have been put forward, all of them trying to find the most important underlying dimensions in the three-way matrix. The three methods that will be discussed here are PARAFAC (Harshman, 1970; Bro, 1997), Tucker-1 and Tucker-2 (Tucker, 1964). The Tucker-1 method is also discussed in the PCA chapter since it can be considered as a PCA of an unfolded matrix of data.

The model for the Tucker-1 for sensory data (see Figure 2.1) can be written as

$$\mathbf{Y}_i = \mathbf{T}_i \mathbf{P}^T + \mathbf{E}_i \tag{17.3}$$

or as

$$\mathbf{Y}_i = \mathbf{T} \mathbf{P}_i^T + \mathbf{E}_i \tag{17.4}$$

depending on whether one is interested in the common loadings (**P**) or the common scores (**T**) variant respectively. In both models, the **E**'s represent the components not extracted and are thought of as noise (Chapter 14). The solution for the model (17.3) is obtained by using PCA on the unfolded matrix with attributes as columns and products*assessors as rows (Chapter 5). For model (17.4), the same technique is used, but now for the unfolded matrix with products as rows and attributes*assessors as columns (Chapter 3). In both cases, the columns of the unfolded matrix are mean centred before calculations.

The Tucker-2 model for assessor i is defined by

$$\mathbf{Y_i} = \mathbf{TW_iP^T} + \mathbf{E_i} \tag{17.5}$$

where now the $\mathbf{W_i}$ matrix is a low-dimensional matrix representing the individual differences among the individual assessors. Note that this is a model with both common scores ($\mathbf{T}$) and common loadings ($\mathbf{P}$). The method allows for individual differences through the low dimensional matrix $\mathbf{W_i}$. Also for Tucker-2, mean centring of each slice $\mathbf{Y_i}$ is the most natural.

The PARAFAC (Bro, 1997; Harshman, 1970) model can be formulated as the Tucker-2 model with diagonal $\mathbf{W_i}$ matrices. The PARAFAC model can also be written as

$$y_{ijk} = \sum_{a=1}^{A} t_{ja} w_{ia} p_{ka} + e_{ijk} \tag{17.6}$$

for each of the elements in the $\mathbf{Y}$-cube. Here the index a represents the number of components. As can be seen, the method is symmetric in the three 'ways'. Mean centring can if wanted be done as for Tucker-2.

The models for the three methods can be put in a hierarchical order with the Tucker-1 as the most flexible and PARAFAC as the most rigid for the same number of components. For Tucker-1 there is no restriction on the scores and loadings and it is therefore the most flexible. Tucker-2 can be considered a Tucker-1 with the restriction that the individual scores $\mathbf{T_i}$ are restricted to be equal to $\mathbf{TW_i}$ or that the individual loadings $\mathbf{P_i}$ are restricted to be equal to $\mathbf{W_iP}$ (depending on which of the Tucker-1 models is used). As mentioned above PARAFAC can be considered a Tucker-2 with the restriction that the $\mathbf{W_i}$ matrices are diagonal.

The concrete solutions for the three methods can all be considered as LS solutions (or approximate) using the Frobenious norm (see above). In all cases, the goal is to minimise the differences between the slices $\mathbf{Y_i}$ of the data table and the corresponding model values in a LS sense. This is also true for Tucker-1, but here the solution is easier to find by regular PCA. Efficient algorithms have been developed for Tucker-2 and PARAFAC (see Smilde *et al.*, 2004).

For the Tucker-1 model the scores and the loadings are plotted in scatter plots, either separately or together as discussed in Chapter 3 and Chapter 14. For the Tucker-2 model, the common loadings and the common scores are also plotted in regular scatter plots. Plotting of the elements of $\mathbf{W_i}$ is useful for understanding the individual differences (see Næs and Kowalski, 1989; Dahl and Næs, 2009). For PARAFAC, separate scatter plots are produced for the assessors, the products and the attributes (see Bro *et al.*, 2008).

All three methods have been used in sensory analysis. We refer to Bro (1997), Bro *et al.* (2008) for applications of PARAFAC, to Brockhoff *et al.* (1996), Næs and Kowalski (1989) and Dahl and Næs (2009) for Tucker-2 applications and to Dahl *et al.* (2008) for Tucler-1 applications. It is clear from all these publications that the methods can be useful both for detecting individual differences in sensory analysis and for interpreting them.

The $\mathbf{W_i}$ values for a Tucker-2 model based on an example from sensory analysis of cheese (see Dahl and Næs, 2009) are given in Figure 17.2. The plot shows the rows of the

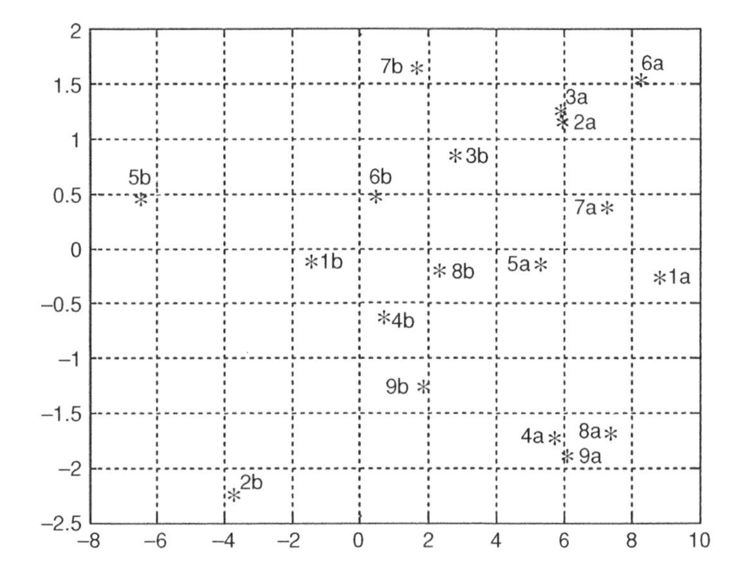

__Figure 17.2__ __individual differences from Tucker-2.__ The individual differences matrices $\mathbf{W}_i$ from Tucker-2 plotted for a cheese data set. The point "ia" represents the elements 1 and 2 in the first row of $\mathbf{W}_i$. The second row is here presented by the symbol "ib". Reprinted from Food Quality and Preference, 20, Dahl and Næs, Identifying outlying assessors in sensory profiling using fuzzy clustering and multi-block methodology, 287–294, 2009, with permission from Elsevier.

$\mathbf{W}_i$ matrices, all of them being 2*2 matrices. The point 'ia' represents the elements 1 and 2 in the first row of $\mathbf{W}_i$. The second row is here presented by the symbol 'ib'. As can be seen there are large differences between the assessors marked by numbers 1–9. A couple of more specific observations can be made: One of them is that assessors 4 and 8 are very similar to each other both for the first row and second row. This means that they are very similar in their assessment of the samples and attributes as compared to the rest of the assessors. Another interesting observation is that assessor number 1, only has a clear contribution for the first element of the first row. This means that this assessor only assesses variation along the first axis and has no sensitivity at all for the second axis; this is a one-dimensional assessor.

17.1.3 Generalised Canonical Analysis (GCA)

Another method which can be used for comparing sensory assessors is the GCA method which is a direct generalisation of regular canonical correlation analysis (see Carroll, 1968). This is a technique which is similar to the Tucker-1 method above in the sense hat it seeks linear combinations of the original variables to describe the important common structures in the data. The difference between GCA and Tucker-1 is that GCA optimises the correlation between linear combinations of all the individual matrices while Tucker-1 focuses on describing variability in the data table as well as possible. Both methods provide consensus information for the individual data matrices. The GCA is a useful method, but has a tendency of overfitting if there are many variables and few objects. It can not be used if the number of variables is larger than the number of objects. If GCA is used, it needs to be validated properly. If this is not done, one can end up with overoptimistic results regarding relations between the data sets.

If we let $\mathbf{V} = (\mathbf{v_1}, \mathbf{v_2}, \ldots, \mathbf{v_A})$ denote the consensus matrix of the A first consensus components, the criterion for optimisation can be described by

$$\max_{\mathbf{h},\mathbf{v_a}} \sum_{\mathbf{i}} \text{corr}(\mathbf{Y_i h_{ia}}, \mathbf{v_a})^2 \tag{17.7}$$

under the restriction that $\mathbf{V^T V} = \mathbf{I}$. Here the $\mathbf{h}$'s are direction vectors for the different components. In Dahl and Næs (2006) it was shown that Tucker-1 and GCA can be put within the same framework by introducing an extra parameter. The parameter can then be varied to find the best compromise between Tucker-1 and GCA for the interpretation of the data (see Chapter 4).

17.1.4 The STATIS Method

Within the area of sensory analysis the STATIS (Structuration des Tableuax À Trois Indices de la Statistique, see e.g. L'Hermier des Plantes (1976), Lavit *et al.* (1994) and Schlich (1996)) method is used to find a compromise between the individual assessor matrices, thus having more or less the same scope as Procrustes analysis. The STATIS method ends up with a weighted average of the individual cross-product matrices $\mathbf{Y_i Y_i^T}$ (using centred $\mathbf{Y_i}$) with weights computed from a matrix of RV coefficients. The RV coefficient is a generalised correlation measure which in this case is used to measure the similarity between two cross-product matrices. The RV coefficient is computed for each combination of assessor matrices and put into a matrix. The first eigenvector $\mathbf{u} = (u_1, u_2, \ldots, u_i)^T$ of this matrix is then used as the weights for the individual cross-product matrices $\mathbf{Y_i Y_i^T}$. The components of $\mathbf{u}$ measure how similar the actual assessor is to the panel average. The weighted average can then be written as

$$\mathbf{V} = \sum_{i=1}^{I} u_i \mathbf{Y_i Y_i^T} \tag{17.8}$$

The strongest weight is put on the assessors with the best agreement with the panel. The $\mathbf{V}$ matrix is typically analysed by computing the first 2 or 3 eigenvectors (essentially principal components of the consensus) and by plotting them in regular scatter plots.

17.1.5 Multiple Factor Analysis (MFA)

This is a technique which is closely related to Tucker-1. It is based on unfolding of the three-way structure such that the samples become the rows and the columns correspond to all the variables for all the data blocks (short and fat matrix). Then each of the blocks is divided by the eigenvalue of its own first principal component. This means that multiple factor analysis is essentially a Tucker-1 after a particular scaling. The same interpretation tools as for the Tucker-1 can be used. For further properties and details about the method we refer to Pages and Husson (2005) and Pages and Tenenhaus (2001).

17.2 Relating Three-Way Data to Two-Way Data

The most obvious way of relating three-way sensory data to two-way data (for instance chemistry) is to first use either PARAFAC or one of the Tucker methods (see Chapter 6) for the three-way data and then relate the common scores to the two-way data using PCR or regular PLS regression. It is, however, also sometimes possible to do the two operations simultaneously using a joint model. One such method is the Tucker-2 approach based on substituting the common scores in the Tucker-2 model by a linear function of the external data, here called **X** (Kroonenberg and De Leeuw, 1980; Van der Kloot and Krooenberg, 1985). The model then becomes

$$\mathbf{Y_i} = \mathbf{XBW_iP^T} + \mathbf{E_i} \tag{17.9}$$

where **B** is the matrix of regression coefficients. The estimation of unknown values is done according to a simple eigenvector calculation. The individual differences, $\mathbf{W}_i$, can be interpreted in the same way as in Figure 17.2. An application of the method in sensory science can be found in Næs and Kowalski (1989).

If the prediction equation is reversed, the multi-block PLS based on unfolding and direct use of PLS regression can be useful. An alternative is the N-PLS method (Bro, 1996; Bro *et al.*, 2001) based on a PARAFAC-like decomposition. The former is the simplest since this is based on simple use of regular PLS regression.

17.3 Path Modelling

Path modelling methods are suitable for finding relations between a number of blocks that are related in a structured way. For instance, one may be interested in understanding how a preference pattern relates to demographic variables, attitudes and habits and also how the demographic variables are related to the habits and attitudes. Path modelling can be used to analyse whole systems of relations simultaneously without splitting up the estimation process.

Usually, it is assumed that each block can be represented by a latent variable. The manifest variables or the measurements are assumed to be linear functions of the underlying latent variables. The latent variables are then linked according to the dependence structure assumed. This is called structural equation modelling or modelling of the inner relations. For linear relations, the structural equation model can be written as

$$\eta = \mathbf{B}\eta + \boldsymbol{\Gamma}\xi + \varepsilon \tag{17.10}$$

where the η is the symbol used for the output variables and ξ the symbol used for the input variables. The ε represents random errors. Since some of the blocks may represent both input and output information, the η is allowed to be represented on both sides in the equation. Blocks that are only considered as input blocks are called exogenous while blocks that are predicted by some other blocks are called endogenous. Typically, some of the elements in the matrix **B** are known to be equal to 0.

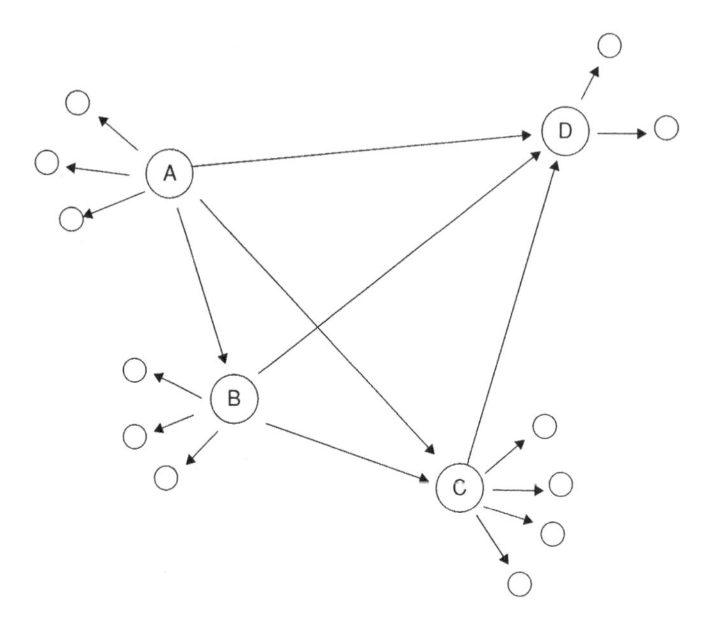

Figure 17.3 An example of a path model diagram. *In this case the block A represents for instance demographic variables, the block B the attitude variables, block C the habits and block D the preference pattern for a certain product category.*

An example of a dependence structure suitable for path modelling is given in Figure 17.3. In this case there are four blocks of information. In two of the blocks (A and B) there are three manifest variables, in one of the blocks (C) there are 4 manifest variables while in the last (D) there are only 2 manifest variables. The block D depends on all the three other blocks as indicated by the arrows. Likewise, block C depends on A and B, block B depends on block A while block A is exogenous. This means that in this case there is one ξ variable and three η's.

The joint fitting of the inner relations and the measurements models can be done in various ways. The most well-known technique is the maximum likelihood (ML) method (Jøreskog, 1977; Kaplan, 2000). This is based on computing the covariance matrix of the measurements using the assumed model and then fitting all the parameters in the model by ML assuming multinormality. This method has the drawback that it can not be used if the number of variables is higher than the number of objects. Another method which solves this problem is the PLS path modelling (see Wold, 1982; Martens *et al.*, 2007; Tenenhaus *et al.*, 2005 for overviews) which is based on iterative fitting of the inner and outer relations using simple linear regression steps. From a theoretical point of view the method is less well developed, but has a number of other advantages. In addition to the fact that it can be used for more variables than samples, the method has good convergence properties and produces both model parameters and scores which can be used for diagnostic purposes. In this book, path modelling will not be described in any further detail, but it is referred to in a few occasions in Chapter 8 and 9.

17.4 MDS-Multidimensional Scaling

In some cases, the input data for a sensory or consumer study come from pairwise comparisons of two products. Instead of being asked to rank or rate samples, the consumers are asked to quantify differences between the products. The data for each assessor can be collected in a distance matrices $\mathbf{D}_i$. In some cases, the distances may be averaged over assessors, but in other cases it may be more natural to consider them independently.

If the data are given as regular scaling data, as is done for instance in sensory profiling, it is possible to compute distances between two products *h* and *j* based on the formula

$$d_{hj} = \sqrt{(\mathbf{y_h} - \mathbf{y_j})^T(\mathbf{y_h} - \mathbf{y_j})} \tag{17.11}$$

The MDS technique is a method for finding a low dimensional representation of the data that fits as well as possible to the distances given. In other words, the MDS tries to find a low-dimensional configuration which gives distances that are as close as possible to the observed distance values using the criterion

$$\sum_{h,j} (d_{hj} - \hat{d}_{hj})^2 \tag{17.12}$$

Here the $\hat{d}_{hj}$ are the values obtained by computing the distances between objects in a low-dimensional space

$$\hat{d}_{hj} = \sqrt{(\mathbf{t_h} - \mathbf{t_j})^T(\mathbf{t_h} - \mathbf{t_j})} \tag{17.13}$$

Note that no loadings will come out of the MDS, only scores. It can be shown that if distances are computed from rating data, the PCA scores of the rating data will be the same as the scores from the MDS. In this sense MDS is a kind of PCA with focus on scores only, but MDS can also be used for other types of data.

When individual differences are present and incorporated in the methods, a possible model to use is

$$\hat{d}^i_{hj} = \sqrt{(\mathbf{t_h} - \mathbf{t_j})^T \mathbf{W_i} (\mathbf{t_h} - \mathbf{t_j})} \tag{17.14}$$

where the $\mathbf{W}_i$'s represent the individual differences. Usually the $\mathbf{W}_i$ are assumed diagonal as they are for the INDSCAL method (Carroll and Chang, 1970). If we assume diagonal $\mathbf{W}_i$ this model is strongly related to PARAFAC described above.

For more information about the MDS methods we refer to Mardia *et al.* (1979), Carroll and Chang (1970) and to Popper and Heymann (1996). Nonmetric versions of the MDS are also available (see Mardia *et al.*, 1979).

17.5 Analysing Rank Data

Rank data may come from actual ranking task protocols or from transformations of rating scale data. Such data are most often analysed by nonparametric methods (Lehmann, 1975); for instance the sign test is used as an alternative to the paired t-test, the Mann Whitney test as alternative to the independent samples t-test, the Kruskal-Wallis test as alternative to the independent samples one-way ANOVA, the Friedman test as alternative to the randomised block two-way ANOVA, Spearman rank order correlation as alternative to usual Pearson Correlations, etc. The advantage of these approaches is that no distributional assumptions are made about the data, and for someone concerned about the ability of assessors/consumers to use the scale properly, these are good alternatives. A downside of the methods is the lack of a natural estimation/quantification of effects, although some of the methods do provide the possibility of some kind of post-hoc comparisons of factor levels. Another disadvantage is that the methods are somewhat less intuitive and they are not so easy to generalise.

In Rayner *et al.* (2005) a general approach for extended analysis of ranked data in all the classical situations is given. The approach is based on a particular decomposition of the χ^2-tests of the resulting tables of counts linking this with the standard statistical tests mentioned above (which appears as the first component in the decompositions) and again with a further identification of potential dispersion effects.

Another method based on ranks is the Eggshell plot (Chapter 3, Hirst and Næs, 1994) introduced as a technique to investigate (rank) agreement between assessors in sensory profiling. The method is simple and easy to use, but is difficult to generalise.

Rank data and ordinal data may also in some cases be analyzed by parametric models; see for instance the class of generalised linear models described in Chapter 15. For rank data models based on the Bradley-Terry approach for paired comparisons we refer to Courcoux and Semenou (1997). In Vigneau *et al.* (1999) this is extended to partial ranking patterns and combined with latent class modelling for the purpose of simultaneous segmentation of consumers. Latent class regression models for general response patterns based on the generalised linear models have been used extensively in market research, see e.g. Wedel and DeSarbo (1995) (see also Chapter 10 and Chapter 16).

17.5.1 Optimal Scaling for Rank Data

The optimal scaling (OS) is another technique that can be used for analysing rank data, but can also be used to find good transformations of rating data if linearity is not satisfactory. The method is based on transforming the data and fitting the model at the same time within the same algorithm.

Let us as an example consider a situation with a set of explanatory numerical variables $\mathbf{x}$ and response data y based on ranking. Since rank data are not suitable for direct use in regression analysis, a possibility is to consider the data as based on an underlying continuous process and then try to reveal or estimate this underlying process in such a way that it fits the best possible way to the model assumed. A possibility is to estimate both the transform and the regression coefficients that fit the best using for instance minimisation of the criterion

$$(f(y) - \mathbf{X}\beta)^T(f(y) - \mathbf{X}\beta) \tag{17.15}$$

This can be done by iterative fitting of the regression coefficients given the data $f(y)$ and fitting of the transform $f(y)$ for given estimates of the regression coefficients. This is the principle of alternating least squares (ALS). The coefficients are found the usual way (see Chapter 15). The transform is found under the restriction that the initial ranking is preserved.

Since estimation of the transform is a very flexible process, this method can easily overfit for small data sets and should always be used with care. The method can possibly be useful for handling ranking data within for instance the framework of conjoint analysis. We refer to Young (1981) for a thorough description of the method (see also Kruskal, 1965; Gifi, 1991; Green, 1973).

17.6 The L-PLS Method

The L-PLS (Martens *et al.*, 2005) method is a technique for combining three matrices which are linked according to the scheme in Figure 9.1. A typical situation where this is appropriate is in a preference mapping situation where the two horizontal matrices are the sensory data and the consumer ratings while the matrix on top is the matrix of additional consumer attributes. The L-PLS method (Martens *et al.*, 2005) uses the SVD on products of the three matrices involved and provide two-dimensional scatter plots of four different types. The first type represents the samples, the next represents the sensory variables, the third the consumer hedonic scores and the last the additional consumer attributes. All these pieces of information can be plotted separately or in the same plot and interpreted the same way as for regular bi-plots. The method is an elegant extension of regular PLS regression. For alternative, but similar approaches, we refer to Lengard and Kermit (2006) and Endrizzi *et al.* (2009).

17.7 Missing Value Estimation

In some situations there are missing values in large data sets. If single elements are missing at random, there are techniques available for replacing them by estimated or imputed values. Some of the techniques, for instance PLS and PCA solved by the NIPALS method, have this as an inbuilt feature, at least in many available implementations.

Walczak and Massart (2001) provides a discussion of various ways of solving the general problem of replacing missing values by estimates. A possible procedure that seems to be used frequently is the iterative algorithm where one starts out by initial estimates of missing values (obtained by for instance averaging), then one calculates the SVD (or PCA) before reconstructing the **Y** by using a pre-defined number of components. The missing values are then replaced by the reconstructed or predicted ones. The procedure iterates until convergence. Multivariate regression analysis of the missing values vs. related variables in the data set is another possible strategy. For an overview we refer to Donders *et al.* (2006) (see also Grung and Manne, 1998).

When there are systematic tendencies in the missing value pattern, one should always be careful when using these methods. Many ANOVA models allow for imbalance and also incompleteness in the data set, and these methods should be used when possible. Incomplete data with missing cells are, however, very difficult to handle in ANOVA if interactions are involved. Note that if the number of missing values is high, there will always be problems with the validity of the imputation results.

References

Arnold, G.M., Williams, A.A. (1987). The use of generalised procrustes techniques in sensory analysis. In J.R. Piggott (ed.), *Statistical Procedures in Food Research*. London: Elsevier Science Publishers, 244–53.

Bro, R. (1996). Multiway calibration. Multilinear PLS. *Journal of Chemometrics* 10, 47–61.
Bro, R. (1997). PARAFAC. Tutorial and applications. *Chemometrics and Intelligent Laboratory Systems*, 38, 149–71.
Bro, R. Smilde, A.K., de Jong, S. (2001). On the difference between low-rank and subspace approximations: improved model for multilinear PLS regression. *Chemometrics and Intelligent Laboratory Systems* 58, 3–13.
Bro, R., Qannari, E.M, Kiers, H.A., Næs, T., Frøst, M.B. (2008). Multi-way models for sensory profiling data. *J. Chemometrics* 22, 36–45.
Brockhoff, P, Hirst, D., Næs, T. (1996). In T. Næs, E. Risvik (eds), *Multivariate Analysis of Data in Sensory Science*. Amsterdam: Elsevier.
Carroll, J.D. (1968). Generalisation of canonical analysis to three or more sets of variables. *Proceedings of the 76th Convention of the American Psychological Association*, vol. 3. 227–8.
Carroll, J.D., Chang, J.J. (1970). Analysis of individual differences in multidimensional scaling via n-way generalization of Eckhart-Young decomposition. *Psychometrika* 35, 283–319.
Courcoux, P., Semenou M. (1997). Preference data analysis using a paired comparison model, *Food Quality and Preference* 8(5–6), 353–8.
Dahl, T., Næs, T. (2004). Outlier and groups detection in sensory panels using hierarchical cluster analysis with the Procrustes distance. *Food Quality and Preference* 15, 195–208.
Dahl, T., Tomic, O. Wold, J.P., Næs, T (2008). Some new tools for visualising multi-way sensory data. *Food Quality and Preference* 19, 103–13.
Dahl, T., Næs, T. (2009). Identifying outlying assessors in sensory profiling using fuzzy clustering and multi-block methodology, *Food Quality and Preference* 20(4), 287–94.
Dijksterhuis, G. (1996). Procrustes analysis in sensory research: In T. Næs, E. Risvik (eds), *Multivariate Analysis of Data in Sensory Science*, pp. 185–217. Amsterdam: Elsevier.
Donders, A.R.T., Geert, J.M.G., van der Heijden, Stijnen, T., Moons, K.G.M. (2006). Review: A gentle introduction to imputation of missing values. *Journal of Clinical Epidemiology* 59, 1087–91.
Endrizzi, I., Gasperi, F., Calo, D.G., Vihneau, E. (2009). Two-step procedure for classifying consumer in a L-structured data context. *Food Quality and Preference* (in press).
Gifi. A. (1991). *Nonlinear Multivariate Analysis*. New York: John Wiley & Sons, Inc.
Gower, J.C. (1975), Generalized Procrustes analysis, *Psychometrica* 45(1), 3–24.
Gower, J.C., Dijksterhuis, G. (2004). *Procrustes Problems*. Oxford: Oxford University Press.
Green, P.E. (1973). On the analysis of interactions in marketing research data. *Journal of Marketing Research* 10, 410–20.
Grung, B., Manne, R. (1998). Missing values in principal components analysis. *Chemometrics and Intelligent Laboratory Systems* 42, 125–39.
Harshman, R.A. (1970). Foundations of the PARAFAC procedure; models and conditions for an 'explanatory' multi-modal factor analysis. UCLA Working Papers in Phonetics, 16, 1–84.
Hirst, D. d Næs, T. (1994). A graphical technique for assessing differences among a set of rankings. *Journal of Chemometrics* 8, 81–93.
Jøreskog, K.G. (1977). Structural equation models in the social sciences: Specifications, estimation and testing. In P.R. Krishnaiah (ed), *Applications of Statistics*. Amsterdam: North Holland, 265–87.
Kaplan, D. (2000). *Structural Equations Modelling: Foundations and Extensions*. California: Sage Publications Inc.
Kroonenberg, P., DeLeeuw, J. (1980). Principal components analysis of three-mode data by means of alternating least squares algorithms, *Psychometrika* 45, 69–97.
Kruskal, J.B. (1965). Analysis of factorial experiments by estimating monotone transformations of the data. *Journal of Royal Statistical Society*, Series B, 27, 251–63.
Langron, S.P., Williams, A.A., Collins, A.J. (1984). A comparison on the consensus configuration from a genealised Procrustes analysis with the untransformed panel mean in sensory profile analysis. *Lebensmittel-Wissenschaft und Technologie* 17, 296–8.
Lavit, C., Escoufier, Y., Sabatier, R., Traissac, T. (1994). The ACT (STATIS method). *Computational Statistics and Data Analysis* 18, 97–119.
Lehmann, E.L. (1975). *Nonparametrics*. San Francisco: Holden-Day, Inc.
Lengard, V., Kermit, M. (2006). 3-way and 3-block regression in consumer preference analysis. *Food Quality and Preference* 17, 234–42.

L'Hermier del Plantes (1976). Structuration des tableaux a trios indices de la statistique. Thesis, Montepellier II.
Mardia, K.V., Kent, J.T., Bibby, J.M. (1979). *Multivariate Analysis*. London: Academic Press.
Martens, H., Anderssen, E., Flatberg, A., *et al.* (2005). Regression of a data matrix on descriptors of both rows and of its columns via latent variables: L-PLSR. *Comp. Stat. and Data Analysis* 48, 103–23.
Martens, M., Tenenhaus, M, Vinzi, V.E., Martens, H. (2007). The use of partial least squares methods in new food product development. In MacFie, H. (ed.), *Consumer-led Food Products Development*. Cambridge: Woodhead, 492–523.
Næs, T., Kowalski. B. (1989). Predicting sensory profiles from external instrumental measurements. *Food Quality and Preference* (4/5), 135–47.
Pages, J. and Tenenhaus, M. (2001). Multiple factors analysis combined with path modelling. Application to the analysis of relationships between physicochemical variables, sensory profiles and hedonic judgements. *Chemometrics and Intelligent Laboratory Systems* 58, 261–73.
Pages, J., Husson, F. (2005). Multiple factors analysis with confidence ellipses: a methodology to study the relationships between sensory and instrumental data. *J. Chemometrics* 19, 138–44.
Popper, R., Heymann, H. (1996). Analyzing differences among products and panellists by multidimensional scaling. In T. Næs, E. Risvik (eds), *Multivariate Analysis of Data in Sensory Science*. Amsterdam: Elsevier, pp. 159–84.
Rayner, J.C.W., Best, D.J., Brockhoff, P.B., Rayner, G.D. (2005). *Nonparametrics for Sensory Science: A More Informative Approach*. Ames, USA: Blackwell Publishing.
Smilde. A., Bro, R., Geladi, P. (2004). *Multi-Way Analysis*. Chichester: John Wiley & Sons, Ltd.
Tenenhaus, M., Vinzi, V.E., Chatelin Y-M., Lauro, C. (2005). PLS path modelling. *Computational Statistics and Data Analysis*, 48, 159–205.
Tucker, L.R. (1964). The extension of factor analysis to three-dimensional matrices. In Frederiksen, N. and Gulliksen, H. (eds), *Contributions to Mathematical Psychology*. New York: Holt, Rinehart & Winston, pp. 110–82.
Van der Kloot, Q.A., Kroonenberg, P.M. (1985), External analysis with three-mode principal components models. *Psychometrika* 50(4), 479–94.
Vigneau E., Courcoux P., Semenou M. (1999). Analysis of ranked preference data using latent class models, *Food Quality and Preference* 10(3), 201–7.
Walczak, B., Massart, D.L. (2001). Dealing with missing data, Part I. *Chemometrics and Intelligent Laboratory Systems* 58, 15–27.
Wedel, M., DeSarbo, W.S. (1995). A mixture likelihood approach for generalized linear models. *Journal of Classification* 12(1), 21–55.
Wold, H. (1982). Soft modelling: The basics and some extensions. In K.G. Jøreskog, H. Wold (eds), *Systems under Indirect Observation*. Amsterdam: North Holland.
Young, F. (1981). Quantitative analysis of qualitative data. *Psychometrika* 46(4), 357–88.

Nomenclature, Symbols and Abbreviations

Nomenclature and Distinctions

Assessor Member of a sensory panel. *See also* **Judge**. May also be used for a consumer.

Acceptance Degree of acceptance for a product, either a food product or a combination of various extrinsic attributes. Usually given as a score between a lower and an upper limit (for instance between 1 and 9). The term is used for rating studies while the term preference is used when a consumer chooses one of several alternatives. The term preference mapping is generally used for acceptance studies and represents as such a different use of the term. For most cases, the distinction between the two terms has no influence on the understanding of the concepts presented.

A posteriori segmentation Segmentation done based on acceptance or preference data. The term post hoc is also sometimes used.

A posteriori use of consumer attributes The consumer attributes are used after the primary analysis of the acceptance or preference data.

A priori segmentation Segmentation done before the acceptance data are analysed, based on for instance demographic variables.

A priori use of consumer attributes The consumer attributes are used in the primary analysis of the data.

Balanced design An experiment where the number of observations is the same for all possible levels of all factors involved.

Calibration Concept sometimes used for the process of estimating regression parameters.

Choice test When a consumer is asked to choose the product he prefers or likes best among several, this is a choice test.

Conjoint analysis Method for analysing the effect of a number of designed factors, for instance packaging and information factors, on consumer acceptance or choice.

Consumer Person selected from a larger population to participate in a hedonic or discrimination test.

Statistics for Sensory and Consumer Science Tormod Næs, Per B. Brockhoff and Oliver Tomic

Consumer attribute Aspect related to the consumer, demographic information or information related to attitudes, values and habits.

Descriptive sensory analysis Sensory analysis based on a trained panel of assessors. Results in intensity score vales for a number of sensory attributes.

Experiment Either one single experiment or a series of experimental runs (in an experimental design).

Experimental design A series of experimental runs constructed from statistical principles.

Experimental run One single element in a full experimental design.

Extrinsic/intrinsic attributes Intrinsic attributes are attributes related to the food product itself, while the extrinsic attributes are other properties related to the product, such as for instance health information and brand name.

Frobenius norm Norm for a matrix defined as the square root of the sum of squares of all elements.

Greek letters Used for describing parameters in models.

Incomplete data set A data set from for instance a sensory study where some of the cells (combinations of factors) are empty.

Interval scale data Numerical data. Distances between numbers make sense. No fixed zero point. Typical for sensory and consumer liking data. An interval scale with a fixed zero point is called a ratio scale.

Intrinsic/extrinsic attributes Intrinsic attributes are attributes related to the food product itself, while the extrinsic attributes are other properties related to the product, such as for instance health information and brand name.

Liking Generic term for describing consumers' degree of pleasure. Sometimes used instead of the concept acceptance.

Judge Member of a sensory panel. *See also* **Assessor**.

Missing values Values that by some (random or other) reason have been left out of the data set.

Nominal scale data Categorical data without ordering. For instance blue, red, white etc.

Object Physical product to be tested by some assessors, also used as a generic term for rows in a data table.

Ordinal scale data Categorical data with ordering. Rank data are ordinal.

Preference Used in more or less the same way as choice. A consumer chooses a product among several, i.e. the one he/she prefers that product. *See also* **Acceptance**.

Product Physical object to be tested by consumers. *See also* **Sample**; **Object**.

Profile Generic name of a combination of attributes. *See also* **Product**.

Ranking If the consumer is asked to rank the samples according to for instance liking, this gives ranking data, i.e. ordinal data.

Repeatability Degree of similarity between sensory replicates of the same sample. More generally used as a measure for similarity between replicated measurements taken of the same sample.

Replicate If a product is tested several times by for instance a sensory panel, these measurements are sensory replicates. If the samples are tested two times, we say that there are two replicates. This is also called duplicates. If there are three replicates, this is sometimes called triplicates. There are different types of replicates.

Reproducibility Used here as similarity between assessors in a sensory panel. More generally used when comparing full replicates, not only measurement replicates (repeated measurements).

Rating This is used in situations where the consumers are asked to rate the products, i.e. to give scores between a lower and an upper limit (for instance between 1 and 9). *See also* **Acceptance**.

Sample Physical object to be tested. A few places the word is also used for a set of observations, a statistical sample (see Chapter 11). When not described explicitly, the former use of the word is applied in this book.

Sensory analysis In this book used mainly as synonymous to descriptive sensory analysis. *See also* **Sensory science** which is used in a broader context.

Sensory panel A group of trained humans used to judge properties or differences between products.

Sensory profiling *See* **Descriptive sensory analysis**.

Sensory science General term used for all studies where the human senses are used. Can be used both for descriptive sensory analysis and consumer tests.

Scores Used for the values given by the sensory assessors (or by the consumers). Also used as a specific term related to scatter plots of the product information in PCA.

Self-explicated test This is a type of consumer test where the consumers are asked directly about which aspects of the product they like best. This is different from conjoint analysis which is based on an experimental design.

Use of Lowercase, Italics, Boldface

Boldface capital letter Matrix, for instance **X**.

Boldface lower-case letter Vector, when presented without transpose (T), it will always be a column vector, for instance **x**.

Lowercase, italics Scalar, for instance x.

Uppercase, italics Used as symbol for number of columns, number of observations etc. Also used in expressions like Y-variable, Y-value etc.

Indices Lower case when counting, upper case for the total number. For instance x_i, $i=1,..,I$.

Symbols

α **and** β Symbols used to describe regression coefficients. Used both for scalars and vectors. Also used as symbols for effects in ANOVA models.

E and F Matrices of residuals for **X** and **Y** respectively (in PLS and PCR).

ε Residual term in a regression model (used in a similar way as f).

γ Symbol used for effect in ANOVA model.

Hat, for instance $\hat{\alpha}$ Used for indicating estimated value. See **P**, **T** and **Q** below for how the hat is used in connection with PCA,. PCR and PLS.

H Transformation matrix.

i, I Number of assessors in sensory profiling.

j, J Number of samples/products tested in sensory profiling. J also used as symbol for criterion to be optimised in fuzzy clustering.

k, K Number of variables in sensory analysis. Also used as a generic term for columns of a matrix and as a generic term for the number of variables in a data matrix. Number of factors in a design.

m, M Number of additional consumer attributes.

n, N Number of consumers in a consumer test. Also used a generic term for rows in a matrix. Also number of independent observations.

P, $\hat{\mathbf{P}}$ Matrix of loadings for PCA and X-loadings for PLS. Without hat when used in a model, with hat when estimated.

Q, $\hat{\mathbf{Q}}$ Symbol for Y-loadings in PCR and PLS. Without hat when used in a model, with hat when estimated.

ρ Population correlation coefficient.

r, R Replicates.

r Empirical correlation coefficient.

R^2 Squared correlation coefficient – explained variance.

s, s^2 Empirical standard deviation and variance.

θ Symbol used for effect in ANOVA model.

T, $\hat{\mathbf{T}}$ Matrix of scores for PCA and PLS. Without hat when used in a model, with hat when estimated.

T Transpose of a matrix or vector.

w Symbol for weight in weighted average.

X Uses to describe explanatory, input or independent variables. Can be sensory, consumer, chemistry variable etc. depending on application. Only used in situations with input variables. Can be used as matrix, vector or scalar.

Y Output or response variables. Can be sensory, consumer, chemistry variables etc. depending on application. Also used as a generic term, when only one matrix of data is available. Can be used as matrix, vector or scalar.

Abbreviations and Acronyms

AIC	Akaike's information criterion.
ALS	Alternating least squares.

ANOVA	Analysis of variance.
ANCOVA	Analysis of covariance.
ASCA	ANOVA-simultaneous component analysis.
BIBD	Balanced incomplete block design.
BIC	Bayesian information criterion.
CCA	Canonical correlation analysis.
CCD	Central composite design.
CPCA	Consensus PCA.
CV	Cross-validation.
CVA	Canonical variate analysis.
CVANOVA	ANOVA used for cross-validated results.
DF	Degrees of freedom.
DoE	Design of experiments.
EM	Algorithm – expectation minimisation algorithm.
FA	Factor analysis.
FCM	Fuzzy clustering method.
FCP	Free choice profiling.
GCA	Generalised canonical correlation analysis.
GLM	Generalised linear model.
GPA	Generalised procrustes analysis.
IIA	Independent of irrelevant alternatives.
INDSCAL	Method for individual MDS
LDA	Linear discriminant analysis.
LS	Least squares.
LS-PLS	Method that combines LS and PLS regression.
LSD	Least significant differences.
L-PLSR	PLS for L-type of data.
MANOVA	Multivariate ANOVA.
MDS	Multidimensional scaling.
MFA	Multiple factor analysis.
ML	Maximum likelihood.
MLR	Multiple regression.
MS	Mean square.
MSC	Multiplicative signal correction.
MSE	Mean square error.
MSEP	The square of RMSEP.
NIPALS	Algorithm for PCA.
NIR	Near infrared.
N-PLS	N-way partial least squares regression.
OS	Optimal scaling.
OVAT	One variable at the time.
PARAFAC	Parallel factor analysis.
PCA	Principal component analysis
PCR	Principal component regression.
PLS	Partial last squares.
PLSR	Partial least squares regression.

Prefmap and MdPref	Names used for two different methods for preference mapping.
QDA	Quadratic discriminant analysis.
REML	Restricted maximum likelihood.
RMSEP	Root means square error of prediction. Used to describe prediction ability of an equation.
RSM	Response surface methodology.
RV	Coefficient – criterion for comparison of matrices.
SD	Standard deviation.
SE	Standard error.
SS	Sums of squares.
STATIS	Structuration des Tableuax À Trois Indices de la Statistique.
SVD	Singular value decomposition.
VAF	Variance accounted for.

Index

Page numbers in italics refer to figures.

Statistics for Sensory and Consumer Science Tormod Næs, Per B. Brockhoff and Oliver Tomic

Printed and bound by CPI Group (UK) Ltd, Croydon, CR0 4YY
16/04/2025
14658385-0003